PHILOSOPHY *of* MATHEMATICS

PHILOSOPHY of MATHEMATICS

Structure and Ontology

STEWART SHAPIRO

New York Oxford
Oxford University Press
1997

Oxford University Press

Oxford New York
Athens Auckland Bangkok Bogota Bombay Buenos Aires
Calcutta Cape Town Dar es Salaam Dehli Florence Hong Kong
Istanbul Karachi Kuala Lumpur Madras Madrid Melbourne
Mexico City Nairobi Paris Singapore Taipei Tokyo Toronto Warsaw

and associated companies in
Berlin Ibadan

Published by Oxford University Press, Inc.
198 Madison Avenue, New York, New York 10016

Library of Congress Cataloging-in-Publication Data
Shapiro, Stewart, 1951–
Philosophy of mathematics : structure and ontology / Stewart
Shapiro.
p. cm.
Includes bibliographical references and index.
ISBN 0-19-509452-2
1. Mathematics—Philosophy. I. Title.
QA8.4.S533 1997
510'.1—dc20 96-31722

9 8 7 6 5 4 3 2 1

Printed in the United States of America
on acid-free paper

For Beverly

Numbers . . . are known only by their laws, the laws of arithmetic, so that any constructs obeying those laws—certain sets, for instance—are eligible . . . explications of number. Sets in turn are known only by their laws, the laws of set theory . . . arithmetic is all there is to number . . . there is no saying absolutely what the numbers are; there is only arithmetic.

Quine [1969, 44–45]

If in the consideration of a simply infinite system . . . set in order by a transformation . . . we entirely neglect the special character of the elements; simply retaining their distinguishability and taking into account only the relations to one another in which they are placed by the order-setting transformation . . . , then are these elements called *natural numbers* or *ordinal numbers* or simply *numbers*.

Dedekind [1888, §73]

Contents

PHILOSOPHY *of* MATHEMATICS

Introduction

This book has both an old topic and a relatively new one. The old topic is the ontological status of mathematical objects: do numbers, sets, and so on, exist? The relatively new topic is the semantical status of mathematical statements: what do mathematical statements mean? Are they literally true or false, are they vacuous, or do they lack truth-values altogether? The bulk of this book is devoted to providing and defending answers to these questions and tracing some implications of the answers, but the first order of business is to shed some light on the questions themselves. What is at stake when one either adopts or rejects answers?

Much contemporary philosophy of mathematics has its roots in Benacerraf [1973], which sketches an intriguing dilemma for our subject. One strong desideratum is that mathematical statements have the same semantics as ordinary statements, or at least respectable scientific statements. Because mathematics is a dignified and vitally important endeavor, one ought to try to take mathematical assertions literally, "at face value." This is just to hypothesize that mathematicians probably know what they are talking about, at least most of the time, and that they mean what they say. Another motivation for the desideratum comes from the fact that scientific language is thoroughly intertwined with mathematical language. It would be awkward and counterintuitive to provide separate semantic accounts for mathematical and scientific language, and yet another account of how various discourses interact.

Among philosophers, the prevailing semantic theory today is a truth-valued account, sometimes called "Tarskian." Model theory provides the framework. The desideratum, then, is that the model-theoretic scheme be applied to mathematical and ordinary (or scientific) language alike, or else the scheme be rejected for both discourses.

The prevailing model-theoretic semantics suggests realism in mathematics, in two senses. First, according to model-theoretic semantics, the singular terms of a mathematical language denote objects, and the variables range over a domain-of-discourse. Thus, mathematical *objects*—numbers, functions, sets, and the like—exist. This is what I call *realism in ontology*. A popular and closely related theme is the Quinean dictum that one's ontology consists of the range of the bound variables in properly regimented discourse. The slogan is "to be is to be the value of a variable." The second sense of realism suggested by the model-theoretic framework is that each well-formed, meaningful sentence has a determinate and nonvacuous truth-value, either truth or falsehood. This is *realism in truth-value*.

We now approach Benacerraf's dilemma. From the realism in ontology, we have the existence of mathematical objects. It would appear that these objects are *abstract*, in the sense that they are causally inert, not located in space and time, and so on. Moreover, from the realism in truth-value, it would appear that assertions like the twin-prime conjecture and the continuum hypothesis are either true or false, independently of the mind, language, or convention of the mathematician. Thus, we are led to a view much like traditional Platonism, and the notorious epistemological problems that come with it. If mathematical objects are outside the causal nexus, how can we know anything about them? How can we have any confidence in what the mathematicians say about mathematical objects? Again, I take it as "data" that most contemporary mathematics is correct. Thus, it is incumbent to show how it is possible for mathematicians to get it right most of the time. Under the suggested realism, this requires epistemic access to an acausal, eternal, and detached mathematical realm. This is the most serious problem for realism.

Benacerraf [1973] argues that antirealist philosophies of mathematics have a more tractable line on epistemology, but then the semantic desideratum is in danger. Here is our dilemma: the desired continuity between mathematical language and everyday and scientific language suggests realism, and this leaves us with seemingly intractable epistemic problems. We must either solve the problems with realism, give up the continuity between mathematical and everyday discourse, or give up the prevailing semantical accounts of ordinary and scientific language.

Most contemporary work in philosophy of mathematics begins here. Realists grab one horn of the trilemma, antirealists grab one of the others. The straightforward, but daunting strategy for realists is to develop an epistemology for mathematics while maintaining the ontological and semantic commitments. A more modest strategy is to argue that even if we are clueless concerning the epistemic problems with mathematics, these problems are close analogues of (presumably unsolved) epistemic problems with ordinary or scientific discourse. Clearly, we do have scientific and ordinary knowledge, even if we do not know how it works. The strategy is to link mathematical knowledge to scientific knowledge. The ploy would not solve the epistemic problems with mathematics, of course, but it would suggest that the problems are no more troublesome than those of scientific or ordinary discourse. The modest strategy conforms nicely to the seamless interplay between mathematical and ordinary or scientific discourse. On this view, everyday or scientific knowledge just is, in part, mathematical knowledge.

For a realist, however, the modest strategy exacerbates the dichotomy between the abstract mathematical realm and the ordinary physical realm, bringing the problem of *applicability* to the fore. The realist needs an account of the relationship between the eternal, acausal, detached mathematical universe and the subject matter of science and everyday language—the material world. How it is that an abstract, eternal, acausal realm manages to get entangled with the ordinary, physical world around us, so much so that mathematical knowledge is essential for scientific knowledge?

Antirealist programs, on the other hand, try to account for mathematics without assuming the independent existence of mathematical objects, or that mathematical statements have objective truth-values. On the antirealist programs, the semantic desideratum is not fulfilled, unless one goes on to give an antirealist semantics for ordinary or scientific language. Benacerraf's observation is that some antirealist programs have promising beginnings, but one burden of this book is to show that the promise is not delivered. If attention is restricted to those antirealist programs that accept and account for the bulk of contemporary mathematics, without demanding major revisions in mathematics, then the epistemic (and semantic) problems are just as troublesome as those of realism. In a sense, the problems are equivalent. For example, a common maneuver today is to introduce a "primitive," such as a modal operator, in order to reduce ontology. The proposal is to trade ontology for ideology. However, in the context at hand—mathematics—the ideology introduces epistemic problems quite in line with the problems with realism. The epistemic difficulties with realism are generated by the richness of mathematics itself.

In an earlier paper, Benacerraf [1965] raises another problem for realism in ontology (see also Kitcher [1983, chapter 6]). It is well known that virtually every field of mathematics can be reduced to, or modeled in, set theory. Matters of economy suggest that there be a single type of object for all of mathematics—sets. Why have numbers, points, functions, functionals, and sets when sets alone will do? However, there are several reductions of arithmetic to set theory. If natural numbers are mathematical objects, as the realist contends, and if all mathematical objects are sets, then there is a fact concerning *which* sets the natural numbers are. According to one account, due to von Neumann, the natural numbers are finite ordinals. Thus, 2 is $\{\phi, \{\phi\}\}$, 4 is $\{\phi, \{\phi\}, \{\phi, \{\phi\}\}, \{\phi, \{\phi\}, \{\phi, \{\phi\}\}\}\}$, and so $2 \in 4$. According to Zermelo's account, 2 is $\{\{\phi\}\}$, 4 is $\{\{\{\{\phi\}\}\}\}$, and so $2 \notin 4$. Moreover, there seems to be no principled way to decide between the reductions. Each serves whatever purpose a reduction is supposed to serve. So we are left without an answer to the question of whether 2 is really a member of 4 or not. Will the real 2 please stand up? What, after all, *are* the natural numbers? Are they finite von Neumann ordinals, Zermelo numerals, or other sets? From these observations and questions, Benacerraf and Kitcher conclude that numbers are not objects, against realism in ontology. This conclusion, I believe, is not warranted. It all depends on what it is to *be an object*, a matter that is presently under discussion. Benacerraf's and Kitcher's conclusion depends on what sorts of questions can legitimately be asked about objects and what sorts of questions have determinate answers waiting to be discovered.

The philosophy of mathematics to be articulated in this book goes by the name "structuralism," and its slogan is "mathematics is the science of structure." The subject matter of arithmetic is the *natural-number structure*, the pattern common to any system of objects that has a distinguished initial object and a successor relation that satisfies the induction principle. Roughly speaking, the essence of a natural number is the relations it has with other natural numbers. There is no more to being the natural number 2 than being the successor of the successor of 0, the predecessor of 3, the first prime, and so on. The natural-number structure is exemplified by the von Neumann finite ordinals, the Zermelo numerals, the arabic numerals, a sequence of distinct moments of time, and so forth. The structure is common to all of the reductions of arithmetic. Similarly, Euclidean geometry is about Euclidean-space structure, topology about topological structures, and so on. As articulated here, structuralism is a variety of realism.

A natural number, then, is a place in the natural-number structure. From this perspective, the issue raised by Benacerraf and Kitcher concerns the extent to which a place in a structure is an *object*. This depends on how structures and their places are construed. Thus, in addition to providing a line on solving the traditional problems in philosophy of mathematics, structuralism has something to say about what a mathematical object is. With this, the ordinary notion of "object" is illuminated as well.

The first part of the book, "Perspective," takes a broad approach and examines general matters in philosophy of mathematics. Its purpose is to set the stage for the detailed elaboration and argument in the rest of the book. I propose that the varieties of realism and antirealism be understood not so much in terms of rhetoric or slogans but as competing, sober *programs*. They look for answers to philosophical questions in different places. All of the programs have prima facie difficult problems to solve. It may be that two competing programs can both succeed, each by its own lights, but this remains to be seen.

Chapter 1, "Mathematics and Its Philosophy," deals with the relationship between the practice of mathematics and the philosophy of mathematics. Although the disciplines are interrelated in important ways, they are roughly autonomous. One cannot "read off" the correct way to do mathematics from the true philosophy, nor can one "read off" the true ontology, epistemology, or semantics from mathematics as practiced.

My orientation demurs from two extremes, one old and one relatively new. For some time, philosophers and mathematicians held that ontology and other philosophical matters determine the proper *practice* of mathematics. Accordingly, the philosopher must first figure out what we are talking about, by describing or discovering the metaphysical nature of mathematical entities. For example, are they objective or mind dependent? Only on completion of this philosophical chore do we know what to say about mathematical objects, in mathematics itself. Call this the *philosophy-first principle*. This orientation, which goes back at least to Plato, underlies many modern debates over items like the law of excluded middle, the axiom of choice, impredicative definition, and the extensional nature of functions and sets. Antirealists rejected such items, claiming that they presuppose the independent existence of mathematical objects. However, philosophy-first is not true to the history of mathematics. Classical logic, impredicative definition, and so on, are so thoroughly entrenched in the contemporary practice of mathematics that practitioners are often unaware of when those principles are applied. An advocate of philosophy-first would think that the realists won. However, at no time did the mathematical community decide that mathematical objects really exist, independent of the minds of mathematicians, and *for that reason* decide that it was all right to engage in the erstwhile questionable methodologies. The techniques and inferences in question were accepted because they were needed for mathematics. It is not much of an exaggeration to claim that mathematicians could not help using the techniques and inferences and, with hindsight, we see how impoverished mathematics would be without them. Of course, to point out that philosophy-first is not true to practice is not to refute it as a normative claim, but the normative matter is not on the present agenda.

The opposite of philosophy-first is a view that philosophy is irrelevant to mathematics. From this perspective, mathematics has a life of its own, independent of any philosophical considerations. Call this the *philosophy-last-if-at-all principle*. Sadly, this has something to recommend it. Most mathematicians are not in the least interested in philosophy, and it is mathematicians, after all, who practice and articulate their field. Some philosophers have expressed sentiment for philosophy-last. A popular view today is naturalism, characterized by Quine [1981, 72] as "the abandonment of first philosophy" and "the recognition that it is within science itself . . . that reality is to be identified and described." The first clause is the rejection of philosophy-first, and the second may be read as an endorsement of philosophy-last-if-at-all. Some philosophers (not Quine) do lean in this direction, applying naturalism to mathematics and declaring that it is and ought to be insulated from any inquiries that are not resolved by mathematicians qua mathematicians.

I suspect that philosophy-last is unhealthy for mathematics, but this weak normative line is not pursued here. Even if philosophy-last is correct, and should be correct, as far as the health of mathematics is concerned, it does not follow that philosophy of mathematics is worthless. The interests of philosophers are sometimes contiguous with those of their colleagues in other departments, but they are different interests. The philosopher must say something about mathematics, something about mathematical language, and something about mathematicians. How do we manage to do mathematics? How is it possible for humans to know and teach mathematics, and how is mathematics applied in the sciences? These important questions are not entirely mathematical matters.

The second chapter, "Object and Truth: A Realist Manifesto," articulates the aspirations and main problems of realism as a philosophical program concerning mathematics. Several "grades" of realism are distinguished. I define a *working realist* to be someone who uses or accepts the inferences and assertions suggested by traditional realism, items like excluded middle, the axiom of choice, impredicative definition, and general extensionality. As such, working realism is not a philosophical view. It is a statement of how mathematics is done, or perhaps a statement of how mathematics ought to be done, but there is no attempt to answer the important philosophical questions about mathematics. Working realism, by itself, has no consequences concerning the semantics, ontology, and epistemology of mathematics, nor the application of mathematics in science. The strongest versions of working realism are no more than claims that mathematics can (or should) be practiced *as if* its subject matter were a realm of independently existing, abstract, eternal entities. Working realism does not go beyond this "as if." Indeed, it is consistent with antirealism. Anyone who is not out to revise contemporary mathematics on philosophical grounds is probably a working realist at some level.

Philosophical realism, or *realism simpliciter*, is a more full-fledged philosophical program. It is a plan for structuring part of the ship of Neurath—the part concerning mathematics—and showing how that part relates to the rest of the ship. Traditionally, of course, realism is the view that mathematical objects exist independently of the mathematician and that mathematical truth is objective, holding (where it does) independent of the mind and language of the mathematical community. This would be the conjunction of realism in ontology and realism in truth-value. The program of realism starts here, asserting the independent existence of a realm of mathematical objects and the objectivity of mathematical truth. But this is only a start. The realist must go on to say something about the nature of this realm and how we manage to obtain knowledge about it.

From these characterizations, one may take philosophical realism to be the default philosophy, or the first guess, of the working realist. The straightforward explanation of why *X* happens as if *Y* is that *Y* is in fact the case. Working realism characterizes how the philosophical realist would practice mathematics. However, the proof is in the philosophical pudding. Without plausible answers to the philosophical questions, realism is no more than an empty promissory note.

In line with Benacerraf's desideratum, I propose that model-theoretic semantics is the central frame of philosophical realism. In model theory, one specifies a range of the variables of a mathematical discourse—an ontology—and then one specifies extensions for the predicates and relations. This determines satisfaction conditions for the complex formulas of the language and truth conditions for the sentences, via the familiar program. The point here is that if realism is correct, then model theory provides the right picture, or "model," of how mathematical languages describe mathematical reality. According to realism, the relationship between language and reality is analogous to the relationship between a formal language and a model-theoretic interpretation of it. Of course, how we manage to "specify" the domains and the various extensions, without vicious circularity, is a major problem with realism.

This chapter provides a context to illuminate several Quinean themes, as they apply to philosophical realism. I have in mind naturalism, ontological relativity, and the thesis that ontology is marked by the range of bound variables. Each of these doctrines amounts to something different at the levels of working realism and philosophical realism.

Part II has the same title as the view to be articulated, "Structuralism." Chapters 3 and 4 develop the positive philosophical account, executing part of the realist program. I try to say what mathematics is about, how we come to know mathematical statements, and how we come to know about mathematical objects. I show that at least some traditional problems for realism are dispelled. Other problems are no more intractable than other puzzles whose solution is usually presumed, if not known. Along the way, some of the general items of philosophy are illuminated, both as applied to mathematics and in general. Chapter 5 sketches some themes in the history of modern mathematics, showing a trend toward mathematics as the science of structure.

Chapter 3, "Structure," articulates structuralism, with focus on ontological matters. As noted, the structuralist holds that the subject matter of, say, arithmetic, is a single abstract structure, the natural-number structure. Natural numbers are the places of this structure. Structures are prior to places in the same sense that any organization is prior to the offices that constitute it. The natural-number structure is prior to "6," just as "baseball defense" is prior to "shortstop" or "U.S. Government" is prior to "vice president."

Even though a structure is a reified "one-over-many" of sorts, structures are not mobilized to play the kind of explanatory or justificatory roles for which universals have been invoked. Structuralists do not claim, for example, that the reason the system of finite, von Neumann ordinals is a model of arithmetic is that it exemplifies the natural-number structure, nor do we argue that the fact that this system exemplifies this structure is what justifies us in using the system as a model of the natural numbers. If anything, it is the other way around.

Two interrelated groups of ontological questions arise, one concerning structures themselves and the other concerning places in structures. In line with a partial analogy with universals, there are several ontological views concerning structures. One can be a Platonist, an Aristotelian, a nominalist, and so on. Without pretending to be exhaustive, I develop three alternatives. The first takes structures, and their places, to exist independently of whether there are any systems of objects that exemplify them. The natural-number structure, the real-number structure, the set-theoretic hierarchy, and so forth, all exist whether or not there are systems of objects structured that way. I call this *ante rem* structuralism, after the analogous view concerning universals.

The second option is a more in re approach. Statements of arithmetic, like "2 + 3 = 5," are not about specific *objects* denoted by "2," "3," and "5." Rather, each such statement is a generalization over all natural-number *systems*: "2 + 3 = 5" comes to "in every natural-number system, the object in the 2 place added to the object in the 3 place yields the object in the 5 place." When understood like this, a seemingly bold ontological claim, like "0 exists," comes to the innocuous "every natural-number system has an object in its 0 place." In other words, on this in re view, statements whose surface grammar indicates that they are about numbers are actually generalizations. Advocates may deny that the natural-number structure exists, or else they will insist that there is no more to the natural-number structure than natural-number systems. There is no more to the natural numbers than countably infinite collections with a distinguished initial object, and so on. Destroy all the systems and the natural-number structure is itself destroyed. This view is called *eliminative structuralism*. It is a structuralism without structures.

Notice that eliminative structuralism requires a background ontology to fill the places of the various structures. Suppose, for example, that there are only finitely many objects in the universe. Then there are no natural-number systems, and every sentence in the language of arithmetic turns out to be true. For example, the above rendering of "2 + 3 = 5" is true because there are no natural-number systems, but the renderings of "2 + 3 = 0" and "2 + 3 ≠ 5" are also true. If the background ontology is not big enough, then mathematical theories will collapse into vacuity. In particular, statements of arithmetic are vacuous unless the universe is infinite; statements of real analysis are vacuous unless the universe contains a continuum of objects; and set theory is vacuous unless the universe contains a proper class—or at least an inaccessible number of objects. Because there are probably not enough *physical* objects to keep some theories from being vacuous, the eliminative structuralist must assume there is a large realm of abstract objects. Thus, eliminative structuralism looks a lot like traditional Platonism.

Our third option, *modal structuralism*, patches the problem of vacuity. It is an eliminative program with a wrinkle. Instead of taking a statement of arithmetic to be about all natural-number systems, the theorist asserts that the statement is about all *possible* natural-number systems. Our example, "2 + 3 = 5," comes to "in every possible natural-number system, the object in the 2 place added to the object in the 3 place yields the object in the 5 place." On this option, there is an attenuated threat of vacuity. According to the modal structuralist, arithmetic is vacuous unless it is possible for there to be a countably infinite system of objects, and real analysis is vacuous unless it is possible for there to be a system the size of the continuum. Modal eliminative structuralism is elaborated and defended in Hellman [1989].

The other group of ontological questions concerns the status of individual mathematical objects—places in structures. Notice first that on the eliminative and modal options, natural numbers, for example, are not individual objects. Singular terms that seem to denote numbers, such as numerals, are disguised bound variables ranging over objects in systems. In contrast, the *ante rem* structuralist holds that natural numbers exist as bona fide objects. These objects are the places of the natural-number structure. In sum, numbers are offices rather than officeholders. The same goes for real numbers and members of the set-theoretic hierarchy. So *ante rem* structuralism is a realism in ontology. Eliminative structuralism and modal structuralism are antirealist in ontology. All three are realist in truth-value.

In ordinary language, there are different ways that positions in patterns are treated. Most of the time, the positions of a pattern are offices, which can be occupied by various sorts of objects and people. In many ways, words that denote offices act like predicates. For example, we speak of different people who have held the office of Speaker of the House, different people who have played shortstop, and different pieces of wood and plastic that have played the role of white queen's bishop. I call this the *places-are-offices* perspective. In other contexts, however, we treat positions of a pattern as objects in their own right. When we say that the Speaker presides over the House and that a bishop moves on a diagonal, the terms "Speaker" and "bishop" are singular terms, at least grammatically. Prima facie, they denote the offices themselves, independent of any objects or people that may occupy the offices. This is the *places-are-objects* perspective. The *ante rem* structuralist holds that natural numbers are places in the natural-number structure, and the theory of arithmetic treats these numbers from the places-are-objects perspective. Numerals are genuine singular terms, semantically, as well as grammatically.

Clearly, there is an intuitive difference between an object and a place in a structure—between an office and an officeholder. The *ante rem* structuralist respects this distinction but argues that it is a relative one. What is an office from one perspective is an object—and a potential officeholder—from another. In arithmetic, the natural numbers are objects, but in some other theories natural numbers are offices, occupied by other objects. Thus, in set theory we say that the finite von Neuman ordinals and the Zermelo numerals both exemplify the natural-number structure. In one system, $\{\phi, \{\phi\}\}$ occupies the 2 place, and in the other $\{\{\phi\}\}$ occupies that place. Sometimes, we take the places of the natural-number structure from both the places-are-offices and the places-are-objects perspectives in the same breath. We say, for example, that the even natural numbers exemplify the natural-number structure. The *ante rem* structuralist interprets this as follows: the even natural numbers, construed from the places-are-objects perspective, are organized into a system, and this system exemplifies the natural-number structure. In the system in question, 0 occupies the 0 place, 2 occupies the 1 place, 4 occupies the 2 place, and so on.

In effect, the notion of "mathematical object" is relative to a structure. What is an object from one perspective is a place in a structure from another. Some structures are exemplified by places from other structures, and there are structures of structures. The various dichotomies and notions of relativity are related to others prevalent in the literature, most notably the abstract/concrete dichotomy, ontological relativity, and the inscrutability of reference.

In sum, the *ante rem* structuralist interprets statements of arithmetic, analysis, set theory, and the like, at face value. What appear to be singular terms are in fact singular terms that denote bona fide objects. Moreover, *ante rem* structuralism accommodates the freestanding nature of mathematical structures. Anything at all can occupy the places of the natural-number structure, including natural numbers themselves. Thus, I hold that *ante rem* structuralism is the most perspicuous account of contemporary mathematics.

Nevertheless, there is a sense in which *ante rem* structuralism is equivalent to eliminative structuralism over a sufficiently large ontology, and to modal structuralism with sufficiently robust "possible-existence" assumptions. There is a straightforward "translation" between the frameworks, such that anything said by an advocate of one of the views corresponds to statements acceptable to advocates of each of the other two views.

The chapter closes with a brief account of some connections between structuralism and functionalism in the philosophy of mind. I suggest that many functionalist theses can be understood in the structuralist terms of this chapter. Functionalism is an eliminative structuralism, of sorts.

Chapter 4, "Epistemology and Reference," takes up epistemological and semantical matters. Because the problems of epistemology are among the most serious and troublesome for realism, this chapter is central to the case made in this book.

Several different strategies are applied to the epistemological problems, depending on the size and complexity of the structures in question. Small, finite structures are apprehended through abstraction via simple pattern recognition. A subject views or hears one or more structured systems and comes to grasp the structure of those systems. Of course, we do not have direct causal contact with structures, because they are abstract. The idea is that we grasp some structures through their systems, just as we grasp character types through their tokens.

To be sure, pattern recognition represents a sticky problem for psychology and cognitive science. There is no consensus among scientists as to how it works. Nevertheless, humans clearly can recognize at least some patterns. The philosophical problem is to show how pattern recognition can lead to knowledge about freestanding, *ante rem* structures.

Even if this strategy were successful, it would not go very far. To grasp a structure via simple pattern recognition, one must see or hear a system that exemplifies it. This limits the technique to small, finite structures. The next task is to extend the epistemological scheme to larger finite structures. For this, I propose a few extensions of pattern recognition, beyond simple abstraction. As above, the subject, perhaps a small boy, comes to grasp small, finite structures via pattern recognition. He grasps the 2 pattern, the 3 pattern, and the 4 pattern. The subject then sees that these small, finite structures themselves come in sequence, one after another, and he extends the sequence to structures he has not seen exemplified. Clearly, we do have some sort of grasp of the 14,792 pattern and the 1,000,000 pattern, even if we have never seen systems that exemplify these structures. Again, the philosophical problem is to see how this extended pattern recognition can lead to knowledge of the structures themselves.

Moving on, our subject, no longer a child, grasps the concept of a *finite cardinal structure* as such, and he realizes that these structures themselves exemplify a structure. This is the natural-number structure, our first infinite pattern. Notice that I do not claim that the natural numbers *are* the finite cardinal structures, that 4 is the 4 pattern, for example. The natural numbers are places in the natural-number structure, and the finite cardinal patterns exemplify this structure. The subject may also notice that the natural-number structure is exemplified by the system of numerals and other sequences of abstract objects.

From here, the subject can apprehend other countably infinite structures as variations on this one. Conceivably, we get to the real numbers by considering sets of natural numbers, and perhaps we can proceed through the levels of type theory. However, these techniques are unnatural at the more advanced levels. We surely do not apprehend the complex-number structure this way. Moreover, even our extended pattern recognition is limited. We are nowhere near the set-theoretic hierarchy. Our other epistemic techniques make a direct appeal to language and language acquisition. A subtheme of this book is that many metaphysical and epistemological matters are closely tied to language.

The next strategy employs an abstraction similar to one developed in Kraut [1980] and Wright [1983] (see also Hale [1987]). It begins with a version of the Leibniz principle of the identity of indiscernibles. If two items cannot be distinguished—if anything true of one is true of the other—then they can be or should be identified. The items are one, rather than two. A relativity of sorts emerges when we add Kraut's observation that indiscernability depends on available resources. Some items can be distinguished in one framework but not in another.

Our final epistemic strategy is *implicit definition*, a common and powerful technique of modern mathematics. For present purposes, implicit definition is the most speculative of all the strategies, and it yields knowledge of very large structures.

Typically, the theorist gives a collection of axioms and states that the theory is about any system of objects that satisfies the axioms. As I would put it, the axioms characterize a structure or a class of structures, if they characterize anything at all. The subject matter of the theory is that structure or class of structures.

An implicit definition characterizes a structure or class of structures by giving a direct description of the relations that hold among the places of the structure. The second-order Peano axioms characterize the natural-number structure, the second-order axioms of real analysis characterize the real-number structure, and the axioms of group theory characterize the class of group structures.

A purported implicit definition characterizes *at most* one structure if it is *categorical*—if any two models of it are isomorphic to each other. A purported implicit definition characterizes *at least* one structure if it is *coherent*, but here things are not so clear. Unlike isomorphism, coherence is not a rigorously defined mathematical notion, and there is no noncircular way to characterize it. Because the envisioned background metatheory is second-order, it will not do to define coherence as deductive consistency. When it comes to structures, consistency does not imply existence, contra Hilbert. Some consistent second-order theories have no models (see Shapiro [1991, chapter 4]). Surely, such theories are not coherent. A better explication of coherence is satisfiability, but now the circularity is apparent. Satisfiability is defined in set theory, and we structuralists take set theory to be about a particular structure—the set-theoretic hierarchy. I develop the various circularities and argue that they are not vicious. Satisfiability is a model of coherence, not a definition of it.

To complete the picture, then, if an implicit definition is coherent and categorical, then a single structure is characterized. According to *ante rem* structuralism, the variables of the theory range over the places of that structure, the singular terms denote places in that structure, and the relation symbols denote the relations of the structure.

The chapter closes with a brief account of reference and other semantic matters. I show that the schematic, model-theoretic notions of reference and satisfaction are particularly suited to mathematics, construed as the science of structure. A general understanding of language acquisition and use is sufficient to understand an implicit definition and sufficient to grasp reference to the places of the structure. I thus make good on the suggestion that model theory provides the central framework of philosophical realism. The present framework of *ante rem* structuralism also fulfills Benacerraf's desideratum that there be a uniform semantics for mathematical and ordinary or scientific languages—to the extent that model-theoretic semantics is appropriate for ordinary or scientific languages.

Chapter 5, "How We Got Here," provides some historical sketches and precursors to the idea that mathematics is the science of structure. Originally, the focus of geometry was space—matter and extension—and the subject matter of arithmetic was quantity. Geometry concerned the continuous, whereas arithmetic concerned the discrete. The nineteenth century saw a gradual liberation of both fields from these roots.

The early sections of this chapter give a bare sketch of the complex transition from geometry as the study of physical or perceived space to geometry as the study of freestanding structures. One early theme was the advent of analytic geometry, which served to bridge the gap between the discrete and the continuous. Projective geometry was a response. Another theme is the attempt to accommodate ideal and imaginary elements, such as points at infinity. A third thread is the assimilation of non-Euclidean geometry into mainstream mathematics and into physics. These themes contributed to a growing interest in rigor and the detailed understanding of deduction as independent of content. Structuralism is little more than a corollary to these developments.

The liberation of geometry from physical and perceived space culminated around the turn of the twentieth century, when Poincaré and Hilbert each published an account of geometry. Both authors argued that concepts like "point," "line," and "plane" are defined only in terms of each other, and they are properly applied to any system of objects that satisfies the axioms. Despite their many philosophical differences, these two mathematicians thus approached an eliminative structuralism (Hilbert more than Poincaré). Each found spirited opposition from a different logicist who maintained the dying view that geometry essentially concerns space or spatial intuition. Russell disputed Poincaré, and Frege disputed Hilbert. The two debates illustrate the emerging idea of mathematics as the science of structure.

Although neither Frege nor his interpreters speak in structuralist terms, the development of arithmetic in Frege [1884] (as articulated by Dummett [1981], [1991]; Wright [1983]; and Hale [1987]) goes some way toward structuralism. Frege approached full *ante rem* structuralism, not the eliminative variety suggested by Hilbert. Frege held that numerals are genuine singular terms that denote natural numbers. We can at least motivate structuralism by combining this with some of his other theses, such as the context principle. The envisioned view avoids the infamous pitfalls of Frege's logicism. Of course, structuralism is far from Frege's own philosophy of mathematics—by any stretch of the exegetical imagination.

A direct forerunner of *ante rem* structuralism is another logicist, Dedekind. His development of the notion of continuity and the real numbers, in [1872], his presentation of the natural numbers via the notion of Dedekind infinity, in [1888], and some of his correspondence constitute a structuralist manifesto, as illustrated by the passage from [1888 §73] that appears as an epigraph to this book: "If in the consideration of a simply infinite system . . . set in order by a transformation . . . we entirely neglect the special character of the elements, simply retaining their distinguishability and taking into account only the relations to one another in which they are placed by the order-setting transformation . . . then are these numbers called *natural numbers* or *ordinal numbers* or simply *numbers*. . . ."

Part III of this book, "Ramifications and Applications," extends the scope of structuralism to other aspects of mathematics and to science and ordinary language. Along the way, I evaluate other views in the philosophy of mathematics in light of structuralism.

Chapter 6, entitled "Practice: Construction, Modality, Logic," concerns an interesting and illuminating gap between the practice of mathematics and its current philosophical and semantic formulations. Mathematicians at work speak and write as if they perform dynamic operations and constructions. Taken literally, this language presupposes that mathematicians envision creating their objects, moving them around, and transforming them. In contrast to this dynamic picture, the traditional realist in ontology, or Platonist, holds that the subject matter of mathematics is an independent, *static* realm. In a deep metaphysical sense, this mathematical realm is eternal and immutable, and so the universe *cannot* be affected by operations, constructions, or any other human activity.

To belabor the obvious, then, the traditional Platonist does not take dynamic language literally. In contrast, the traditional intuitionists do take the constructional language literally, and they deny the static elements of mathematical practice. Intuitionists hold that this perspective requires revisions in the practice of mathematics. They demur from the law of excluded middle and other inferences based on it. In present terms, intuitionists are not working realists.

We can bring intuitionism and classical dynamic language under a common framework by thinking in terms of an imaginary, idealized constructor who can perform various operations, such as drawing lines and applying functions. Neither the intuitionistic constructor nor his classical counterpart has bounds on his lifetime, attention span, and so on. The classical constructor is idealized further, in that he has a greater omniscience and has the ability to complete infinite tasks. There is an intimate three-way connection between the logic of a mathematical system, the metaphysical nature of its objects, and the moves allotted to the ideal constructor.

In this framework, structuralism provides a single umbrella for the various static/dynamic, classical/intuitionistic systems. Structuralism has the resources to assess and compare the different systems, by using a common measure.

Dummett [1973], [1977] provides another, semantic route to intuitionistic logic. He argues that reflections on the learnability of language and its role in communication demand that an assertabilist semantics replace a truth-valued semantics. Following Heyting [1956], sentences should be understood in terms of their "proof conditions" rather than their truth conditions. Dummett's argument concludes that this perspective requires revisions in the practice of mathematics.

A philosopher of mathematics might find a Heyting-style, assertabilist semantics attractive, because "proof" or "assertability" may prove more tractable than "mathematical truth" on the sticky epistemic front. It may be more natural (or more realistic) to speak of the "assertion abilities" of the ideal constructor, taking this to be an extension of the abilities of human mathematicians. The idea is to use this notion in the semantics. However, revisionism is a steep price to pay, moving us back to the philosophy-first principle. The question before us is whether an assertabilist semantics is compatible with working realism.

Dummett's path from assertabilism to revisionism depends on how the idealized modal notions of provability or assertability are to be understood. One is forced to revisionism only if assertabilist semantics is coupled with a certain pessimism about human epistemic powers (or the epistemic powers attributed to the ideal constructor). A more optimistic viewpoint leads to classical logic, and allows the antirealist to speak as if mathematical assertions have objective truth-values. However, under this plan, the notion of "provable sentence" of, say, arithmetic, is just as complex and semantically intractable as classical "truth." So if this alternative is taken, there is no gain on the epistemic front.

Chapter 7, "Modality, Structure, Ontology," concerns programs of antirealism in ontology, realism in truth-value. The defenders of these programs maintain that mathematics is not about an independently existing realm of abstract objects, but once mathematical assertions are properly interpreted, they have nonvacuous truth conditions that hold or fail independent of the mathematician. In the end, mathematics—or a surrogate—is substantially correct. The advocates of such programs constitute the loyal opposition.

I do not attempt an exhaustive survey of the relevant antirealist literature. I focus on authors whose strategy is to reduce ontology by introducing (or invoking) some ideology, typically a modal operator. The authors include Hartry Field [1980], Charles Chihara [1990], and Geoffrey Hellman [1989], as well as some nominalists who invoke George Boolos's [1984], [1985] account of second-order quantifiers as plural quantifiers. Hellman [1989] is especially relevant to the present project, because he develops a modal structuralism.

It is common, both now and throughout the history of philosophy, to interpret modal operators as quantifiers. This is an analysis of modality in terms of ontology. The programs under study here reverse this trend, reducing ontology by invoking modality, or some other ideology. None of the authors envision a realm of possible worlds, a realm of possibilia, or a model theory to explicate the notions in question. The notions are just used, without apology—as mathematical notions are used without apology by us realists.

The ontological antirealist programs have promising beginnings, because the epistemology of the various notions may be more tractable than an epistemology of abstract objects like sets. However, the contention of this chapter is that, when applied to mathematics, the epistemological problems with antirealist programs are just as serious and troublesome as those of realism. There is no real gain on the most intractable problems in the philosophy of mathematics.

In each case, I show that there are straightforward translations between the set-theoretic language of the realist and the language with added ideology. The translations preserve warranted belief, at least, and probably truth (provided, of course, that we accept both viewpoints, at least temporarily). The contention is that, because of these translations, advocates of one of the systems cannot claim a major epistemological advantage over advocates of the other. Any insight that antirealists claim for their system can be immediately appropriated by realists and vice versa. The problem, however, lies with the "negative" consequences of the translations. Epistemological *problems* with realism get "translated" as well. The prima facie intractability of knowledge of abstract objects indicates an intractability concerning knowledge concerning the "new" notions, at least as they are developed in the works in question here.

To be sure, the notions invoked by our antirealists do have uses in everyday language, and competent speakers of the language do have some pretheoretic grasp of how they work. The antirealists are not inventing the ideology. However, the pretheoretic grasp of the notions does not support the extensively detailed articulations that are employed in the antirealist explications of mathematics. We do in fact have a decent grasp of the extensively articulated notions, but this understanding is not pretheoretic. It is mediated by mathematics, set theory in particular.

As I see it, set theory and modal notions like logical consequence illuminate each other. It is unfair to reject set theory, as our antirealists do, and then claim that we have a pretheoretic grasp of modal notions that, when applied to mathematics, exactly matches the results of the model-theoretic explication of them.

Several important philosophical issues lie near the surface of the critical remarks. One general matter is the issue of ontology and ontological commitment. The prevailing criterion, due to Quine, is that the ontology of a theory is the range of its bound variables. This is plausible enough, but recall that Quine insists on a fixed, and very austere ideology. Anything beyond the resources of extensional, first-order formal languages is out of bounds. When this constraint is relaxed, as our present antirealist authors (rightly) propose, things get murky. How are the disputes to be adjudicated? Each side proposes to eliminate or reduce the most basic notions *used* by the other. Typical remarks about judging the matter on holistic grounds are not very helpful, unless these grounds are elaborated and defended. Structuralism has the resources to formulate a further articulation of—or an alternative to—the Quinean criterion. Instead of speaking of the "ontology" of a theory, the philosopher should speak of the strength of the ontology and ideology combined. The notion of the equivalence of two structures plays a central role in this criterion. The ontology/ideology criterion completes the case for the contention, from chapter 3, that the three articulations of structuralism—*ante rem*, eliminative, and modal—are equivalent.

The final chapter 8, "Life outside Mathematics: Structure and Reality," is a brief account of some extensions of structuralism beyond mathematics, to science and to ordinary discourse. As noted earlier, one major problem for philosophy of mathematics is to account for the application of mathematics to the material world. Structuralism has a line on a partial account. Put simply, mathematics is applied when the theorist postulates that a given area of the physical world exemplifies a certain structure. In nearly all scientific theories, the structures of physical systems are modeled or described in terms of mathematical structures.

One consequence of this perspective is a blurring of the boundary between mathematics and science. Typically, (pure) mathematics studies structures as such, independent of whether they are exemplified or not. The focus is on the structures themselves. Empirical science is usually concerned with which structures are exemplified where. The focus is on the structur*ed*. There are many intermediate perspectives, and there is no need to draw a sharp boundary. The same goes for the boundary between mathematical and ordinary discourse. There is no sharp distinction between the mathematical and the mundane. To speak of objects at all is to impose structure on the material world, and this is to broach the mathematical.

One recurring theme in the book is that many of the standard tools that the contemporary philosopher uses to approach a subject matter were developed and honed with mathematics in view. Examples include the usual model-theoretic semantics and the concomitant notions of reference and logical consequence. The tools work best with mathematical languages precisely because mathematical objects are structural. It is in the context of these tools that such theses as "to be is to be the value of a bound variable" are formulated. Model-theoretic semantics is also a good model for ordinary discourse, because, for many purposes, ordinary objects can be viewed as structural. However, the match between model-theoretic semantics and ordinary language is not perfect. An attempt to overextend the analogy and the blind application of model theory results in some artificial puzzles.

This brings us back to where we started, the Benacerraf [1973] desideratum that there should be continuity between the semantic accounts of mathematical and ordinary or scientific languages. To some extent, the desired continuity depends on the extent to which ordinary objects resemble mathematical ones. Structuralism has a lot to say about what *mathematical* objects are—they are places in structures. The book closes with some speculative remarks on the extent to which ordinary objects can be construed as places in structures.

I am greatly indebted to many friends, teachers, colleagues, and students who read and criticized early incarnations of various parts of this book. Given their generous effort, I can only conclude that the remaining errors are due to my own stubbornness. I especially thank Michael Resnik, who twice acted as referee for Oxford University Press for this project. One other referee can only be acknowledged anonymously. Penelope Maddy and Jill Dieterle provided extensive comments on the bulk of the book, for which all readers should be grateful. I also thank the University of St. Andrews for allowing me to run a seminar based on the penultimate version of this book. Many improvements resulted. The main participants were Roy Dyckhoff, Janet Folina, Michele Friend, Stephen Ferguson, Ben Geis, Stephen Read, Penny Rush, and Crispin Wright. An incomplete list of others who have graciously given of their time and expertise includes Paul Benacerraf, George Boolos, Charles Chihara, Peter Clark, Roy Cook, John Corcoran, David Galloway, Bob Hale, Geoffrey Hellman, Catherine Hyatt, Peter King, Jeffrey Koperski, Robert Kraut, Timothy McCarthy, David McCarty, Colin McLarty, Pierluigi Miraglia, Gregory H. Moore, Calvin Normore, Diana Raffman, Joe Salerno, Michael Scanlan, George Schumm, Mark Silcox, Neil Tennant, Michael Tye, Alan Wescoat, and Mark Wilson. I apologize for omissions from this list. This project has taken a long time to come to fruition. I gave papers on related topics to the Eastern Division of the American Philosophical Association, The Center for Philosophy of Science at the University of Pittsburgh, and the Philosophy Research Seminar at the University of St. Andrews. Thanks to the audiences, who listened carefully and provided invaluable feedback. Robert Thomas, editor of *Philosophia Mathematica*, gave permission to use some material originally published there. I thank Angela Blackburn, Cynthia Read, and Lisa Stallings of Oxford University Press for encouraging me to pursue this project and guiding it through the publication process.

Words cannot express my appreciation for my wife, Beverly Roseman-Shapiro, for her loving support and tolerance. Her impatience for the excesses of analytic philosophy constantly saves my humanity. I dedicate this book to her.

Part I

PERSPECTIVE

1

Mathematics and Its Philosophy

For a long time, many philosophers and some mathematicians believed that philosophical matters, such as metaphysics and ontology, determine the proper *practice* of mathematics. Plato, for example, held that the subject matter of mathematics is an eternal, unchanging, ideal realm. Mathematical objects are like the Forms. In Book VII of the *Republic*, he chided mathematicians for not knowing what they are talking about and, consequently, doing mathematics incorrectly: "[The] science [of geometry] is in direct contradiction with the language employed by its adepts. . . . Their language is most ludicrous, . . . for they speak as if they were doing something and as if all their words were directed toward action. . . . [They talk] of squaring and applying and adding and the like . . . whereas in fact the real object of the entire subject is . . . knowledge . . . of what eternally exists, not of anything that comes to be this or that at some time and ceases to be." The geometers of antiquity did not take Plato's advice, as witnessed by just about every source of ancient geometry, Euclid's *Elements* included. They use constructive, dynamic language: lines are drawn, figures are moved around, and so on. If Plato's philosophy is correct, this makes no sense. Eternal and unchanging objects are not subject to construction and movement. According to Proclus [485], the "problem" of dynamic language occupied those in Plato's Academy for some time.[1] Proclus tried valiantly to reconcile mathematical practice with the true philosophy. The task, as he saw it, was to figure out "how we can introduce motion into immovable geometric objects."

At first glance, one might think that the one-way dispute between the Platonists and the geometers concerns little more than terminology. Euclid wrote that between any two points *one can draw* a straight line, and Hilbert [1899] made the Platonistically correct assertion that between any two points *there is* a straight line. Presumably,

1. Proclus [485, 125] credits a nephew of Plato, named Speusippus, and a pupil of Eudoxus, named Menaechmus, with noteworthy accomplishments on this issue.

Hilbert and Euclid said the same thing, if their languages are properly understood. However, the situation is not this simple on either mathematical or philosophical grounds. Prima facie, the long-standing problems of trisecting an angle, squaring a circle, and doubling a cube are not questions of *existence*. Did ancient and modern geometers wonder, for example, whether *there is* an angle of twenty degrees, or was it a question of whether such an angle *could be drawn* and, if so, with what tools? It is not easy to formulate problems like this in static language.[2]

In the twentieth century, debates over intuitionism provide another clear and straightforward example of a philosophical challenge to mathematics as practiced. The traditional intuitionists held that mathematical objects are mental constructions, and mathematical statements must somehow refer to mental constructions. Brouwer [1948, 90] wrote, "The . . . point of view that there are no non-experienced truths . . . has found acceptance with regard to mathematics much later than with regard to practical life and to science. Mathematics rigorously treated from this point of view, including deducing theorems exclusively by means of introspective construction, is called intuitionistic mathematics. . . . [I]t deviates from classical mathematics . . . because classical mathematics believes in the existence of unknown truths." And Heyting [1931, 53], [1956, 1, 2, 10]:

> [W]e do not attribute an existence independent of our thought, i.e., a transcendental existence, to the integers or to any other mathematical objects. . . . [M]athematical objects are by their very nature dependent on human thought. Their existence is guaranteed only insofar as they can be determined by thought. They have properties only insofar as these can be discerned in them by thought. . . . Faith in transcendental . . . existence must be rejected as a means of mathematical proof.

> Brouwer's program . . . consisted in the investigation of mental mathematical construction as such. . . . In the study of mental mathematical constructions, "to exist" must be synonymous with "to be constructed". . . . In fact, mathematics, from the intuitionistic point of view, is a study of certain functions of the human mind.

Heyting expresses a subjectivism toward mathematical *objects*, holding that they do not exist independently of the mind. *Esse est fingi.* Brouwer seems more concerned with mathematical *assertions*, stating that they do not have mind-independent truth conditions.[3] Contemporary intuitionists, like Dummett [1973], [1977] and Tennant [1987], begin with reflections on language acquisition and use, drawing conclusions about the truth conditions for mathematical assertions.

The intuitionists contend that the philosophy has consequences concerning the proper practice of mathematics. Most notably, they reject the law of excluded middle and other inferences based on it. These methodological principles are symptomatic of faith in the transcendental existence of mathematical objects or the transcendental truth of mathematical statements. Let P be a property of numbers. For an intuition-

---

2. See chapter 6 (and Shapiro [1989]) for more on the static/dynamic dichotomy in mathematical language.

3. Despite the subjectivist tone of these passages, Brouwer indicates a Kantian perspective in which mathematics is mind-dependent but still objective in some sense (see Posy [1984]).

ist, the content of the expression $\neg\forall xPx$ is that it is refutable that one can find a construction that shows that P holds of each number; the content of $\exists x\neg Px$ is that one can construct a number x and show that P does not hold of x. The latter expression cannot be inferred from the former because, clearly, it is possible to show that a property cannot hold universally without constructing a number for which it fails. Heyting notes that a realist, someone who does hold that numbers exist independently of the mathematician, will accept the law of excluded middle and related inferences (but see Tennant [1987]). From the realist's perspective, the content of $\neg\forall xPx$ is simply that it is false that P holds universally, and $\exists x\neg Px$ means that there is a number for which P fails. Both formulas refer to numbers themselves; neither has anything to do with the knowledge-gathering abilities of mathematicians, or any other mental feature of them. From the realist's point of view, the two formulas are equivalent.

Before Brouwer, Kronecker held that at least some mathematical entities do not exist independently of the mathematician, as indicated by his much-quoted slogan, "God made the integers, everything else is the work of man." He proposed that this ontological view suggested, indeed demanded, revision of mathematical practice (see Stein [1988] and Edwards [1988]). After Heyting, there is Bishop [1975, 507] and constructivism: "There is a crisis in contemporary mathematics, and anybody who has not noticed it is being willfully blind. The crisis is due to our neglect of philosophical issues."

There are several other methodological battlegrounds that were once thought to turn on philosophical considerations. A definition of a mathematical entity, such as a particular number, is *impredicative* if it refers to a collection that contains the defined entity. Poincaré launched a systematic attack on the legitimacy of impredicative definitions in set theory (beginning in [1906] and continuing throughout his career; see Goldfarb [1988] and Chihara [1973]). The critique was based on the idea that mathematical objects, such as sets, do not exist independently of the mathematician. Poincaré rejected the actual infinite, insisting that the only sensible alternative is the potentially infinite. Accordingly, there is no static set of, say, all real numbers, determined prior to the mathematical activity. From this perspective, impredicative definitions are circular. One cannot construct an object by using a collection that already contains it. From the other side, Gödel [1944] is an explicit defense of impredicative definition, against Russell's "vicious circle principle." The argument is tied to Gödel's *philosophical* perspective, realism:

> [T]he vicious circle . . . applies only if the entities are constructed by ourselves. In this case, there must clearly exist a definition . . . which does not refer to a totality to which the object defined belongs, because the construction of a thing can certainly not be based on a totality of things to which the thing to be constructed belongs. If, however, it is a question of objects that exist independently of our constructions, there is nothing in the least absurd in the existence of totalities containing members, which can be described (i.e., uniquely characterized) only by reference to this totality. . . . Classes and concepts may . . . be conceived as real objects . . . existing independently of us and our definitions and constructions. It seems to me that the assumption of such objects is quite as legitimate as the assumption of physical bodies and there is quite as much reason to believe in their existence. (p. 456)

According to Gödel's realism, a definition does not represent a recipe for constructing, or otherwise creating, an object. Rather, it is a way to characterize or point to an already-existing thing. Thus, an impredicative definition is not viciously circular, no more so than using the expression "the old fart" to refer to the oldest member of the faculty.

Russell himself did not hold a constructivist philosophy of mathematics. His rejection of impredicative definitions was due to metaphysical views on the nature of propositions and to his assimilation of classes to propositions and propositional functions (see Goldfarb [1989]).

Another example is the *axiom of choice*. One version says that for every set A of nonempty sets, there is a function whose domain is A and whose value, for every $a \in A$, is a member of a. The existence of the function does not depend on a mechanical method for picking out, constructing, or even uniquely characterizing a member of each member of A. Today, virtually every mathematician accepts the axiom of choice and, in fact, most are not explicitly aware of cases in which it (or a weaker version) is applied.[4] But it was not always like this. When the axiom was explicitly formulated in Zermelo's [1904] proof of the well-ordering theorem, it was opposed by many, probably most, leading mathematicians, notably Baire, Borel, and Lebesgue. Their opposition often focused on fundamental philosophical considerations that concern the nature of cardinality, functions, and sets. Typically, the opponents adopted antirealist perspectives. Lebesgue [1971], for example, divided mathematicians into two camps: "empiricists," who admit only the existence of real functions that are (uniquely) definable; and "idealists," who admit the existence of functions that are not definable, presumably, functions that exist independently of us and our abilities to define mathematical objects. Lebesgue and fellow "empiricists" reject the axiom of choice, whereas "idealists," like Cantor and Zermelo, accept it (see Moore [1982]).

Lebesgue's terminology is nonstandard, of course, but his battle lines do seem correct. His "idealists" are realists, holding that the universe of sets exists independently of the mathematician. These people do not require an algorithm or other method for uniquely characterizing a choice set before they will believe that the set exists. Indeed, from the realist perspective, the axiom of choice is an obvious truth, and thus its designation as an "axiom." Lebesgue's "empiricists" are antirealists, perhaps constructivists.[5]

A closely related item is the extensional notion of *function*, as an arbitrary correspondence between collections. The same goes for the notion of set, as an arbitrary collection. In both cases, one might think that traditional philosophical issues are

4. An anecdote: My first year in graduate school in mathematics, I made a nuisance of myself by asking, almost constantly, where choice principles are employed. My teachers did not always know, but to their credit, they would patiently figure it out (or ask someone else) and tell me later. The exception was an annoyed visiting logician, who told me to always assume that choice was employed.

5. The axiom of choice holds in some constructivist contexts, but that is because the antecedent is understood in constructivist terms. The principle is "if one can construct a set of nonempty sets, then one can construct a choice function on the set." The issue of choice in constructive mathematics is rather complex. In intuitionistic real analysis, the axiom of choice undermines Brouwer's theorem that every function is continuous.

involved. If mathematical objects are mind-dependent, as an antirealist might contend, then it would seem that the notions of "set" and "function," like all mathematical notions, should be closely tied to some *characterization*, perhaps an algorithm or a description in a particular language. The mind does not have direct access to infinitary objects. Thus, our antirealist would not speak directly of functions and sets themselves but of function and set presentations (see Shapiro [1980]). This would make the notions of "function" and "set" inherently *intensional* and quite different from their contemporary, extensional treatment. A realist, on the other hand, would consider functions and sets themselves, as divorced from any descriptions, algorithms, or names of them. The descriptions *describe something*.

The orientation suggested by these examples is that philosophy determines and thus precedes practice in some sense. One first describes or discovers what mathematics is all about—whether, for example, mathematical entities are objective or mind-dependent. This fixes the way mathematics is to be done. One who believes in the independent existence of mathematical objects is likely to accept the law of excluded middle, impredicative definitions, the axiom of choice, extensionality, and arbitrary sets and functions. As we have seen, many of these methodological principles have been resisted on antirealist grounds and subsequently defended on realist grounds. Let us call the perspective here the *philosophy-first principle*. The idea is that we first figure out what it is that we are talking about and only then figure out how to talk about it, and what to say about it. Philosophy thus has the noble task of determining mathematics.

Despite the above examples, the philosophy-first principle is not true to the history of mathematics. Classical logic, impredicative definition, the axiom of choice, extensionality, and arbitrary functions and sets, are thoroughly entrenched in the practice of modern mathematics. Substantially, the battles are over (at least in mathematics). But at no time did the mathematical community don philosophical hats and decide that mathematical objects—numbers, for example—really do exist, independently of the minds of mathematicians, and *for that reason* decide that it is all right to engage in the erstwhile questionable methodologies. If anything, it is the other way around.

The history of the axiom of choice makes an interesting case study (again, see Moore [1982]). Some of its early proponents, like Zermelo and Hilbert, pointed out that the axiom codifies a principle often used in mathematics. This observation turned out to be insightful, and decisive. The first half of this century saw an intensive study of the role of the axiom of choice in such central fields of mathematics as analysis, algebra, and topology. It was learned that the principle is essential to the practice of these branches as they had developed at the time. Ironically, implicit, but necessary, uses of choice principles permeate the work of Baire, Borel, and Lebesgue, the major *opponents* of the axiom of choice.[6] In short, the principle of choice was not accepted

6. According to Taylor [1993], Zermelo (and Hilbert) held that the fact that choice principles are entrenched in mathematical practice and are applied without thought is evidence that the axiom of choice is self-evident. The self-evidence of choice is not automatically refuted by the fact that many intelligent theorists balked at the principle once it was explicitly formulated. If Taylor is correct, then we have here a substantive argument over whether something is self-evident. This is a much richer notion of self-evidence than the bland one of obviousness.

because realism sanctions it, but because it is needed. In a sense, mathematicians could not help using it, and with hindsight, we see how impoverished mathematics would be without it.

Similarly, the contemporary extensional notion of function as an arbitrary correspondence and the notion of set as an arbitrary collection did not emerge from blatant philosophical considerations. The notions developed from their intensional ancestors because of internal pressures within mathematics, mathematical physics in particular. First principles were not involved. The extensional notions made for a smoother and more useful theory. The tie to language, or to intensional items like propositional functions, proved to be artificial and unproductive.[7]

The same goes for the other disputed principles, such as excluded middle and impredicative definition. The extensive mathematical study of logical systems that lack these tools shows just how different mathematics would be without them. Many subtle distinctions must be made, definitions must be constantly checked for constructive or predicative pedigree, and the mathematician must pay close attention to language. Crucially, many important results must be given up. Mathematicians do not find the resulting systems attractive.[8] The reasons for this may not be clear, but it is not tied to traditional philosophy. It is noteworthy that most of the metatheory used to study intuitionistic and predicative systems is itself nonconstructive.

The opening paragraph of Dedekind's [1888] treatise on the natural numbers explicitly rejects the constructivist perspective. Then there is a footnote: "I mention this expressly because Kronecker not long ago . . . has endeavored to impose certain limitations upon . . . mathematics which I do not believe to be justified; but there seems to be no call to enter upon this matter with more detail until the distinguished mathematician shall have published his reasons for the necessity or merely the expediency of these limitations" (§2). Of course, Kronecker did state his reasons, but they were philosophical. Dedekind apparently wanted to know why the mathematician, as such, should restrict his methods. Philosophy, by itself, does not supply these reasons.

Bernays [1935] is another interesting illustration of the present point. Although his title is "Sur le platonisme dans les mathématiques" there is little concern with traditional philosophical issues. Bernays notes at the outset that some people speak of a "foundational crisis" in mathematics, but it "is only from the philosophical point of view that objections have been raised." This is an interesting contrast to the passage quoted earlier, from Bishop [1975, 507], which was written forty years later (in another context, of course). Both speak of "crisis" in mathematics. For Bishop, the problem was the unchecked use of nonconstructive methods. He thought that the "crisis" was real but mostly unnoticed because of the neglect of philosophy. For Bernays, on the other hand, the situation was the aftermath of various antinomies

7. See Stein [1988] and Wilson [1993a] for illuminating discussions of this historical development, particularly through the work of Dirichlet.

8. An anecdote: An intuitionist once told me that he enjoys teaching graduate-level logic courses. He praised the elegance and power of the Henkin proof of the completeness theorem but then added, with a sigh, "too bad I can't believe it." Too bad indeed.

and formal contradictions. However, Bernays saw no real crisis. In short, Bishop saw an unnoticed but actual emergency, whereas Bernays saw a noticed but merely apparent one.

The primary subject matter of Bernays [1935] is the dispute over impredicative definitions and principles like excluded middle. As we have seen, the methodological principles would be endorsed by realism. Bernays wrote, "The tendency of which we are speaking consists in viewing the objects as cut off from all links with the reflecting subject. Since this tendency asserted itself especially in the philosophy of Plato, allow me to call it 'platonism'" (p. 259). The principles occur almost everywhere: "[T]he application is so widespread that it is not an exaggeration to state that platonism reigns today in mathematics." This is not a statement of the ontological views of mathematicians in 1935, at least not directly. It is a statement of how mathematics is done. Bernays concedes that the principles in question have led to trouble *within* mathematics. For example, the view that all mathematical objects exist independently of the mathematician and the view that all collections of mathematical objects are themselves mathematical objects suggest an unrestricted comprehension principle. Russell's paradox follows and that, at least, is a real problem. One response would be to reject the techniques sanctioned by Platonism and, in effect, to adopt constructivism. This, however, is a "most radical" solution, because if it were followed, mathematics would be crippled: "[M]athematicians generally are not at all ready to exchange the well-tested and elegant methods of analysis for more complicated methods unless there is an overriding necessity for it" (p. 264). Philosophical arguments against realism do not amount to "overriding necessity," nor does Russell's paradox. For Bernays, the real moral of the antinomies is to avoid the tendency to draw rigid philosophical and methodological conclusions from the "Platonistic" mathematics: "We have set forth . . . a restricted platonism which does not claim to be more than . . . an ideal projection of a domain of thought. But the matter has not rested there. Several mathematicians and philosophers interpret the methods of platonism in the sense of conceptual realism, postulating the existence of a world of ideal objects containing all the objects and relations of mathematics. . . . It is this absolute platonism which has been shown untenable by the antinomies. . . . We must therefore give up absolute platonism. But . . . this is almost the only injunction which follows from the paradoxes" (p. 261). In short, there is a middle ground between intuitionism and "absolute Platonism," or there should be: "[T]he characteristic feature of intuitionism is [that of being founded] on the relation of the reflecting and acting subject. . . . This is an extreme methodological position. It is contrary to the customary manner of mathematics, which consists in establishing theories detached as much as possible from the thinking subject. . . . For even if we admit that the tendency away from the [thinking] subject has been pressed too far under the reign of platonism, this does not lead us to believe that the truth lies in the opposite extreme" (pp. 266–267). Note the constant attention to methodology.

Even in Gödel's philosophical writing, the philosophy-first principle is not a dominant theme. The purpose of Gödel [1944] is to *respond* to an explicit, philosophically based attack on mathematical principles. Gödel's argument is that methodological criticisms are based on a philosophy that one need not adopt. Other philosophies support other principles. Gödel did not argue for realism on the grounds of first principles, prior to practice. His philosophical papers [1944], [1964] are lucid articulations of realism, together with arguments that realism conforms well to the practice of mathematics and, perhaps, arguments that realism provides a good guide to practice. Gödel is noted (and notorious) for the argument that the case for the existence of mathematical objects is an exact parallel of the case for the existence of physical objects. As I see it, Gödel's point is that we draw both conclusions on the basis of articulated and successful (mathematical and physical) theories. This is far from philosophy-first.

It would probably beg the present question to reject the philosophy-first principle just because it is not true to mathematical practice, or to the history of mathematics. One can always concede the "data" of practice and history, while maintaining a normative claim that mathematics ought to be dominated by philosophy and, with Plato, Bishop, and others, be critical of mathematicians when they neglect or violate the true philosophical first principles. To pursue this normative claim, the philosopher might formulate a telos for mathematics and then argue either that mathematicians do not accept this telos but should, or else that mathematicians implicitly accept it but do not act in accordance with it. We may be off on a regress, or it may come down to a verbal dispute over what gets to be called "mathematics." In any case, the matter of normativity is a central and baffling item on the agenda of contemporary philosophy. I do not know how this presumed dialectic should continue.

Let us briefly examine a theme opposite to philosophy-first, the thesis that philosophy is irrelevant to mathematics. On this perspective, mathematics has a life of its own, quite independent of any philosophical considerations. A view concerning the status of mathematical objects or statements is at best an epiphenomenon that has nothing to contribute to mathematics, and is at worst a meaningless sophistry, the rambling and meddling of outsiders. If philosophy of mathematics has a job at all, it is to give a coherent account of mathematics as practiced up to that point. Philosophers must wait on the mathematician (perhaps in two senses) and be prepared to reject their own work, out of hand, if developments in mathematics come into conflict with it. Call this the *philosophy-last-if-at-all principle*.

It must be admitted that philosophy-last has something to recommend it. The fact is that many mathematicians, perhaps most, are not in the least interested in philosophy, much less in specific questions of ontology or semantics; and it is mathematicians, after all, who practice and further articulate their field. For better or for worse, the discipline carries on quite independently of the musings of us philosophers.

It might be added that some mathematicians who do "don philosophical hats," as a hobby, present views at odds with their own practice. Cantor's notorious psychologism is but one example. Hersh [1979] suggests that it is typical for a mathematician to be a Platonist during the week, when *doing* mathematics, and a formalist "on Sunday," when there is leisure to think *about* mathematics. Carnap [1950] uses a similar metaphor: "A physicist who is suspicious of abstract entities may perhaps try to declare a certain part of the language of physics as uninterpreted and uninterpretable, that part which refers to real numbers, . . . functions, limits, etc. More probably he will just speak about all these things like anybody else, but with an uneasy conscience, like a man who in his everyday life does with qualms many things which are not in accord with the high moral principles he professes on Sundays" (p. 241). This suggests that it is conducive to mathematics as such to treat, say, numbers *as if* they are part of an eternal, mind-independent realm, even if traditional Platonism, as an articulated philosophy, causes discomfort. For example, the mathematician will wonder how it is possible to *know* anything about this eternal realm. After failing to solve this standard problem with Platonism, our Sunday philosopher is led to views like formalism or psychologism, which are at odds with practice. The practice is resumed Monday morning, as if nothing had happened. Nothing did happen, really.

As indicated by this quote, there is some sentiment for philosophy-last from philosophers. The writings of the Vienna Circle contain pronouncements against traditional philosophical questions, especially those of metaphysics. For Carnap, questions concerning the real existence of mathematical objects are sometimes declared to be "external" to the mathematical language and, for this reason, they are mere "pseudoquestions."[9]

A popular view today is naturalism, characterized by Quine [1981, 72] as "the abandonment of first philosophy" and "the recognition that it is within science itself . . . that reality is to be identified and described" (see also Quine [1969]). The first clause is, of course, the rejection of philosophy-first. The second may be read as an endorsement of philosophy-last-if-at-all, but Quine himself does not go this far. He regards science and philosophy as a seamless web. If he is right, our question loses much of its force, if not its sense. Moreover, Quine's naturalism focuses on science, not mathematics. For Quine, the modern empiricist, the goal of the science/philosophy enterprise is to account for and predict sensory experience—irradiations on our nerve endings. *Science* has the only plausible line on this. *Mathematics* is accepted only to the extent that it is needed for the scientific/philosophical enterprise (perhaps with a little more mathematics thrown in, for "rounding things out"). The parts of mathematics, such as advanced set theory, that go beyond this role are not accepted (as true). Moreover, Quine himself makes proposals *to mathematicians*, based on this overall philosophy of mathematics and science. He suggests, for example, that V = L be accepted in set theory because it makes for a cleaner theory. Presumably, we are to ignore the fact that most set theorists are skeptical of V = L. This is in the spirit of philosophy-*first* with respect to mathematics, even if it is science/philosophy-first.

Some philosophers, like Burgess [1983] and Maddy [1990], [1997], do lean toward philosophy-last with respect to mathematics. They apply naturalism to mathematics and thereby declare that mathematics is and ought to be insulated from much traditional philosophical inquiry, or any other probes that are not to be resolved by mathematicians qua mathematicians. Philosophy-last follows, if one adds that only considerations of methodology matter.

With characteristic wit, and with considerable professional modesty, David Lewis [1993, 15] also explicitly rejects any thought of philosophically based revisions to

9. Carnap's perspective is considered further in chapter 2.

mathematical practice: "I laugh to think how *presumptuous* it would be to reject mathematics for philosophical reasons. How would *you* like to go and tell the mathematicians that they must change their ways . . . ? Will you tell them, with a straight face, to follow philosophical argument wherever it leads? If they challenge your credentials, will you boast of philosophy's other great discoveries: That motion is impossible, . . . , that it is unthinkable that anything exists outside the mind, that time is unreal, that no theory has ever been made at all probable by evidence, . . . , that it is a wide-open scientific question whether anyone has ever believed anything, . . . ? Not me!" Well, me neither. But perhaps this need not be philosophy-last-if-at-all.

The present antirevisionist attitude reflects an important difference between philosophy of mathematics and, say, political theory and philosophy of religion. A philosopher of politics is not to be dismissed for suggesting a moral error in an actual government, like that of Kuwait. We would not expect a philosophical critic to begin "How would *you* like to go down to the palace and tell them they must change their ways?" Or, to the philosopher of religion, "How would *you* like to go to the synagogue and tell them they must change their ways?" In both cases, it is standard practice to recommend that the practitioner "follow philosophical argument wherever it leads," in full knowledge that this suggestion will be ignored, if it is heard at all. Examples like these indicate a difference in attitude (if nothing else) between mathematics/science on the one hand and politics/religion on the other. Many contemporary philosophers, including me, believe that scientists and mathematicians usually know what they are doing, and that what they are doing is worthwhile. I do not attempt to further articulate and defend this deferential attitude here.[10]

On the other hand, in adopting an antirevisionist perspective, I do not mean to worship mathematics and mathematicians. No practice is sacrosanct. As fallible human beings, mathematicians do occasionally make mistakes, even systematic mistakes; and some errors can be, and have been, uncovered by something recognizable as philosophy. The present orientation is that any given principle used in mathematics is taken as correct by default, but not incorrigibly. The correctness of the *bulk* of mathematics is a well-entrenched, high-level theoretical principle.[11]

---

10. To address this problem, one might argue that philosophy of science and philosophy of mathematics are *descriptive*, not normative theories (see Burgess [1992]). Without further elaboration, however, this maneuver does not help. Notice, first, that much of philosophy of religion is descriptive, as well. It deals with actual religions, not (only) what the philosopher thinks religion should be. Moreover, the subject matter of philosophy of mathematics is itself substantially normative, namely, *correct* mathematics, and *good* mathematics. The *logic* of mathematical discourse is an important philosophical topic, and logic—the study of correct reasoning—is normative if anything is (see Shapiro [1997]). A natural rejoinder might be to distinguish a descriptive theory of such norms from a normative theory of norms. The former deals with what the norms are, whereas the latter deals with what the philosopher thinks they should be. I do not pursue this regress further here.

11. The antirevisionist spirit would also suggest that it is unhealthy to use philosophy (alone) to *restrict* oneself or the mathematical community to classical mathematics. If one decides, for whatever reason, that realism is true, it is still a bad move to conclude that there is something toxic about intuitionistic, constructive, or predicative mathematics. There is no reason to oppose a nonclassical program unless it is taken to *replace* classical mathematics. In chapter 6, I join Heyting in urging a more eclectic approach to our subject. Let many flowers try to blossom.

Historically, of course, there has not always been an intellectual wall separating mathematicians and philosophers. Traditionally, most philosophers were aware of the state of mathematics in their time and took it seriously for their work. Parsons [1983, essay 1], for example, notes that rationalism is an attempt to extend the (perceived) methodology of mathematics to all of science and philosophy. Their major opponents, the empiricists, realized that mathematics does not readily fit their mold, and they went to some lengths to accommodate it, sometimes distorting mathematics beyond recognition.

From the other side, recall that many major mathematicians were themselves major philosophers, notably Descartes, Leibniz, perhaps Pascal, and, in more recent times, Bolzano, Russell, Whitehead, Hilbert, Frege, Church, and Tarski. It must be conceded, of course, that some of these thinkers subscribed to a version of philosophy-first. Detailed historical analyses, beyond the present scope, would be needed to determine whether they practiced what they preached in this regard or, indeed, to trace the mutual influence between their philosophy and their mathematics.

In any case, at least some mathematicians were concerned with philosophy and used it at least as a guide to their work. Examples were discussed earlier in connection with the philosophy-first principle. Even if there are no philosophical first principles, philosophy can set the direction of mathematical research. Bernays [1935], for example, can be read as a rejection of philosophy-last, when he wrote that the "value of platonistically inspired mathematical conceptions is that they furnish models [that] stand out by their simplicity and logical strength." In many minds, the current state of mathematics is not good. It is a highly specialized and disoriented discipline, with experts even in related fields unable to understand each other's work. One often hears complaints that mathematics, as a whole, lacks direction. Philosophy, including ontology, might help provide orientation and direction, even if it does not supply first principles.

For a striking example, Gödel claimed that his realism was an important factor in the discovery of both the completeness of first-order logic and the incompleteness of arithmetic. The completeness theorem is an easy consequence of some of Skolem's results. With hindsight, one must do little more than pose the question. Yet Skolem did not do this. The reason can be traced to the different orientations that Skolem and Gödel had toward mathematics, orientations that might loosely be described as philosophical.[12] However, this and other happy cases should be tempered with historical examples of the negative influence of philosophy on mathematics, inspiring researchers to form sects divorced from the mainstream, on the basis of proposals that would inhibit the growth of mathematics.

Once again, someone might attempt to bypass all of this historical analysis and argue for a normative claim, either that mathematics should not be influenced by philosophy, even if it sometimes is, or that mathematics should be influenced by philosophy, even if it usually is not. As before, it is not clear how the normative question is to be resolved. My aim is different. Despite the foregoing antirevisionism,

12. See Gödel's letters to Hao Wang, published in Wang [1974], and the introductions, by Burton Dreben and Jean van Heijenoort, to the completeness results in Gödel [1986]. See also Gödel [1951].

I doubt that the extreme of philosophy-last-if-at-all is healthy, on balance, for either mathematics or philosophy. Whatever one's view on this, however, it does not follow that philosophy of mathematics is a mere handmaiden to mathematics. Philosophers have their own interests, beyond those of their colleagues in other departments, and the pursuit of those interests is interesting and worthwhile. The work of the philosopher of mathematics should dovetail, and sometimes merge, with that of the mathematician, but at least part of it is different work. One job of the philosopher is to give an account of mathematics and its place in our intellectual lives. What is the subject matter of mathematics? What is the relationship between the subject matter of mathematics and the subject matter of science that allows such extensive application and cross-fertilization? How do we manage to do and know mathematics? How can mathematics be taught? How is mathematical language to be understood? In short, the philosopher must say something about mathematics, something about the applications of mathematics, something about mathematical language, and something about ourselves. A tall order.

Although it may be misleading to put it this way, the primary task of philosophy of mathematics is to *interpret* mathematics and thus illuminate its place in the world view. Because much interpretation is linguistic, a prima facie focus is the *language* of mathematics. What do mathematical assertions mean? What is their logical form? What is the best semantics for mathematical language? The answers to these questions determine the terms in which the other questions are to be addressed.

Antirevisionism can be formulated in this framework: it is *mathematics* that is to be interpreted, and not what the philosopher hopes mathematics can be or should be, and not what a prior (or a priori) philosophical theory says mathematics should be. In general, interpretation can and should involve criticism, but here at least, criticism does not come from outside—from preconceived first principles.

It will help illustrate the perspective here to distinguish three types of philosophical issues and tasks. Of course, the borders are rough, even indiscernible in some cases.

First, there are very general philosophical issues, typically concerning all of mathematics. Most of these questions come from general philosophy: matters of ontology, epistemology, and semantics. In this regard, mathematics is a case study for the philosopher, one among many. Some of the problems and issues on the main agenda of contemporary philosophy have remarkably clean formulations when applied to mathematics.

The second group of issues consists of attempts to interpret specific mathematical or scientific results. In recent times, examples come from mathematical logic: the Löwenheim-Skolem theorems, Gödel's completeness and incompleteness theorems, compactness, and the wealth of independence results. To some extent, questions concerning the applications of mathematics fit here, as well. What can a theorem tell us about the natural world studied by science? There are (or were) those who take mathematics to be no more than a meaningless game played with alphanumeric characters, but everyone else holds that mathematics has some sort of independent meaning. But what is this meaning? And when results like the Löwenheim-Skolem theorem and the incompleteness theorems come along, what exactly do they say or imply about the nonmathematical world, about human knowability, and so on? Typically, philosophers cannot go very far in addressing such questions before encountering the more general ones.

The third group of issues consists of attempts to articulate and interpret particular mathematical theories and concepts. Foundational work in arithmetic and analysis fits here. Sometimes, this sort of activity has ramifications for mathematics itself, and thus challenges and blurs the boundary between the disciplines. Interesting and powerful research techniques are often suggested by foundational work that forges connections between mathematical fields. Besides mathematical logic, consider, for example, the embedding of the natural numbers in the complex plane, via analytic number theory. Indeed, whole branches of mathematics have been spawned by foundational activity. Wilson [1992] shows how much of the mathematical foundational activity consists of finding "natural settings" for various structures and concepts—the environment that best illuminates the items under study.

From another perspective, sometimes developments within mathematics lead to unclarities as to what a certain concept is. The example developed in Lakatos [1976] is a case in point. A series of "proofs and refutations" left interesting and important questions over just what a polyhedron is, and this led to foundational work that is at least in part philosophical. Again, consider the task of accommodating or understanding the status of dynamic language in geometry. The attempt to understand the "allowed moves" of compass and straightedge construction led to important insights. For yet another example, recall that work leading to the foundations of analysis led to unclarities over just what a function is, ultimately yielding the modern notion of function as arbitrary correspondence. The questions are at least partly ontological, and so perhaps philosophy can set the direction for at least some of this type of work.

This third group of issues underscores the *interpretive* feature of philosophy of mathematics. We need to figure out what a given mathematical concept *is* and what a stretch of mathematical discourse *says*. The Lakatos [1976] study, for example, begins with a "proof" that consists of a thought experiment in which one removes one of the faces of a given polyhedron, stretches the remainder out on a flat surface, and then draws lines, cuts, and removes the various parts—keeping certain tallies along the way. It is not clear a priori how this blatantly dynamic discourse is to be understood. It is not a lab report of an experiment. What is the logical form of the discourse, and what is its logic? What is its ontology? Much of the subsequent mathematical/philosophical work addresses just these questions.

Similarly, can one tell from surface grammar alone that an expression like dx is not a singular term that denotes a mathematical object, whereas dy/dx may very well denote something (a function, not a quotient)? The history of analysis shows a long and tortuous process of showing just what expressions like this mean.

Of course, mathematics can often go on quite well without this interpretive work, and sometimes the interpretive work is premature and is a distraction at best. Berkeley's famous and logically penetrating critique of analysis was largely ignored among mathematicians—so long as they knew "how to go on," as Wittgenstein might put it. In the present context, the question is whether mathematicians must stop mathematics until they have a semantics for their discourse fully worked out. Surely not.

On occasion, however, tensions within mathematics lead to the interpretive philosophical/semantic enterprise.[13] Sometimes, mathematicians are not all that sure how to "go on as before," nor are they sure just what the concepts are. Moreover, we are never certain that the interpretive project is accurate and complete and that other problems are not lurking ahead.

These considerations temper the antirevisionism urged earlier. Suppose philosophers decide that a certain inference or technique *T* is illegitimate and should not be used in mathematics. Even if they are right, they cannot go on to criticize an actual piece of mathematics unless they know that *T* has been employed in it. However, one typically does not know which principles have been used until at least some of the foregoing interpretive work is complete. Live mathematicians do not speak and write in formal languages, and the underlying logic and ontology of mathematical discourse is not always near the surface. Because lines in mathematical discourse are not usually justified by citing inference rules from a logic text, even attributing a rule of inference to a mathematician involves some interpretation. The Lakatos example, the emerging notion of function, the rigorization of analysis, and the solution of the construction problems in geometry are all cases in point.

To be sure, many mathematicians are interested in the more general philosophical matters, and perhaps one can argue that they should be. Specialization is a good thing, but it can go too far when impenetrable walls are built. It is a question of emphasis and focus. As we have seen, some mathematicians are too quick to accept a philosophy, and some philosophers do not look closely enough at real mathematics.

As far as I can tell, the perspective urged here does not have a preexisting label, and I do not propose to coin one. The word "holism" is so overworked as to be almost meaningless, but I do embrace some holistic elements. Philosophy and mathematics are intimately interrelated, with neither one dominating the other. One cannot "read off" the correct way to do mathematics from the true ontology, nor can one "read off" the true ontology from mathematics as practiced. The same goes for semantics, epistemology, and even methodology. Quine [1960] opens with a quotation from Neurath [1932]: "We are like sailors who have to rebuild their ship on the open sea, without being able to dismantle it in dry dock and reconstruct it from the best components." Quine was right to omit the next sentence in Neurath's text, which is "Only metaphysics can disappear without trace." Metaphysics is an integral part of the "ship" and cannot be exorcized from it. Quine's view, of course, is that there are no sharp borders between philosophy, mathematics, and science, and I agree with this much. Presumably, a philosophical view that is totally divorced from mathematics as practiced should be rejected, and traffic along and across the blurry border is to be encouraged.

So much for my "holism." Notice, on the other hand, that the foregoing considerations are not compatible with many typical "holistic" pronouncements, such as those labeled "epistemological holism" and "meaning holism," nor are they compatible with some of Quine's central theses. Mathematics and philosophy are related and

---

13. I have benefited considerably from discussion with Mark Wilson.

connected, to be sure, but they are not part of a seamless web that is ultimately answerable to (and only to) sensory perception (see Resnik [1997]). There are seams, even if they are fuzzy and even if it sometimes does not matter where they are. More important, sensory perception is not the be-all and end-all of epistemology. Neither the philosopher nor the mathematician faces (only) that tribunal. I agree with the opponents of holism that mathematics and its philosophy are autonomous disciplines, each commanding respect, but I agree with the holist that mathematics and philosophy are interlocked. I propose the metaphor of a partnership or a healthy marriage, rather than a merger or a blending—a stew rather than a melting pot.

As I conceive it, philosophy of mathematics is done by those who care about mathematics and want to understand its role in the intellectual enterprise, in the ship of Neurath. A mathematician who adopts a philosophy of mathematics should gain something by this, an orientation toward the work, some insight into its perspective and role, and at least a tentative guide to its direction—what sorts of problems are important, what questions should be posed, what methodologies are reasonable, what is likely to succeed. The difference between this and first principles is that theses on such matters are defeasible, as mathematics develops and evolves.

2

Object and Truth

A Realist Manifesto

1 Slogans

The topic of this chapter is the matter of realism and antirealism in philosophy of mathematics. What is at stake when one adopts or rejects a view on these matters? Things have gotten murky lately, despite a wealth of illuminating activity in the philosophy of mathematics.

Traditionally, a realist about numbers is someone who earnestly says, or explicitly or implicitly believes, that numbers exist. The antirealist opposition says or believes "Oh no, they don't." Passions run high over this, about as high as they get in contemporary philosophy.[1] Curiously, there is not much attention directed at the questions themselves. It is as if everyone already understands what "existence," "object," and "objectivity" come to here. Armed with this understanding, we are ready to take up arms over the matter. What are we fighting about?

I propose that realism and antirealism be understood not so much in terms of slogans and rhetoric but as types of competing, sober *programs*. Their proponents look for answers to philosophical questions in different places. Both programs begin with the same questions, those enumerated in the previous chapter. What is the semantics, and logical form, of mathematical statements? What is mathematics about (if anything)? How can it be known? How is mathematics applied in the study of the physical world? There are also problems unique to each program. The deepest problems for realism are on the epistemic front: how is it possible for humans to know anything about an eternal (or timeless), abstract, mathematical realm? Conceivably, a realist program and an antirealist program can both succeed, each by its own lights.

1. An anecdote: I once found myself chatting with a logician about a student of his who had just completed a dissertation that defended antirealism in set theory. As luck would have it, the student walked in a few minutes later. After introductions, I suggested that we get together for an argument sometime before he left town. He growled at me, and we never did discuss the matter.

That is, each program might provide good answers to all outstanding problems, answers at least acceptable to its advocates. From a neutral perspective, there would be a standoff. In the meantime, the rhetoric might be submerged and adjudication postponed until the programs have been further executed. Let all of the flowers try to blossom.

Even at the level of slogans, there are two different realist themes. The first is that mathematical *objects* exist independently of mathematicians, and their minds, languages, and so on. Call this *realism in ontology*. The second theme is that mathematical *statements* have objective truth-values independent of the minds, languages, conventions, and so forth, of mathematicians. Call this *realism in truth-value*.

Realism in ontology is sometimes called "platonism," with a lowercase "p." Hellman [1989] dubs the view "objects platonism." The connection with Plato suggests a quasi-mystical connection between humans and an abstract and detached realm, a connection denied by most realists in ontology. Here, I prefer the more cumbersome expression "realism in ontology." On the contemporary scene, realism in truth-value is sometimes just called "realism."

The traditional battles in philosophy of mathematics focused on ontology. Ontological realism stands opposed to idealism, nominalism, and the like. Kreisel is often credited with shifting attention toward realism in truth-value, proposing that the interesting and important questions are not over mathematical *objects*, but over the *objectivity* of mathematical discourse.

A survey of the recent literature reveals no consensus on any logical connection among the different realist theses or their negations. Each of the four possible positions is articulated and defended by established and influential philosophers of mathematics. Once the key notions are developed a little further, this book articulates a variety of realism in both ontology and truth-value. In many respects, my views are a variant of those of Resnik [1981], [1982]. Maddy [1990] and Gödel [1944], [1964] also accept both forms of realism. Chihara [1990] and Hellman [1989] each develop programs best characterized as antirealism in ontology, realism in truth-value. Their idea is to cast mathematics in a language that contains modal operators, with the result that mathematics has no distinctive ontology. However, statements with the modal operators have objective, nonvacuous truth-values (see chapter 7). Dummett [1973], [1977] and the traditional intuitionists are antirealists in both ontology and truth-value (although as we saw in chapter 1, Heyting spoke more of antirealism in ontology, with Brouwer and Dummett defending antirealism in truth-value). I am aware of only one example of a realist in ontology, antirealist in truth-value, Tennant [1987].[2]

It is time to move beyond slogans. The next section characterizes a methodological position, called "working realism," which has been confused with realism in ontology. Working realism is a description of how mathematics is done, but there is little attempt to answer the questions that motivate philosophy of mathematics. Working realism, by itself, has few or no consequences for semantics, ontology, and

2. As intuitionists, these truth-value antirealists demand changes in the way mathematics is done (see chapter 1). In chapter 6, I explore the possibility of truth-value antirealism without revisionism.

how mathematics is applied in the sciences. The strongest versions of working realism amount to claims that mathematics can (or should) be pursued *as if* its subject matter were a realm of independently existing, abstract, and eternal (or timeless) entities. But that is all. Working realism is consistent with antirealism in ontology and with antirealism in truth-value. Anyone who is not out to revise current mathematics is probably a working realist at some level.

Section 3 articulates a more robust program, called "philosophical realism" or "realism" simpliciter, a combination of realism in ontology and realism in truth-value. Philosophical realism is a plan for understanding the structure of part of the ship of Neurath, the part concerning mathematics. The idea is to start the philosophical process by asserting the objectivity of mathematical truth and the independent existence of a realm of mathematical objects. As such, philosophical realism is only a start.

Section 4 is a brief account of antirealist programs, for purposes of contrast. In section 5, some Quinean concepts and ideas on the agenda of contemporary philosophy are woven into the fabric of realism so construed. These include the relativity of ontology, the inscrutability of reference, and the thesis that ontology is determined by the range of bound variables. The closing, section 6, further elaborates the contrast between working realism and philosophical realism by relating that distinction to some formally similar contrasts in the literature on metaphysics and philosophy of science, specifically Carnap's internal/external questions, Arthur Fine's natural ontological attitude, and Putnam's internal realism.

2 Methodology

I define a *working realist* to be a person (or community) who uses or accepts the mathematical inferences and assertions suggested by traditional realism and rejected by others on antirealist grounds. The methodological principles in question were discussed in chapter 1: impredicative definitions, the axiom of choice, general extensionality, arbitrary functions and sets, and classical logic. Working realism is a view concerning mathematical practice or, to be precise, a view concerning how practice is to be described. Resnik [1980, 162] calls this view "methodological platonism."

There are, to speak (very) roughly, several levels of working realism. Each level includes those below it, and the boundaries between them are not sharp. The first, and weakest, level applies to those mathematicians whose practice can be characterized as conforming to the aforementioned principles and inferences. This working realism is purely descriptive and, moreover, such mathematicians themselves may or may not accept the description, were it offered to them. One is a working realist in this sense if one seems to *use* excluded middle, the axiom of choice, and the like, uncritically, even if one does not acknowledge or otherwise admit that one's practice adheres to these principles. We might call this "third-person working realism," because the characterization of practice may be disputed by those who engage in the practice. To continue a point from chapter 1, the analysts Baire, Borel, and Lebesgue were working realists in this weak sense, at least in part. Choice principles, or equivalents, occurred subtly but essentially throughout their work, even though these mathematicians explicitly *rejected* the axiom of choice once it was explicitly formulated. Certainly, they would not accept the characterization of themselves as working realists.

The next level of working realism occurs when mathematicians acknowledge that their current work (more or less) conforms to the items in question. This is "first-person working realism." Because there may be no commitment that their future work will so conform, this working realism is still descriptive. Such theorists agree that, so far, they have been using excluded middle, the axiom of choice, and so on, and all is well; but who knows what the future may bring? This is first-person working realism.

The third level of working realism is normative. The principles and inferences in question are accepted as templates to guide further research and as grounds for criticizing others. These mathematicians hold, for whatever reason, that mathematics *should* conform to classical logic, impredicative definition, choice, and so forth. As we have seen, Dedekind, Bernays, and Gödel were self-acknowledged working realists in this third sense. Baire, Borel, and Lebesgue were not.[3]

There are, perhaps, borderline cases between the various levels of working realism. A "cautious working realist" might make full use of excluded middle, the axiom of choice, and the like, but will keep track of such uses, just in case, and may devote effort to eliminating them, in the interests of security. The normative working realist may take such activity to be a waste of time, energy, and talent.[4] In any case, we do not need this fine a taxonomy here.

As noted in chapter 1, philosophy of mathematics is an *interpretive* enterprise. One must engage in at least some interpretation in order to arrive at working realism. Mathematicians do not speak or write in the formal language of mathematical logic, and the underlying logical forms of real mathematical assertions are not right on the surface. Neither are the rules of inference. The theorist must interpret a given stretch of mathematical discourse *as* a sequence of well-formed formulas. Even after a piece of mathematics has been rendered in a formal language, it may not be clear

3. The differences between the levels of working realism echoes a distinction in cognitive science between "rule-describable behavior" and "rule-following behavior." The former is activity performed *as if* the subject were following a particular rule, with no position on how the activity is actually performed (or should be performed). In the present context, the "rules" are classical logic, impredicative definition, and the like. Advocates of the weaker forms of working realism hold that the practice of mathematics is rule-describable, whereas the strong normative version takes mathematics to be rule-following activity. Of course, there are a number of distinctions to be made here: rules can be explicit or implicit, conscious or unconscious; and there are Wittgensteinian criticisms of rule-following to be dealt with.

4. Questions about what is interesting and worthy of pursuit have ramifications for methodology and for the development of mathematics. In part, this is because such matters weigh on editors at least as much as (if not more than) correctness. Moore [1982, 215] reports an interesting anecdote. In the early 1920s, Tarski proved that the axiom of choice is equivalent to the statement that for every infinite cardinal κ, $\kappa^2 = \kappa$. He sent the report to Lebesgue, asking him to submit it to a journal of the Paris Academy of Science. Lebesgue rejected the note, on the grounds that he opposed the axiom of choice. Trying to be helpful, Lebesgue suggested that Tarski send the note to Hadamard. When Tarski followed this advice, Hadamard also rejected the note, saying that because the axiom of choice is true, there is no point in proving it from the equation.

which principles and rules of inference have been invoked, nor do we automatically know what caveats and restrictions the principles and rules are supposed to have. Logic teachers usually require students to justify each line with an explicit, previously articulated rule, but mathematicians do not write that way. Typically, the rules themselves are not part of the discourse being interpreted. The philosopher or logician understands the discourse *as* an instance of excluded middle, countable choice, and so on.

The situation here is a matter of matching theory to data—curve fitting. The "theory" is working realism and the "data" is actual (correct) mathematical discourse. Kreisel [1967] remarks that it is rare for mathematicians to dispute *individual inferences*, as they occur in practice. The serious and sustained disagreements are usually at the level of *axioms* and *rules* of inference. Kreisel's historical claim may be an overstatement, but it does highlight the important distinction between arguments over the correctness of a given piece of mathematical discourse and arguments over how that discourse is to be described and codified. Working realism is a global position in the latter debate, and as an interpretive project, it is in part philosophical.

Nevertheless, working realism, even the bold normative version of it, is not a full philosophy, at least not by itself. It is a scant beginning to a philosophical program. Working realism is a theory of how mathematics is done and, in the bold articulation, a theory of how mathematics ought to be done, but so far there is no attempt to answer the philosophical questions about mathematics that motivate our enterprise in the first place. There is no ontology, no epistemology, no semantics, and no account of application. Working realism, at all levels, is silent on these matters. The normative version is a statement of how mathematics should be pursued, but nothing yet is added as to why it should be pursued that way. Working realism may give the norms, but it does not give the telos.

To use an overworked term, working realism is more or less "internal" to mathematics, limited to methodology, and it does not do what philosophy of mathematics is supposed to do. We do not get any perspective on the place of mathematics in the ship of Neurath. If there is a coherent antirealist program that does not require revisions in how mathematics is to be done (see chapter 6), working realism is consistent with antirealism. For that matter, it may not even be compatible with realism if that philosophical program is thwarted. When I stated that working realism is only a scant beginning to a program of philosophy of mathematics, this holds for realism and the various forms of antirealism alike.

One might ask about the orientation of working realism toward statements, like the continuum hypothesis and various determinacy principles, which are independent of set theory. In less foundational terms, the issue concerns statements that are logically independent of whatever is held by consensus of mathematicians working in certain fields.

Notice, first, that no one is troubled by the fact that the commutativity of multiplication is independent of the axioms of group theory. This is because, on all accounts, group theory is not about a single structure that is unique up to isomorphism. Rather, group theory is about a class of related structures. The same goes for field theory, topology, and so on. I call such fields "algebraic." The status of independent statements arises (if it does) only in fields like arithmetic, analysis, and perhaps set theory. Mathematicians sometimes call fields like arithmetic and analysis "concrete," but that term has other uses in philosophy. The idea is that each such field is about a single structure, or isomorphism type. For lack of a better term, I call the fields "nonalgebraic."

Of course, I just put the distinction between algebraic and nonalgebraic fields in terms of intended subject matter. This raises matters of ontology, and thereby goes beyond working realism. The problem is to account for the difference between algebraic and nonalgebraic fields in terms of mathematical practice alone.

The orientation toward independent statements is potentially important for the practice of mathematics. If I hold (for whatever reason) that a given statement has a determinate truth-value, then I may devote some effort to figuring out this truth-value (never mind how). If, on the other hand, I do not think the statement has a truth-value, and yet I take the statement to be significant, then I will assert it or not, depending on which option produces the best, the most convenient, the most useful, or even the most elegant theory. By way of analogy, in algebraic areas, one may study theories without the "disputed" item, theories with it, and theories with its negation. At this point, it is not clear whether the criteria one mathematician would use for determining the "truth" of the disputed item are the same as those another would adopt for determining what is best or most convenient.[5]

Again, a working realist is one who accepts (at some level) principles and inferences like the law of excluded middle, impredicative definition, and the axiom of choice. So far, this carries no commitment to there being a fact of the matter concerning, say, the continuum hypothesis (CH). Such mathematicians may assert that CH has a truth-value, that it does not, or even that they do not know (or care) whether CH has a truth-value. Because our working realists do accept the law of excluded middle, they will accept "CH or not-CH," but this is different from the assertion that CH has a determinate truth-value. To pursue our analogy, notice that excluded middle holds in algebraic fields. For example, it follows from the axioms of group theory that either multiplication is commutative or it is not:

$$\forall x \forall y(xy = yx) \vee \exists x \exists y(xy \neq yx),$$

but there is no group-theoretic fact of the matter whether multiplication is commutative. It is commutative in some groups and not in others. In semantic terms, excluded middle and bivalence are not equivalent.

Of course, semantic bivalence does follow from excluded middle together with the main principles of Tarskian semantics. That is, if we take the so-called "T-sentences" as premises and we define F(A) as T($\neg A$), then $A \vee \neg A$ is equivalent to T(A) $\vee$ F(A) (in intuitionistic logic). The latter seems like a good (object-language) rendering of the statement that A has a truth-value. This suggests that the distinction between algebraic and nonalgebraic fields may be articulated in terms of semantics, but here again we go beyond working realism. I conclude that work-

5. See Maddy [1990] for a lucid account of the methodological role of independent statements. I am indebted to Maddy for pressing this issue.

ing realism, as articulated so far, does not entail that independent statements have a truth-value.

Traditional Platonists often remark that, on their view, mathematical objects and ordinary physical objects are the same kind of thing. Resnik [1980, 162], for example, defines an "ontological platonist" to be someone who holds that ordinary physical objects and, say, numbers are "on a par." For working realism there is some truth to this slogan, but it is potentially misleading. In the first place, working realism is not a view about the ontological status of mathematical objects. Working realism is limited to methodology. So, the phrase "on a par" must be restricted to analogies between how mathematical discourse and reasoning is best described and how ordinary discourse and reasoning is best described. The working realist may contend that nouns and variables function much the same way in both contexts. This would depend on whether ordinary discourse conforms to the items in question: extensionality, arbitrary functions and sets, impredicative definition, choice, and classical logic.

When it comes to ordinary language, most of the items are straightforward. Clearly, there are extensional contexts in ordinary discourse, and there is no problem with arbitrary functions and sets. Moreover, impredicative definitions are allowed without controversy in ordinary discourse, as noted in chapter 1. If, for example, someone defines the "old fart" to be "the oldest member of the faculty," there would be no raised eyebrows, or at least none motivated by philosophy. In ordinary discourse, the axiom of choice is also moot, but even in mathematics there is no problem with choice principles when all the domains are finite. Outside of mathematics, one rarely wonders about choice functions for infinite collections. Indeed, if such a question did come up in, say, physics, one would be inclined to think that it is really mathematics that is being done (even if we agree that there is no sharp border between mathematics and physics).

To finish the analogy between ordinary and mathematical discourse (for now), the major remaining theme of working realism is that the *logic* of mathematics is classical. The logic of ordinary discourse, however, is not at all straightforward. Logic teachers quickly become adept at showing beginning students how classical logical connectives differ from their counterparts in ordinary language. The largest gap is probably that between the truth-functional conditional and the various conditionals of, say, English. The two statements, "If Nixon had been more belligerent in Viet Nam, the Communists would have backed down" and "If Nixon had been more belligerent in Viet Nam, a world war would have started" are not both true, and certainly not both true just because the antecedents are false. Of course, those too steeped in elementary logic may think otherwise. To some extent, every truth-functional connective differs from its counterpart in ordinary language. Classical conjunction, for example, is timeless, whereas the word "and" often is not. The inference, from A & B infer A, is not refuted by pointing out that "Larkin is at first and there is a ground ball up the middle" does not entail "Larkin is at first." Similarly, "Socrates runs and Socrates stops" is not equivalent to "Socrates stops and Socrates runs."[6]

6. A previous referee pointed out that temporality can be invoked without mentioning conjunction. The locution "Socrates stops. Socrates runs" is not equivalent to "Socrates runs. Socrates stops,"and

Here, perhaps, we can *make* classical logic fit ordinary discourse by introducing time parameters. "Larkin is at first at t_1, and there is a ground ball up the middle at t_2" does clearly entail "Larkin is at first at t_1." The conditional is not handled so easily, however. A casual survey of the literature reveals no consensus on the logic of conditional sentences. Moreover, ordinary language has features like ambiguity, vagueness, and modality, and again there is no agreement that we can make everything fit classical logic.

There are several possible diagnoses for this situation, but for present purposes they all point in the same direction. First, one may hold that classical connectives and classical logic somehow capture the "deep structure" of both ordinary and mathematical discourse. If so, it is closer to the "surface" in the case of mathematics. A second diagnosis is that classical connectives are highly simplified mathematical models of their counterparts in both ordinary and mathematical discourse. On this view, the logic of *both* contexts is only "more or less classical." Even so, the approximations are much better in the case of mathematics. Statements of pure mathematics, for example, are not usually inflected for tense, even implicitly (if we ignore dynamic language, Turing-machine talk, etc., see chapters 1 and 6). Furthermore, mathematical predicates are not usually vague, and modality is not invoked in a nontrivial, substantive manner. In contrast, one must do a considerable amount of contentious philosophical/semantic analysis in order to get ordinary discourse even approximately into the mold of classical logic. A third diagnosis is that the connectives of both ordinary and mathematical discourse are "stronger" than the truth-functional ones, but this extra strength does not matter in the case of mathematics. The idea is that the semantics of the conjunction always has a temporal component, but this component makes no difference in the case of mathematics, because there are no time-bound mathematical events. Or perhaps the truth of a conditional always requires some sort of connection between antecedent and consequent, but this connection is moot in the case of mathematics. Accordingly, the semantics of both mathematical and ordinary discourse is nonclassical, but the logic of mathematics *looks* classical, because the nonclassical elements play no role there.[7]

Under each of the diagnoses, then, classical logic describes mathematical discourse *better* than it does ordinary discourse. Thus, it is not helpful, and is potentially misleading, to characterize working realism as the linguistic, or logical, treatment of mathematical objects "on a par with" ordinary physical objects. In effect, the orientation of working realism is that mathematical objects are to be treated more like the

---

"Larkin is at first. There is a ground ball up the middle." does not entail "Larkin is at first." This suggests another mismatch between traditional logic and reasoning in ordinary, nonmathematical contexts. In ordinary contexts, the *order* of the premises sometimes matters. The practice of taking premises to be unordered *sets* may work reasonably well in modeling mathematical reasoning, but for studying of ordinary discourse, it may be better to structure premises into ordered sequences.

7. I thank a previous referee for the last suggestion. The major tools of both proof theory and model theory were developed in large part by paying attention to mathematical language (from the perspective of working realism) and not by focusing on ordinary language. This theme recurs throughout this book.

kinds of things traditional Platonists say they are. That is, mathematical objects are to be treated *as if* they are timeless, eternal, and independent of the mathematician; and as if mathematical properties and relations are absolutely precise. This orientation serves to eliminate the complicating features of ordinary discourse, such as temporality and vagueness, and it sanctions impredicative definition, classical logic, and so on.[8] But, as Hans Vaihinger [1913, 73] reminds us, to treat mathematical objects as if they are a certain way is not to hold that they are that way: "We are dealing with a closely woven net, a fine tissue of . . . concepts in which we envelop reality. We achieve a passable success; but that does not mean that the content must take the form of the net woven around it." Once again, adopting working realism may not rule out antirealism.

3 Philosophy

There is a more full-fledged version of realism, which I call *philosophical realism*, or *realism* simpliciter. In line with the conclusions of chapter 1, this philosophy does not provide first principles for mathematics, nor is it a foundation in bedrock, nor does it provide the ultimate justification for mathematics. Rather, realism is a type of program for structuring part of the ship of Neurath, the part concerning mathematics. It provides a framework for addressing philosophical questions about mathematics: to characterize its subject matter (if it has one); to account for how mathematics is learned, communicated, and extended; and to delineate the place of mathematics in our overall intellectual lives, science in particular.

As above, realism has traditionally been characterized as the view that mathematical objects exist independently of the mathematician or the view that mathematical statements have objective, nonvacuous truth conditions also independently of the mathematician. These views, which I call "realism in ontology" and "realism in truth-value" respectively, are a good place to start. Our philosophical realist attacks the philosophical questions by postulating an independently existing realm of mathematical objects and suggesting that mathematical discourse is objective.

There is a natural, if not inevitable, link between working realism and philosophical realism. In endorsing classical logic, and the like, the working realist proposes that the formal languages of now-standard logic texts give something like the deep structure of mathematical discourse—at least for the purposes of describing the methodology of mathematics. Philosophical realism is just the attempt to take these formal languages at face value. Taken literally, the variables of mathematical discourse range over a domain of discourse. This domain is the ontology, or at least a gesture in that direction. Standard semantics suggests genuine, bivalent, nonvacuous truth conditions for the sentences in the language. Thus, philosophical realism is the received view or the first guess or the default philosophy suggested by the combination of working realism and standard logical theory.

---

8. To pursue this characterization of working realism, one can state that the mathematician treats independent statements as if they have determinate truth-values.

To reiterate the obvious, then, the philosophical realist does believe that mathematics has a subject matter. It is about the items in the range of the variables of the formal languages. The first item on the agenda of this program is an account of the nature of the postulated objects. What, after all, are these numbers, sets, and so on? The account would presumably yield truth conditions for the statements of mathematics. For the program to succeed, we also need to be told how a language with these truth conditions can be learned and how humans manage to exchange information about the mathematical realm. How do mathematicians make true assertions most of the time? Clearly, philosophical realism needs an epistemology. That is the second item on the agenda.

At just this point, the Benacerraf [1973] dilemma raises its head. It looks like we cannot accomplish both jobs and remain realists. How can there be a plausible epistemology in which living, breathing organisms come to know anything substantial about an atemporal realm of mathematical objects, all of which are outside the causal nexus? I take it that some humans do have mathematical *knowledge*. How can this knowledge be squared with the ontological claims of philosophical realism? How can we say anything about mathematical objects and have any confidence that what we say is true?

To briefly summarize familiar material, these questions have been developed into outright objections to philosophical realism by arguing for an epistemological claim that it is impossible for thoroughly biological organisms to get information about an acausal realm. The premier exemplar of this claim is the so-called causal theory of knowledge, a thesis that one cannot know about a particular kind of object unless there is some causal link between the knower and at least samples of the objects. Ordinary objects, it is claimed, are known via causal links, but according to realism mathematical objects cannot be. Even though the causal theory is not the only, or even the most prominent, epistemology nowadays,[9] the Benacerraf challenge remains. How can we have mathematical objects and know about them? I take the development of an epistemology to be the central problem facing the program of philosophical realism.

Another major item on the philosopher's agenda is a plausible account of how mathematics is applied in science. After all, mathematics is an important ingredient in most of our best efforts to study physical and human reality. Thus, our realist needs to show us how the mathematical objects relate to the physical and human world, so that mathematical truth can figure in scientific truth, and mathematical knowledge can figure in scientific knowledge.

For realism, the question of applications is a double-edged sword, supporting realism and yet indicating a deep problem for it. First, it is widely agreed that the connection with science is a motivation for realism. Mathematics is thoroughly entrenched in the practice of science, to the point that many hold that there is no sharp border between them. Thus, it is natural to attempt a uniform semantics for both mathemati-

9. See Dieterle [1994, chapter 2] for an application of "reliabilist" epistemology to a realism rather similar to the present one.

cal and scientific languages. If our best interpretation of science points toward realism, then the desire for a uniform semantics would suggest realism in mathematics, as well.

A much-discussed argument for realism, attributed to Quine and Putnam, focuses on the connections between mathematics and science (see, for example, Putnam [1971]). Because mathematics is indispensable for science and because the basic principles of science are (more or less) true, most of the basic principles of mathematics are true as well. This is realism in truth-value. Moreover, if we take the language of mathematics, as reformulated in the idiom of mathematical logic, at face value, then we are committed to the *existence* of numbers, sets, and so forth, and have endorsed realism in ontology. This is a variant of our path from working realism to philosophical realism.

The conclusion here is tentative, at best. Pressure can be applied at either of two joints. First, the desire for a uniform semantics can be fulfilled by developing a similar *antirealism* for both mathematics and science. The philosophical literature contains a number of antirealist proposals in the philosophy of science. Second, a philosopher can reject the uniformity and develop separate semantics for mathematical and scientific languages. Then one would show how the two languages "link up." In other words, the semantics can be complementary, if not uniform (see Tennant [1987], [1997]).

Thus, it is much too cozy to leave things at the Quine–Putnam indispensability argument. We encounter the other edge of the sword. The realist must provide an account of exactly *how* mathematics is applied in science. Indeed, from the perspective of philosophical realism, the main premise of the Quine–Putnam indispensability argument is a big mystery. What does an abstract, acausal, atemporal realm of numbers and sets have to do with the physical world studied in science? We cannot honestly sustain the conclusion of the Quine–Putnam indispensability argument until we know this. It is not enough for the philosopher to note the apparent indispensability and then draw conclusions that spawn more questions than they answer.

In sum, the program of philosophical realism is a tall order. These considerations, and others, have led some philosophers to reject the program and attempt antirealism. In chapters 6 and 7, we will look at some samples to see if there is progress on the common problems, and if there are other problems with these programs that are just as intractable as those of realism.

Because, as noted in chapter 1, philosophy of mathematics is primarily an interpretive enterprise, semantics plays a central role. Philosophical realism is well served by a bivalent, model-theoretic framework, sometimes called "Tarskian."[10] To execute the program, one must somehow specify an ontology, which is to be the range of the variables of suitably formalized mathematical discourse. Presumably, one would specify extensions (if not intensions) for the predicates, relations, and functions of the language. This, in effect, is a model-theoretic interpretation of the language. In

---

10. See Tarski [1933], [1935]. There is some controversy over Tarski's role in the historical emergence of model theory in logic. See, for example, Etchemendy [1988]; Sher [1991]; and Shapiro [1997].

the now-familiar manner, the interpretation determines Tarskian satisfaction conditions for the formalized sentences.

The role of model-theoretic semantics here is not a deep one. Model theory was developed to study (or model) various logical notions, like consequence and consistency. Its central notion is "truth in a model." The only point here is that according to philosophical realism, the relationship between mathematical language and mathematical reality—between word and world—is analogous to the relationship between a formal language and a model-theoretic interpretation of it. The conditions for truth in a model match the ways that meaningful (mathematical) statements get their truth-values. In other words, the conditions for truth in a model represent truth conditions. As Hodes [1984] elegantly puts it, truth in a model is a model of truth; at least for a philosophical realist it is.

The focus on model-theoretic semantics highlights the main problems with philosophical realism. Although the notion of "satisfaction" and the concomitant notion of "reference" are the crucial elements of the semantic picture, model-theoretic semantics has very little to say about these concepts. Model theory provides only the bare structure of how the semantic notions relate to each other. There is a general problem of explaining how mathematical languages are understood—perhaps as a case study for how any language is understood. There is also a problem of understanding how formal languages relate to the natural languages of mathematics. Model theory itself does not address these questions. Thus, even if model-theoretic semantics is the central framework for philosophical realism, by itself it is an empty framework. Model theory is the structure around which answers are developed. The task of the program of philosophical realism is to say more precisely what the domain of discourse is and how we come to know things about it.

Tarski [1944, 345] himself was aware that semantics alone does not provide grandiose solutions to global philosophical problems: "It is perhaps worthwhile saying that semantics as it is conceived in this program . . . is a sober and modest discipline which has no pretensions of being a universal patent medicine for all the ills and diseases of mankind. . . . You will not find in semantics any . . . illusions of grandeur. . . . Nor is semantics a device for establishing that everyone except the speaker and his friends is speaking nonsense." Tarski claimed only that his system provides a method to overcome some traditional difficulties with semantic notions and the "possibility of a consistent use of semantic concepts." Here the framework is part of a philosophical research program, not a solution in and of itself.

In contemporary philosophy, Tarskian semantics has been put to a number of different uses, and it is worth sorting some of these out.[11] There is, first, the Davidson project of giving a theory of meaning for a natural language, like English, in terms of a notion of truth. On this program, "truth" is presupposed as unproblematic. In contrast, Tarski's purpose was to give a coherent, rigorous definition of "truth" and other semantic notions, because such notions had fallen into disrepute (due to the

11. I thank a referee for this point, as well as some of the detail that follows.

antinomies). Tarski's work focused on formal languages. Of course, the "convention T" is common to both programs. Consider one clause:

$$\Phi \vee \Psi \text{ is true (or true in } M\text{) if either } \Phi \text{ is true (in } M\text{) or } \Psi \text{ is true (in } M\text{).}$$

On the Davidson project, this is part of what one must grasp in order to understand the notion of disjunction, truth being presupposed. On the Tarski project, this is part of what one must grasp in order to understand the notion of truth, or truth in M, disjunction being presupposed.

The present orientation is closer to Tarski's. As a working realist, I assume that formal languages more or less accurately render the languages and logical forms of mathematics. Moreover, I assume that these languages are understood—somehow. The model-theoretic semantics is erected on these languages, and in terms of these languages. Put otherwise, from working realism we have that standard formal languages capture something about real mathematical languages. The philosophical realist takes at least some of these languages to be about the mathematical universe. Tarski's work then provides the relevant semantic notions in a remarkably straightforward and unproblematic manner.

I do not claim that a realist is somehow forced to develop a model-theoretic semantics. If one is interested only in semantic notions, like the concept of truth, then there are interesting alternatives available, and some philosophers eschew semantics altogether. In the spirit of naturalism, some wonder whether the central relations between words and the world—reference and satisfaction—are philosophically occult or otherwise lack scientific respectability. Those who hold that "truth" is "robust" are opposed by those who call themselves "deflationists" and "minimalists" (see, for example, Horwich [1990]; Wright [1992]; and Kraut [1993]). The debate rages on. Here, I have nothing to offer on this question of metaphysics. My present concern is with realism in mathematics, and I make a modest proposal to invoke a certain framework. The notion of truth is a straightforward by-product of this framework. Notice that if one believes that there is a realm of mathematical objects and that (formalized) languages of mathematics are about this realm, then there is no further impediment to developing a model-theoretic, truth-valued semantics. Tarski established this much. One simply adds a predicate for satisfaction to a language that one presumably *already understands and uses*, and then defines the semantic notions in the textbook manner. If the defined notions are deflationary then so be it, and if they are robust then so be it.

It might be added that model-theoretic semantics does not *characterize* realism. The model-theoretic program is so benign that there is nothing to prevent an antirealist from carrying it out. Presumably, the program itself would get an antirealist interpretation. Moreover, model-theoretic semantics, by itself, does not even sanction the inferences and principles of *working* realism, unless those same principles are used in the metatheory. For example, one can develop a model theory for an intuitionistic language within intuitionistic set theory, and it should not be surprising that only intuitionistic logic is sound for this semantics. Of course, such a model theory does not play any role in articulating intuitionistic *philosophy*. No one would take the relationship between language and model to be a significant analogue of how sentences of intuitionistic languages are to be understood.

In model-theoretic semantics, the practice is to separate object language and metalanguage. The former is the formal language we are giving a semantics for, whereas the latter is the informal or semiformalized language in which the semantics is carried out. This distinction may not be necessary or even desirable for characterizing semantic notions,[12] but here it is convenient. The framework, conceptual scheme, or simply the language *in which* a program of philosophical realism is cast need not be the framework *about which* it deals. The latter is mathematics itself, or one of its branches. Philosophical realism, as I have formulated it so far, has a truth predicate for mathematics, or for the branch in question. Thus, if the languages are sufficiently formalized and sufficiently rich, then in light of Tarski's theorem on the undefinability of truth, the metalanguage must properly extend the resources of the object language.

Typically, the object language can be faithfully translated into the metalanguage. Because the envisioned metalanguage of philosophical realism should have the resources to make substantial assertions about the ontology that is attributed to the object language, this metalanguage must also contain a faithful representation of the object language. For example, the philosopher should be able to state that there are infinitely many primes and be able to formulate the Goldbach conjecture and the continuum hypothesis.

This observation underscores the rejection of philosophy-first in chapter 1. Philosophy does not supply first principles for mathematics nor a foundation in anything nonmathematical. Its goal is to interpret the mathematics done by real mathematicians, and, in a sense, the interpretation begins from within. On the other hand, someone might feel cheated by philosophical realism. In a sense, one interprets mathematics and outlines its ontology by using (a facsimile of) that same mathematics. An all-too-cozy and all-too-lazy philosopher might just say that "6" refers to 6, that the variables of arithmetic range over the natural numbers, and leave it at that. But such a philosopher should not pretend to have shed any light on anything. One ought to feel cheated if the realist has done no more than provide a model-theoretic semantics, especially if it is done in much the same terms as the object language. If that is all the philosopher is willing to say, then one can wonder what it means to be a realist. The "all-too-cozy" pronouncements are surely deflationary or minimal and are not of much use on the philosophical questions under study here.[13] One cannot hope to answer every philosophical question about every subject all at once, but one

12. There have been a number of interesting attempts to develop rich, formal languages that contain their own semantical notions (e.g., Kripke [1975]; Gupta and Belnap [1993]). Some of these involve a many-valued semantics and reject the law of excluded middle (and so are prima facie inconsistent with working realism). For purposes of characterizing semantical notions, it may also be possible to get by without a comprehensive truth predicate. For example, for any natural number n, there is a formula of set theory that characterizes "truth for set-theoretic formulas with fewer than n symbols."

13. Wright [1992] is an insightful account of what sense can be made of the realist/antirealist debate in light of such "minimalist" conceptions of truth.

can demand some progress of philosophy and ontology, some dissolution of puzzles. Model-theoretic semantics, by itself, does not provide such progress.

The model-theoretic framework allows a relatively neat distinction between algebraic and nonalgebraic branches of mathematics. A field is nonalgebraic if it has a single "intended" interpretation among its possible models or, more precisely, if all of its "intended" models are isomorphic (or at least equivalent). This, together with excluded middle (in the object-language theory) yields bivalent truth conditions in the usual manner. Every sentence, including those independent of the axioms, has a truth-value. In such branches, "truth" amounts to "truth in the intended models." Of course, the philosophical puzzle now concerns how we manage to apprehend, pick out, or communicate the intended interpretations, and how we manage to know things about it (or them)—of which more later. A field is algebraic if it has a broad class of (nonequivalent) models. Group theory, for example, is about all groups. In practice, however, the mathematician does not use the language of group theory. The group theorist works in a (set-theoretic) metatheory in order to study groups. It is an open question whether this metatheory is itself algebraic or, for that matter, whether there are any nonalgebraic fields.

What of the philosophical metalanguage, the framework in which philosophical realism is developed? One can start the grades-of-realism discussion at this level. In particular, one can be a working realist about the framework, explicitly or implicitly adopting classical logic, impredicative definition, and so on. A more ambitious project, perhaps, would be a program of philosophical realism for the metalanguage, which would include a metasemantics, metaepistemology, and so forth. This would be philosophical realism about the language of philosophical realism. The grades-of-realism discussion can be restarted yet again, this time about the metaphilosophical meta-metalanguage. Alternately, one can adopt a program of antirealism at one of the advanced stages. No doubt, it would seem odd to give an antirealist interpretation to a metalanguage that is itself used for a realist program for an object-language theory of mathematics. One with realist leanings might feel cheated or betrayed by such a move. But this possibility cannot be ruled out a priori.[14]

In any case, the regress of metalanguages cannot go on forever. Life is too short. Eventually, usually rather quickly, one comes to a metalanguage that is just *used*, and used without benefit of more semantics, epistemology, and the like. At some point we stop interpreting. To paraphrase Quine, perhaps this is where theorists "acquiesce" into their "mother tongue." Here, the "mother tongue" is our background mathematical and philosophical vernacular. One can be a *working* realist about this language, of course, even a bold, normative working realist, because working realism is more or less internal. It is just a matter of adopting standard renderings of accepted practice. Moreover, it is not that we cannot go on to pro-

14. Hellman [1983] suggests that realism (in philosophy of science) should be formulated in such a way that antirealists cannot reinterpret everything the realist says ("from within") to their own satisfaction. The situation brings to mind Berkeley's claim that he can accept everything the materialist (or dualist) says about physical objects, suitably (re)interpreted of course. I do not see a need to stop contemporary counterparts of Berkeley in their tracks, and the attempt to do so may be futile.

vide a semantics and epistemology for this framework. If pressed, we might take on a program of philosophical realism at this stage. But we do not. To paraphrase Wittgenstein, surely out of place, one might say that the background language is used without the "justification" of more semantics and more philosophy. We use the language without (more) interpretation, but this is not to say that we use it "without right." We understand the language without interpreting it. In the chosen metaphor of holism, everyone must stand somewhere *on* the ship of Neurath. This metalanguage is where the philosopher stands.

The old formalist Curry (e.g., [1950, 11–12]) puts the role of the "mother tongue" well:

> Whenever we talk about a language it is said that we must do so in a second language. . . . [T]he first language is called the *object language*, the second the metalanguage. If then one talks about the metalanguage, one has to do so in a third, meta-metalanguage and so on. But this way of speaking ignores one particular fact: *viz.*, that any investigation . . . has to be conducted, not in an arbitrary metalanguage but in *the communicative language which is mutually understood by speaker and hearer*. I shall call this language the *U-language*, i.e., the language being used. . . . It is not enough to say that the U-language is English. . . . [The U-language] is a growing thing. As we proceed we shall modify it, add to, and refine it. . . . We can never transcend [the U-language]—whatever we study we study by means of it. It cannot be exhaustively described, and it can lead to contradiction if carelessly used. . . . It follows from the foregoing that there is no such thing as a meta-U-language.

Here, I follow Curry's advice and his terminology. The U-language contains the language of mathematics, and yet we use it to try to make progress on philosophical questions. The circle is not vicious.

This dovetails with the conclusions of chapter 1. Philosophy of mathematics does not supply first principles for mathematics, nor does philosophy operate from a privileged, more secure perspective. Mathematics and philosophy of mathematics are interlocked but autonomous disciplines. In philosophy, we try to interpret the language of mathematics and try to understand its role in the intellectual enterprise.

4 Interlude on Antirealism

From the present, programmatic perspective, let me briefly contrast philosophical realism with the loyal opposition, the various antirealist programs. This is a preview of coming attractions, in chapters 6 and 7. The concern of this book is only with views that take the full range, or almost the full range, of contemporary mathematics seriously. It is a basic datum that the bulk of mathematics is legitimate, whatever this legitimacy may turn out to be, and it is a basic datum that mathematics plays a significant role in our intellectual lives, however that role is ultimately characterized and understood. Mathematics, and its role in our overall theorizing, is something to be interpreted and explained, not explained away.

Given the antirevisionist theme of this book (see chapter 1), I limit my attention to views that accept *working* realism. It is real mathematics that is to be interpreted and explained, not what a philosopher says mathematics should be. Working realism is just the result of accepting the common rendering of mathematical discourse into the idiom of logic books, as least as an attempt to describe the methodology. Our question is what to make of the mathematics. The various programs make different things out of working realism.

Philosophical realism is a type of research program. In like manner, I would describe antirealism as another type of program, one that attempts to answer philosophical questions about mathematics and its place in our intellectual life without postulating that mathematical truth is objective, bivalent, and nonvacuous, and without postulating that the subject matter of mathematics is an objective, mind-independent realm. Often, this involves a non–model-theoretic semantics for mathematical languages. To put it baldly, if the variables of a language of mathematics are not to be understood as ranging over a domain of discourse, then they must be understood in some other way—if, as presumed here, the language is understood at all. This yields antirealism in ontology. And if the bulk of the justified assertions of mathematics are not true, then some other account of which assertions are warranted or legitimate is requisite—if, as presumed, mathematics is significant. This is antirealism in truth-value. Either way, then, a non–model-theoretic semantics is appropriate.

On the contemporary scene, I believe that the main motivation behind antirealist programs is the prima facie difficulty of obtaining an epistemology that squares with its ontology. Some antirealist programs do begin with what appears to be a more tractable line on the epistemological front. Some, like Field [1980], [1984], regard mathematical knowledge as a variant of "logical knowledge," and others, like Hellman [1989] and Chihara [1990], invoke logical modality (see chapter 7). Still others, like Tennant [1987] and Dummett [1973], attempt to replace the notion of mathematical *truth* with proof, or warranted assertability (never mind, for now, that these last two are revisionist programs; see chapter 6). At least prima facie, an epistemology of logical knowledge is not problematic, and warranted assertability may not be either. It depends on what a warrant is. With each program, the goal is to show how its epistemology can underwrite a mathematics rich enough for science and its foundations, and to do so without introducing notions as problematic as those of realism. In short, the goal is to reproduce a rich mathematics while keeping the epistemology tractable.

This, I believe, is a tall order, about as tall as the demands on a program of realism. I submit that it is not wise to adjudicate the simple question of realism versus antirealism without detailed consideration of the specifics of one or another of the programs.

5 Quine

I propose to weave some familiar Quinean themes into the fabric of realism, partly in a non-Quinean way. I have in mind the relativity of ontology, the inscrutability of reference, and the thesis that ontology consists of the range of bound variables (see, for example, Quine [1969]). Each doctrine amounts to something different at the levels of working realism and philosophical realism.

Let us begin with working realism. Quine himself has remarked that his thesis that relates ontology to bound variables is not deep. The idea is only that the existential quantifier of classical logic is a good gloss on the ordinary word for existence (in the U-language). A term denotes something just in case it can be replaced by a variable in existential generalization, and if a variable can be replaced by it in universal instantiation. An object exists, or is in our ontology, just in case it is in the range of a bound variable. Now, working realists surely have a predicate for "natural number," call it N, and there is a term for zero, 0. They hold that zero is a number: $N0$. It follows by logic that $\exists xNx$. Thus, for the working realist, numbers exist. Period.

Of course, things are rarely this simple. Ordinary languages, even those used by mathematicians, are not sufficiently articulated to make it determinate just what the ontology is. For example, there are surely parts of the language that one does not wish to take literally or at face value. Those parts are not to be rendered in a formal language in the syntactically most straightforward way. If I say that Bob Dole has only a few leadership qualities in common with Ronald Reagan, I am not automatically committed to qualities; when Three Dog Night says that one is the loneliest number, they may not be embracing numbers, and certainly not lonely numbers. The criterion for ontology seems designed for languages like those in logic books, and the criterion "works" only for such languages. Quine [1981, 9] himself notes that the notions of ontology and ontological commitment do not comfortably fit ordinary discourse: "The common man's ontology is vague and untidy. . . . We must . . . recognize that a fenced ontology is just not implicit in ordinary language. The idea of a boundary between being and nonbeing is . . . an idea of technical science in the broad sense. . . . Ontological concern is . . . foreign to lay culture, though an outgrowth of it." This is of a piece with Curry's remarks that one cannot precisely describe the U-language.

Quine allows ontologists room to regiment their language, by paraphrasing parts of it and eliminating others. The process of regimentation is to systematically render sentences of natural language into the idiom of logic. Truth be told, ontologists are only supposed to pretend or envision that they have regimented their language, because no one actually speaks in a regimented language. Quine and his followers continue to write in ordinary English. The purposes of the regimentation exercise include elucidating the logic and assessing the ontology. One of the goals is economy. For example, variables that range over numbers can be replaced by variables that range over sets, so we need not hold that both numbers and sets exist, just sets. If we are "ontologically committed" to sets anyway, we need not embrace numbers as well, because some sets will do as numbers.

Ontological relativity and the inscrutability of reference arise at this level because there may be more than one way to regiment the same part of a given language. Because regimentation is akin to translating the language into the idiom of logic, the theses in question are corollaries of the indeterminacy of translation. Famously, real numbers can be thought of as Cauchy sequences of rational numbers (and those interpreted as pairs of integers), or else real numbers can be thought of as Dedekind cuts, certain sets of rational numbers. Thus, it is inscrutable whether "π" in the mathematical vernacular refers to a certain Dedekind cut or to an equivalence class of Cauchy sequences. Ontological relativity relates to the fact that the entire range of the variables can differ in different regimentations, and there is no fact of the matter which is correct. In a well-known illustration, Quine suggests that one can paraphrase an entire language of physics so that all variables range over sets. According to that regimentation, the only things that exist are sets.

All this is at the level of working realism. One need not envision a model-theoretic semantics (in a separate metalanguage or U-language) in order to tighten up a stretch of discourse or render it into the idiom of logic. For example, a decision to take real numbers to be Dedekind cuts can amount to no more than an allowed paraphrase. Although we surely have too many grades of realism on our chart already, this observation suggests one more. *Committed working realists* are theorists who envision that their language, after regimentation, will still have variables ranging over some mathematical objects, typically numbers or sets. Committed working realism is, in effect, a decision to *use* mathematical language and take at least some of it at face value, not to recast the ontology in nonmathematical terms.

Quine insists that a fully regimented language be extensional and first-order and that it employ classical logic. Against this, some theorists favor higher-order logic (Shapiro [1991]), and some adopt primitive modal terminology (e.g., Field [1984], [1991]; Hellman [1989]; Chihara [1990]; see also Shapiro [1985] and chapter 7). Because the U-language (à la Curry) certainly contains such items, the question is whether the disputed terminology is to be used in the envisioned regimentation—in the idiom into which the vernacular is rendered.

Any of these options would alter and complicate the picture of ontology through bound variables, because decisions concerning the resources in the regimented language affect the ontology. Typically, one can reduce ontology by, in effect, increasing ideology. For example, some theorists suggest that a primitive modal operator eliminates a need for mathematical entities. With intuitionistic mathematics, things are not even this simple. Perhaps, with Parsons [1983, essay 1], each kind of quantification marks a different kind of "ontological commitment." Presumably, the trade-offs involved in the various alternatives are to be judged on some sort of holistic grounds and, once again, there is no reason to think that there will always be a clear winner. It is not even clear what the rules of this holistic game are. In any case, the relativity of ontology seems to be the relativity of ontology/ideology.

From the internal perspective of (committed) working realism, one can specify what *kinds* of mathematical entities exist, but there is a limit to the ability to circumscribe the *extent* of these entities or, in other words, there is a limit to one's ability to specify the *range* of the variables. Generally, one can describe ontology for terminology that has been paraphrased into something else. For example, if, through regimentation, terminology for numbers is cast in terms of sets, then one can assert that the extension of "the natural numbers" is the finite ordinals, and one can specify that the extension of "the real numbers" is the Dedekind cuts. There is, however, a limit to the working realist's ability to state the range of variables that are taken at face value, not paraphrased. For example, it is typical for set theory to be taken as the language of foundations. In effect, one envisions that the whole of mathematics is regimented into the language of set theory. The ontology is V. Working realists who accept this can, of course, assert the existence of sets or even particular sets, such as $0^{\#}$ or a measurable cardinal. They can also relate this ontology to certain items within the ontology: "V consists of the empty set, its powerset, the powerset of that, and so on—through the ordinals." But, with Russell, if asked the general question, "which sets exist?" or "which ordinals exist?" our working realist has but one answer, "all of them." If asked "how many sets exist?" or "how many ordinals exist?" the answer is "a proper class." Not very helpful.

Normally, variable ranges are specified in *semantics*. The central component of model theory is the domain of discourse. That is what the language is about. Model theory is typically done in a separate metalanguage or U-language, and we turn to the perspective of *philosophical realism*. To be sure, Quine's own writings are filled with attacks on ordinary semantical notions, like "meaning" and "synonymy." Model-theoretic semantics is, of course, a respectable mathematical enterprise, but even this is not emphasized in Quine's works concerning logic.

Nevertheless, the theses under consideration here—ontology through variables, the relativity of ontology, and the inscrutability of reference—are meaningful in the context of philosophical realism and are substantially correct.[15] The ontology of a (nonalgebraic) theory consists of the range of its variables. That is, the ontology is the domain of discourse of the interpretation of the object language. The relativity of ontology and the inscrutability of reference begin with the fact that one can give different models for the same theory, and in some cases there is no fact of the matter which one is correct. Because a model-theoretic interpretation can be understood as a translation of the object language into the informal language of set theory, the two theses are again seen to be corollaries of the indeterminacy of translation.

In effect, no object-language theory determines its ontology all by itself. In any model-theoretic semantics worthy of the name, isomorphic structures are equivalent (see Barwise [1985]). This is a manifestation of the slogan that logic is "formal." Thus, given any model of a fully regimented language, one can specify a different but equivalent model. It is a simple matter of changing the referents of some of the terms and making straightforward adjustments, a routine exercise for those who have had a course or two in mathematical logic. The best one can achieve is that all models of a theory are isomorphic, in which case the ontology is determined "up to isomorphism," and we get this much only if the domain of discourse is finite, or else the ideology of the object language is stronger than first-order. With first-order object languages and infinite domains, the Skolem paradox rears its (ugly) head at this point. If an object-language theory has at least one infinite model, then, by itself, it cannot even determine the *cardinality* of its ontology, let alone the extension of it.

To pick up a recurring subtheme of this book, theses like the inscrutability of reference and the relativity of ontology may be artifacts of the fact that formal (or regimented) languages of logic books are the philosophical paradigms of natural languages (or U-languages) and that model theory is the philosophical paradigm of

---

15. In chapter 4, I will show that if we construe a language of mathematics as about a structure or a class of structures, then there is no inscrutability. In effect, the insights behind the inscrutability of reference are shared by structuralism, but different conclusions on the nature of objects and reference ensue.

semantics. If ordinary discourse does have mechanisms to fix ontology and to make reference "scrutable," these mechanisms are not registered in formal or regimented languages, nor in model theory. Now, as before, I suggest that formal languages, and mathematical logic in general, were developed with focus on the languages of mathematics and not on ordinary discourse. As I argue in subsequent chapters (3 and 4), in mathematics reference and ontology really are determined only "up to isomorphism." Structure is all that matters. So, in this respect, formal languages and model-theoretic semantics are good models for the U-languages of mathematics. The extent to which this technique for assessing ontology (and logic) apply to ordinary and scientific discourse remains to be seen (see chapter 8).

Potentially, there are differences (and possibly conflicts) in assessing ontology at the various levels of realism. In the framework of philosophical realism, the ontology of an object language is typically specified in a metalanguage (or the U-language). If the metalanguage is sufficiently regimented, then it has its own ontology, the values of *its* variables. There is no a priori reason to insist that the ontology of the metalanguage be the "same" as that of the object language. Presumably, the metatheory must contain at least a rudimentary theory of sets, in order to characterize domains of discourse. Even if the object language also countenances sets, it is possible that the metatheory countenances "more" of them.

To take an example, suppose that an object-language theory contains substantial set theory, say ZFC (Zermelo-Fraenkel set theory plus the axiom of choice). An interesting puzzle arises. By assumption, the regimented object language contains variables that range over sets and, of course, it has a universal quantifier, "for every set, . . ." The working realist who adopts this language will surely say "sets exist." Moreover, from this perspective, the quantifier is taken literally—it ranges over *all* sets. What else is a working realist going to say? This is what it is to *use* the language of set theory, and take it at face value. But suppose now that one becomes a philosophical realist and develops a metatheory. On the present account, a *domain* is specified for the variables of set theory. That is, in the metatheory the domain of ZFC is an object. Russell's paradox indicates that this domain cannot be a member of itself. So the metatheory countenances at least one collection of sets that, according to this same metatheory, is not in the domain of set theory. Indeed, from the perspective of the object-language theory, the domain of the original theory is not a set and, thus, from the perspective of ZFC, the domain *does not exist* (as a set). It is a theorem of ZFC that $\neg\exists x\forall y(y \in x)$. So much for the idea that the ontology of the object-language theory includes every collection.

The moral, I believe, is that in adopting a metatheory, one can expand ontology. In effect, we may require more sets in order to provide a set-theoretic interpretation of the original theory. This is in line with Curry's observation that a U-language is a "growing" thing and that it is impossible to describe it with the precision of a logic text. Of course, one can take the variables of a regimented metalanguage to range over "all sets," for, again, this is just to use the metalanguage, and take it literally. But there is nothing (except boredom or disinterest) to prevent the formulation of a meta-metalanguage with an even greater ontology. One cannot, in the same breath, take set variables to range over all sets and have a fixed domain, a set, in mind as the range of the variables.[16] This is the danger of what Bernays calls "absolute platonism."

Notice, on the other hand, that an increase in ontology with a move from an object language to an adequate metalanguage, or from working realism to philosophical realism, is not inevitable. There are other ways to go about it. Suppose that the object-language theory is regimented as *second-order* set theory with variables ranging over both sets and (proper) classes and relations. As far as ontology goes, a model-theoretic semantics for this language does not require any more than second-order set theory, with the same universe of discourse and the same class of proper classes. On this plan, we introduce a notion of satisfaction defined over proper classes, and so models whose domains are proper classes are countenanced.[17] Notice that this satisfaction relation cannot be formulated in ordinary (object-language) second-order set theory. Indeed, because set-theoretic truth is just "satisfaction by the universe," truth (for the object language) can be defined from "class satisfaction."

Tarski's theorem indicates that a metatheory must have resources beyond those of its corresponding object-language theory. Here we have a choice between an increased ontology (more sets) and an increased ideology (a richer satisfaction relation). It seems that there are many trade-offs in this enterprise, and these must be compared somehow.

6 A Role for the External

I have emphasized that working realism is more or less internal to mathematics. The working realist accepts some erstwhile disputed methodological principles. To arrive at working realism, one must do some philosophical interpretation but not much. There is no attempt to answer philosophical questions about mathematics: what it is about, how it is known, how it relates to science, and so on. The common goal of the programs of philosophical realism and the various antirealisms is to address such questions. Thus, in a sense, those programs are external to mathematics. The questions are not necessarily addressed by mathematicians as such and are not to be decided by mathematical techniques.

Other writers have delimited internal/external contrasts. Typically, the "internal" perspective is internal to something acceptable, whereas the "external" perspective is rejected as illegitimate, or even incoherent. Examples include Carnap's internal questions versus external pseudoquestions, Arthur Fine's natural ontological attitude versus realism, and Putnam's internal realism versus metaphysical realism. It will

16. See Hellman [1989] and Parsons [1983, essays 10–11] for a similar account of the variables of set theory.

17. See Shapiro [1991, especially chapters 5–6]. The present plan works only for the "standard semantics" of second-order languages, in which the predicate variables range over every subclass of the domain of discourse. For nonstandard, Henkin semantics, one needs to specify a range of the predicate (and relation) variables, and in large models this is a class of proper classes, a third-order item.

prove instructive to contrast these distinctions with the present one between working realism and philosophical realism. Of course, I do not claim to do full justice to the detailed and subtle work of these authors.

One important disanalogy between the present perspective and those of the others is that I regard the external perspective to be legitimate. This section serves to link the theme of the present chapter with that of chapter 1. One conclusion there was that mathematics and its philosophy are interlocked but autonomous disciplines. Philosophical realism is a program for one of those disciplines.

It is natural to start with Carnap (e.g., [1950]), despite any disrepute his program may be under. According to Carnap, there is a crucial prerequisite to asking ontological questions like "Do numbers exist?" and "Do electrons exist?" The philosopher, mathematician, or scientist must first formulate an appropriate "linguistic framework" with explicit and rigorous syntax and rules of use. In the case of mathematics, a linguistic framework is similar to a formal language and deductive system. For example, first-order Peano arithmetic can be a framework for arithmetic. For scientific or everyday discourse, empirical rules that involve observation would be added. In a sense, the notion of linguistic framework is a precursor to regimentation in the program of Quine, Carnap's student. Here, too, I presume that it is a question of envisioning a framework, rather than actually using one for serious communication. We do not really get out of Curry's U-language.

Once a framework has been formulated, there are two sorts of ontological questions one might ask; or, better, there are two senses to questions of ontology. Suppose one asks, "Are there natural numbers?" On one level, this may be regarded as a simple version of a question like "Is there a prime number greater than one million?" In this case, the ontological question is *internal* to the natural-number framework and the answer is "yes," on utterly trivial grounds. Carnap says that in the internal sense, the existence of numbers is an analytic truth, because it is a consequence of the rules of the framework. This is similar to the working realist's assertion that numbers exist—although working realism does not make reference to this or that linguistic framework.

In traditional philosophy, the question about the existence of numbers is regarded as anything but trivial. One asks whether numbers *really exist*, or whether numbers exist independently of the mind and independently of any given linguistic framework. In asking the traditional question, we do not wonder whether "numbers exist" is a theorem of a particular formal deductive system but whether it is true. In present terms, the traditional question is whether the envisioned natural-number framework accurately describes an intended domain of discourse. This, Carnap says, is the external question, which he sometimes calls a "pseudoquestion." When considering a framework as a whole, the only legitimate question we may ask is whether to accept or adopt the framework, but this is a pragmatic matter, not susceptible to an all-or-nothing, yes-or-no answer: "The acceptance of a new kind of entity is represented . . . by the introduction of a framework of new forms of expression to be used according to a new set of rules. . . . The acceptance cannot be judged as true or false because it is not an assertion. . . . [T]he acceptance of a linguistic framework must not be regarded as implying a metaphysical doctrine concerning the reality of the entities in question" (Carnap [1950, 249–250]).

To illustrate this program, Carnap envisions a "thing" framework for making ordinary assertions about ordinary objects:

> If someone decides to accept the thing language, there is no objection against saying that he has accepted the world of things. But this must not be interpreted as if it meant his acceptance of a *belief* in the reality of the thing world; there is no such belief or assertion or assumption, because it is not a theoretical question. To accept the thing world means nothing more than to accept a certain form of language, in other words, to accept rules for forming statements and for testing, accepting, or rejecting them. . . . But the thesis of the reality of the thing world cannot be among those statements, because it cannot be formulated in the thing language or, it seems, in any other theoretical language. (Carnap [1950, 243–244])

In other words, the good "theoretical" questions must be *internal* to a more or less explicit linguistic framework. Traditional ontological questions fail this test, miserably. In this regard, an opponent, a metaphysician "might try to explain . . . that it is a question of the ontological status of numbers; the question of whether or not numbers have a certain metaphysical characterization called reality . . . or status of 'independent entities.' Unfortunately, these philosophers have so far not given a formulation of their question in terms of the common scientific language. Therefore . . . they have not succeeded in giving to the external question . . . any cognitive content. Unless and until they supply a clear cognitive interpretation, we are justified in our suspicion that their question is a pseudo-question" (Carnap [1950, 245]).

Let us pretend that there is an explicit, rigorous linguistic framework for mathematics, or for one of its branches. Call the framework *L*. For Carnap, the legitimate pragmatic question is whether to accept *L*. An affirmative answer is a main thesis of *working realism*, and this much is beyond question here. One can still ask whether *L* captures mathematics as practiced, at some level, but even this question is not raised here—especially because *L* is something of a philosopher's fiction. I do not know the Carnapian status of questions of how a given linguistic framework relates to a preformal practice, the time before the rules of the framework were explicitly formulated.[18]

To be sure, the present programs of philosophical realism and the various antirealisms are "external" to mathematics in Carnap's pejorative sense. The philosophical programs are attempts to reach beyond both the internal questions and the pragmatic questions. So this book runs afoul of the Carnap program. Nevertheless, the discrepancy can be tempered and, I think, usefully explained. A closer look at some of Carnap's writing reveals that in attacking external questions, much of his fire is

18. In a collection of notes entitled "What does a mathematical proof prove?" (published posthumously in [1978]), Lakatos distinguishes three stages in the evolution of a branch of mathematics: the preformal stage, the formal stage, and the postformal stage. The issue of how a formal system relates to the practice it formalizes is an inherently *informal* matter in the last stage. See chapter 6 of this book.

directed at what I call the philosophy-first principle, a thesis that philosophical principles *determine* the proper practice of mathematics (see chapter 1). On such a view, the philosopher either provides the ultimate *justification* for mathematics as practiced, or else demands revisions in mathematics, to get it to conform to the true philosophy. Carnap takes external questions to have just this motivation:

> [T]hose philosophers who treat the question of the existence of numbers as a serious philosophical problem . . . do not have in mind the internal question. . . . [I]f we were to ask them: "Do you mean the question as to whether the framework of numbers, *if* we were to accept it, would be found to be empty or not?", they would probably reply: "Not at all; we mean a question *prior* to the acceptance of the new framework." (p. 245)

> Many philosophers regard a question of this kind as an ontological question that must be answered *before* the introduction of the new language forms. The latter introduction, they believe, is legitimate only if it can be justified by an ontological insight supplying an affirmative answer to the question of reality. (p. 250)

> They believe that only after making sure that there really is a system of entities of the kind in question are we justified in accepting the framework. (p. 253)

As in chapter 1, I join Carnap in rejecting the philosophy-first orientation. The program of philosophical realism is not prior to mathematics in any sense. If anything, the metaphor is that philosophical realism comes after mathematics, or during mathematics. However, it does not follow from the rejection of philosophy-first that there is no legitimate perspective from which to ask at least some traditional-sounding philosophical questions. Are the questions themselves out of order, or is the problem with the explanatory and justificatory role that philosophy once took upon itself? On the present view, the philosopher stands somewhere *on* the ship of Neurath (where else?) and tries to shed light on another part. One does not have to embrace philosophy-first in order to regard questions of the ontology, epistemology, and application of mathematics as serious and substantial questions. And they are not entirely mathematical questions.

There has been a lot of criticism of the Carnap program, and I do not wish to add any more (even if I could). Notice, however, that despite Carnap's antimetaphysical attitude, his program does presuppose substantive *answers* to some traditional philosophical questions. Because a linguistic framework has an explicitly formulated syntax and explicit rules of use, it is like a deductive system in a formal language. Natural languages of mathematics do not seem to have such neat rules and, even if they do, the rules are not explicit. Carnap seems to hold that the practice of mathematics resembles—or ought to resemble—moves in an explicit linguistic framework. Indeed, the thesis seems to be that rule-following is of the essence of mathematics. Do we have philosophy-first here? Should mathematicians suspend their comfortable, but sloppy and informal practice in the U-language until that practice is rendered in an explicit linguistic framework? What if the practice resists such treatment? The relationship of mathematics to science is also delimited in Carnap's work (and that of other logical positivists). Their idea is to envision a linguistic framework for science—or a branch of it—and to see how its rules relate to the rules of mathematical frameworks.

Notice also that if the linguistic rules for a framework of mathematics are those of a formal system (e.g., Carnap [1934]), then his views look a lot like formalism or perhaps logicism (see Carnap [1931]). If instead the "rules" are put in a context like model theory (Carnap [1942]), we may find something close to a structuralist philosophy of mathematics (of which more later, of course). Both of these are substantial philosophical positions.

We turn to a more contemporary internal/external contrast, Arthur Fine's [1986] natural ontological attitude (NOA), as contrasted with what he calls "realism" and with what he calls "antirealism." For our purposes, there are interesting similarities with Carnap. Fine's sketch of NOA begins with what he calls the "homely line": "[I]f the scientists tell me that there really are molecules, and atoms . . . then so be it. I trust them and, thus, must accept that there really are such things with their attendant properties and relations. . . . [We] accept the evidence of one's senses and [we] accept, *in the same way*, the confirmed results of science . . . we are being asked not to distinguish between kinds of truth or modes of existence. . . . [We] accept the certified results of science as on a par with more homely and familiarly supported claims" (pp. 127–128). In short, the NOA advocate trusts science to tell us what exists. So do I. I also trust mathematicians. In the present context, NOA counterparts might reason, "if the mathematicians tell me that there really are numbers, then so be it" or at least they would if they paid to mathematics the same homage that the NOA advocates pay to science (and if they forget that mathematicians rarely use the word "really" when speaking of existence). So far, then, NOA resembles *working realism.*

Despite Fine's claim, NOA is not completely internal to science. Like working realism, NOA invokes some interpretation of mathematics/science. In particular, scientists do not speak a language sufficiently regimented to delimit the existential claims. To paraphrase Quine, a "fenced ontology is just not explicit" in the "vague and untidy" language of the scientist. Sure we trust the scientists, but it is not always clear just what they say. Moreover, scientist-mathematicians qua scientist-mathematicians do not claim that their results are *on a par* with ordinary claims and that the two are to be accepted *in the same way*—unless a certain type of semantics and philosophy is part of mathematics/science. The NOA advocate urges that scientific language, once regimented, be *interpreted* the same way as ordinary language (once that is regimented, as well).

Fine himself agrees that a brand of semantics is involved in NOA:

> When NOA counsels us to accept the results of science as true, . . . we are to treat truth in the usual referential way, so that a sentence . . . is true just in case the entities referred to stand in the referred-to relations. Thus, NOA sanctions ordinary, referential semantics and commits us, via truth, to the existence of individuals, properties, relations, processes, and so forth. (p. 130)

> NOA recognizes in "truth" a concept already in use and agrees to abide by the standard rules of usage. These rules involve a Davidsonian-Tarskian referential semantics and they support a thoroughly classical logic of inference. (p. 133)

The main theme of Fine [1986] is that NOA falls short of both realism and antirealism. How is NOA different from realism? It is not that the realist holds that science is, for the most part, true, because the NOA advocate holds that as well; and it is not that the realist holds that theoretical scientific entities exist, because the NOA advocate holds that, too. Fine wonders "whether there is any necessary connection moving us from acceptance of the results of science as true to being a realist" (p. 128). To distinguish NOA from realism, Fine first invokes an auditory metaphor:

> What . . . does [the realist] add to his core acceptance of the results of science as really true? . . . [T]he most graphic way of stating the answer [is] that the realist adds . . . a desk-thumping, foot-stomping shout of "Really!" . . . "There really are electrons, really!" (p. 129)

> [W]hat of the "external world"? How can I [the NOA advocate] talk of reference and of existence claims unless I am talking about reference to things right out there in the world? And here, of course, the realist . . . wants to stamp his feet. (p. 131)

I do not think that the philosophical realist, in the sense of this book, needs to get so excited that he stamps his feet, especially because the program is so daunting.

More seriously, Fine joins Carnap in invoking an internal/external metaphor. Realism and, for that matter, antirealism are external to science: "The realist, as it were, tries to stand outside the arena watching the ongoing game and then tries to judge (from the external point of view) what the point is. It is, he says, *about* some area external to the game" (p. 131). To some extent, the same goes for the present program of philosophical realism. One goal is to say what the point of mathematics is. I would think, however, that an advocate of NOA would accept at least the last sentence of the above passage. Suppose we were to ask an NOA advocate or a working scientist, "Were there molecules before molecular theory was formulated?" or "Had there been no molecular theory, would there still be molecules?" Affirmative answers, quite in line with common sense, seem to make the subject matter of molecular theory "external" to the game of science. Where else are the molecules? It would be helpful to have a more literal reading of this particular internal/external metaphor or perhaps an argument that there is something wrong with the questions themselves.

The fact is that many theses that Fine says are constitutive of realism are simple consequences of NOA. For example, we read, "First, realism holds that there is a definite world; that is, a world containing entities with relations and properties that are to a large extent independent of human acts and agents. . . . Second, according to realism, it is possible to obtain a substantial amount of reliable and relatively observer independent information concerning this world and its features, information not restricted, for example, to just observable features" (pp. 136–137). Again, it is hard to imagine a working scientist, and thus an NOA advocate, denying this (certain interpretations of quantum mechanics aside).

Fine is aware that the realist and the NOA advocate seem to say the same things: "[I]t is not the *form* of a claim held true that marks off realism, it is rather the significance or content of the claim" (p. 139). The difference between NOA and realism, then, seems to lie at the level of semantics. This much accords with a theme of this chapter. We have seen that NOA advocates accept a Tarskian–Davidsonian account of their language. What of Fine's realist? The most common theme that Fine attributes to this character is an "extratheoretic" or "correspondence" notion of truth. The following is a sampling: "[T]he realist wants to explain the robust sense in which *he* takes . . . claims to truth and existence; namely as claims about reality—what is really, really the case. The full blown version of this involves the conception of truth as correspondence with the world, ...The realist adopts a standard, model-theoretic correspondence theory of truth" (pp. 129, 137).

There is, presumably, a difference between the Tarskian–Davidsonian semantics of NOA and the model-theoretic one of realism. The contrast between robust correspondence accounts and deflationary accounts is both deep and complex, and as noted already, I would like to steer clear of it as much as possible. In previous sections, I take a model-theoretic semantics to be the central framework of philosophical realism, but, by itself, it is an empty framework—in line with Tarski's own orientation. To be sure, the central notions of reference and satisfaction (and thus truth)—shared by the NOA advocate and the philosophical realist—do point to interesting philosophical puzzles, but we should not prejudge all resolutions to be metaphysically "robust" in a pejorative sense.

Another recurring theme in Fine [1986] is that the realist and the antirealist both insist on providing some sort of theory, analysis, or *interpretation* of science, whereas the NOA advocate does not. Again, a sample:

> [A] distinctive feature of NOA . . . is [its] stubborn refusal to amplify the concept of truth by providing a theory or analysis (or even a metaphorical picture). (p. 133)
>
> What binds realism and anti-realism together is [that they both] see science as a set of practices in need of an interpretation, and they see themselves as providing just the right interpretation. (pp. 147–148)
>
> The quickest way to get a feel for NOA is to understand it as undoing the idea of interpretation. (p. 149)
>
> [W]e could certainly agree with the physicists that there are quarks without necessarily imposing on the conclusion the special interpretive stance of realism. (We might . . . follow NOA's recommendation and not adopt any special stance at all.) (p. 152)

In light of my characterization of the philosophical enterprise as interpretive, these are fighting words. I have argued that, like working realism, NOA involves *some* (philosophical) interpretation. Fine does not deny this. He only rejects a special *kind* of interpretation. What kind? We are brought back to the internal/external metaphor: "I do not suggest that science is hermeneutic-proof, but rather that in science, as elsewhere, hermeneutical understanding has to be gained *from the inside*" (p. 148n). The question is, inside what? What is inside and what is outside?

If there is a divergence between Fine's perspective and the present one, it can be tempered if we look at Fine's most compelling arguments against realism. We will find out what Fine's realist believes by seeing exactly where Fine thinks the critter goes wrong. Like Carnap, much of Fine's attack is directed at what I call "philosophy-first."[19] His realist is not just out to interpret science but to give some extrascientific justification or grounding to the scientific enterprise. The realist tries to tell us what the scientist means in order to figure out whether the science is true: "For the problem of the external world (so-called) is how to satisfy the realist's demand that we justify the existence claims sanctioned by science (and, therefore, by NOA) as claims to the existence of entities 'out there'" (p. 132). The operative item here, I believe, is not the phrase "out there" in scare quotes, but the word "justify." For Fine's realist, *scientific* justification for molecules is not enough. Fine's realists see themselves as providing the ultimate justification.

The chapter on NOA begins with an intriguing and revealing analogy with the Hilbert program. Fine praises Hilbert for having the correct idea that "Metatheoretic arguments must satisfy more stringent requirements than those placed on the arguments used by the theory in question, for otherwise the significance of the reasoning about the theory is simply moot" (p. 114). The same goes for Fine's realist, who must support realism with arguments *more* stringent than those of science: "Hilbert's maxim applies to the debate over realism: to argue for realism one must employ methods more stringent than those in ordinary scientific practice. . . . [T]he form of argument used to support realism must be more stringent than the form of argument embedded in the very scientific practice that realism itself is supposed to ground" (pp. 115, 116n).

Well, it depends on what one is trying to do. Hilbert's program *was* designed to put mathematics on a firm foundation, and so he was trying to ground mathematics, in a strong epistemological sense. Here is the rallying cry: "The goal of my theory is to establish once and for all the certitude of mathematical methods. . . . No one shall drive us out of the paradise which Cantor has created for us" (Hilbert [1925, 191]). Thanks to Gödel, of course, the program failed—in this purpose. With Fine, Quine, and others, it is now hard to imagine a perspective more secure than that of science and mathematics, taken very generally. Epistemological foundationalism is dead, and if Fine's realist embraces it, then we can agree on a "do not resuscitate" order. But it

19. Fine has other lines, of course. He argues, for example, that the realist's "extratheoretic" correspondence is "unverifiable" and that the realist requires "genuine access to the relation of correspondence." This is reminiscent of arguments against realism in ontology in mathematics, on the ground that no one has access to abstract objects like numbers and sets. In either case, we need more information on what the relation of "access" is and why the various realisms are committed to it. Fine also discusses the relationship between realism and scientific practice, an analogue of the main topic of chapter 1. He suggests that Einstein's early antirealism, under the influence of Mach, played a crucial motivating factor in the development of relativity theory, and Fine argues that Einstein's later adoption of realism was stifling. Fine commends the pioneers of quantum mechanics for ignoring Einstein's philosophical advice. However, he later shows that virtually any (consistent) scientific theory is subject to both a realist and an antirealist interpretation. The upshot is that, as far as practice goes, philosophy is idle. In the end, Fine seems to embrace what I call the philosophy-last-if-at-all view (see chapter 1).

does not follow that there is no role for foundational activity. To borrow the title from my other book, can there be foundations without foundationalism? The philosopher of mathematics or science has interesting and important questions to attack. We can try to make sense of mathematics and its place in our intellectual lives, even if we cannot ground it in something more secure—and even if we have to work on less sturdy ground than that of mathematics/science itself. The only concession we need, so far, is that it may be possible for a program of philosophical realism to make some progress on some worthwhile front.

Our third instance of the internal/external metaphor is Hilary Putnam's "internal realism" (or "internalism") and its ugly cousin "metaphysical realism" (Putnam [1981], [1987]; see also [1980]). This distinction does not match the present one between working realism and philosophical realism and, indeed, much of what Putnam says about internalism resonates well with structuralism.

"Internal" is, of course, a relative term. The present working realism is, roughly, internal to mathematics and its methodology. Carnap's internal ontological questions are internal to a given linguistic framework, and Fine's NOA is internal to science, broadly conceived. In contrast, Putnam's internal realism is internal to our total "conceptual scheme," the entire range of conceptual resources with which we approach the world. In Wittgensteinian terms, internalism is internal to our "form of life." It is hard to imagine a perspective external to *that*, but this, I suppose, is Putnam's point.

According to Putnam, the main thesis of metaphysical realism is that "the world consists of some fixed totality of mind-independent objects. There is exactly one true and complete description of 'the way the world is'" ([1981, 49]). The key term here is "fixed." According to metaphysical realism, the objects, properties, and relations of the universe are determined independently of whatever language and theory we happen to hold. The "one true and complete description" of the world has terms that name those objects and refer to those predicates and relations.

So, according to metaphysical realism, the aim of science/philosophy is to come up with this "one true and complete description," or at least to approach it as an ideal. The view presupposes that there is "a God's Eye view of truth, or, more accurately, a No Eye view of truth—truth as independent of observers altogether" ([1981, 50]). This supposed "view" is the perspective of the single, complete description of the universe.

There is no reason why the present program of philosophical realism must presuppose this God's-eye view of reality. The program involves a metalanguage or, better, a U-language à la Curry, but there is no commitment that this language is "uniquely correct," whatever that might mean. We have already endorsed Curry's metaphor that the U-language is a growing thing. One can, perhaps, wonder whether there is a fixed totality of *mathematical* objects, the cumulative hierarchy V perhaps. One might hold that this totality is uniquely correct. A rejection of this idea is what distinguishes the thorough structuralism articulated here from a more limited version (see chapter 3).

A corollary of Putnam's rejection of the single, correct description, God's-eye view, is an ontological relativity: "[I]t is characteristic of [internal realism] to hold that *what objects does the world consist of?* is a question that it only makes sense to ask *within* a theory or description" ([1981, 49]). In Putnam [1987], internalism is characterized as a "conceptual relativity":

> [W]e cannot say—because it makes no sense— . . . what the facts are independent of all conceptual choices. (p. 33)

> We can and should insist that some facts are there to be discovered and not legislated by us. But this is something to be said when one has adopted a way of speaking, a language, a 'conceptual scheme'. (p. 36)

Putnam temporarily invokes a metaphor of the universe as a rolled-out dough, with conceptual schemes as various cookie cutters. The schemes produce different batches of objects out of the same universe. Even this metaphor is rejected, however, because there is no theory-neutral way of asking what the dough is like or, in particular, what the parts of the dough are—or whatever the literal counterparts of these questions might be.

Like Fine's realist, the metaphysical realist endorses the dreaded "correspondence" account of truth. And like Fine, Putnam speaks of the metaphysician's need for "access" to the truth, in order to figure out just what reference is (see note 19). What is the relationship between "Fido" and Fido? This is not a problem for the internal realist: "[A] sign that is actually employed . . . by a . . . community of users can correspond to particular objects *within the conceptual scheme of those users*. 'Objects' do not exist independently of conceptual schemes. *We* cut up the world into objects when we introduce one or another scheme of description. Since the objects *and* the signs are alike *internal* to the scheme of description, it is possible to say what matches what. . . . But [the metaphysical realist] does not regard such statements as telling us what reference *is*. For him finding out what reference *is*; i.e. what the *nature* of the 'correspondence' between words and things is, is a pressing problem" ([1981, 52]). This much is reminiscent of Carnap. The "internal" question of reference is trivial; the external one appears hard but is really a meaningless pseudoquestion.

As part of the argument, Putnam invokes the fact that in standard model-theoretic semantics, no (formal) language determines its interpretation. At best, a given theory determines its models only up to isomorphism. This, Putnam says, shows that metaphysical realists do not really have any access to the correspondence they postulate:[20] "[C]ouldn't there be some kind of abstract isomorphism, or . . . *mapping* of concepts onto things in the (mind-independent) world? Couldn't truth be defined in terms of such an isomorphism or mapping? The trouble with this suggestion is . . . that *too many* correspondences exist. To pick out just *one* correspondence between words or mental signs and mind-independent things we would have to already have referen-

20. To bolster this point, Putnam [1980] invokes the Löwenheim-Skolem theorem, which suggests an intriguing analogy between his own conceptual relativity and the sort of relativity invoked by Skolem (e.g., [1922]). However, the only thing that Putnam needs is the less fancy fact that isomorphic interpretations are equivalent. The Löwenheim-Skolem theorems apply only to first-order languages, whereas the equivalence of isomorphic models holds in any model-theoretic semantics worthy of the name. See Shapiro [1991].

tial access to the mind-independent things. . . . This simply states in mathematical language the intuitive fact that to single out a correspondence between two domains one needs some independent access to both domains" ([1981, 72–74]).

The model-theoretic semantics invoked by the present philosophical realist seems to involve only the internal perspective that Putnam endorses. What other perspective is there? The present program is carried out in a U-language; the philosopher starts *on* the ship of Neurath. In a sense, structuralism is built around Putnam's insights, pitched at a more local level. We end up with at least an "internal relativity of ontology"—a relativity to a theory or, better, to a structure (rather than to a grand conceptual scheme). The fact that formal, and mathematical, theories characterize their models only up to isomorphism does not make reference impossible to grasp, nor is reference as trivial as Putnam indicates.[21] The question is "reference to what?" Structuralism has interesting ramifications concerning what it is to be a mathematical *object*. It is well past time to turn our attention in that direction.

21. There are important differences between Putnam's internal realism and the present program of philosophical realism, structuralism in particular. The most-discussed aspect of Putnam [1981] is his endorsement of an epistemic account of truth. Truth, Putnam says, "is some sort of (idealized) rational acceptability. . . . [T]o claim a statement is true is to claim it could be justified. . . . Truth is ultimate goodness of fit" (pp. 49, 56, 64). In contrast, this book invokes an ordinary model-theoretic notion of truth. For all we know, some truths may be independent of rational acceptability. Notice, incidentally, that the truth-is-rational-acceptability thesis does not appear in Putnam [1987]. There, he attacks the very dichotomy between parts of discourse that have objective truth conditions and parts that have only conditions for warranted assertion (again, see Wright [1992]; also Kraut [1993]). Another theme of Putnam [1987] is that words like "object" are equivocal: "The device of reinterpretation [recognizes] that one person's 'existence' claim may be another person's something else. . . . The notions of 'object' and 'existence' are not treated as sacrosanct, as having just one possible use . . . there are no standards for the use of even the logical notions apart from conceptual choices" (pp. 34–36). This may be only a matter of terminology, but my view is that once the language is regimented, "object" and "existence" are univocal, but these notions are relative to a conceptual scheme, theory, or structure.

Part II

STRUCTURALISM

3

Structure

1 Opening

Every theory, philosophical or otherwise, must take some notions for granted. The philosopher inherits a fully developed language, a U-language in Curry's sense (see the previous chapter). Nevertheless, in both historical and contemporary philosophy, the most basic concepts of discourse are open to articulation and analysis. Every notion or principle in the inherited U-language is up for scrutiny and perhaps revision, at least in principle—even if one cannot revise every notion, all at once. In Neurath's metaphor, the philosopher is at work rebuilding parts of the floating ship of concepts. The notions of existence, object, and identity occur in just about every philosophical work, usually without further ado. Indeed, it is hard to imagine writing philosophy without invoking and presupposing these notions. Should we conclude that everyone already has clear and distinct ideas of them? Is any attempt to articulate such notions a waste of time and effort?

Presumably, one cannot go about articulating basic notions without presupposing and even using them. We have to start somewhere. This part of the book exhibits what is sometimes called a "dialectical" approach. We begin by using certain notions. As we go, some of these notions get refined and even modified. This tempers some of the very statements we use in getting the procedure off the ground. As the notions get further modified, the statements used to make the modifications themselves get modified. In the end, the original statements should be regarded as first approximations.

Structuralism has interesting consequences for the basic building blocks of ontology. Among other things, structuralists have something to say about what an object is and what identity is, at least in mathematics. Along the way, we speculate about how far the structuralist notion of object carries over to ordinary, nonmathematical contexts (see also chapter 8). The problem, however, is that structuralism cannot be articulated without invoking the notion of object, and so I ask for the reader's dialectical indulgence.

My structuralist program is a realism in ontology and a realism in truth-value—once the requisite notions of object and objectivity are on the table. Structuralists hold that a nonalgebraic field like arithmetic is about a realm of objects—numbers—that exist independently of the mathematician, and they hold that arithmetic assertions have nonvacuous, bivalent, objective truth-values in reference to this domain.

Structuralism is usually contrasted with traditional Platonism. Ultimately, the differences may not amount to much when it comes to ontology, but the contrast is a good place to begin our first approximation. Like any realist in ontology, the Platonist holds that the subject matter of a given nonalgebraic branch of mathematics is a collection of objects that have some sort of ontological independence. The natural numbers, for example, exist independently of the mathematician. As I noted in the previous chapter, Resnik [1980, 162] defines an "ontological platonist" to be someone who holds that ordinary physical objects and numbers are "on a par." Numbers are the same kind of thing—objects—as beach balls, only there are more numbers than beach balls and numbers are abstract and eternal.

To pursue this analogy, one might attribute some sort of ontological independence to the individual natural numbers. Just as each beach ball is independent of every other beach ball, each natural number is independent of every other natural number. Just as a given red beach ball is independent of a blue one, the number 2 is independent of the number 6. An attempt to articulate this idea will prove instructive. When we say that the red beach ball is independent of the blue one, we might mean that the red one could have existed without the blue one and vice versa. However, nothing of this sort applies to the natural numbers, as conceived by traditional Platonism. According to the Platonist, numbers exist necessarily. So we cannot say that 2 could exist without 6, because 6 exists of necessity. Nothing exists without 6. To be sure, there is an epistemic independence among the numbers in the sense that a child can learn much about the number 2 while knowing next to nothing about 6 (but having it the other way around does stretch the imagination). This independence is of little interest here, however.

The Platonist view may be that one can state the *essence* of each number without referring to the other numbers. The essence of 2 does not invoke 6 or any other number (except perhaps 0 and 1). If this notion of independence could be made out, we structuralists would reject it. The essence of a natural number is its *relations* to other natural numbers. The subject matter of arithmetic is a single abstract structure, the pattern common to any infinite collection of objects that has a successor relation with a unique initial object and satisfies the (second-order) induction principle. The number 2, for example, is no more and no less than the second position in the natural-number structure; 6 is the sixth position. Neither of them has any independence from the structure in which they are positions, and as places in this structure, neither number is independent of the other. The essence of 2 is to be the successor of the successor of 0, the predecessor of 3, the first prime, and so on.

Plato himself distinguishes two studies involving natural numbers. *Arithmetic* "deals with the even and the odd, with reference to how much each happens to be" (*Gorgias* 451A–C). If "one becomes perfect in the arithmetical art," then "he knows also all of the numbers" (*Theatetus* 198A–B; see also *Republic VII* 522C). The study called *logistic* deals also with the natural numbers but differs from arithmetic "in so far as it studies the even and the odd with respect to the multitude they make both with themselves and with each other" (*Gorgias 451*A–C; see also *Charmides* 165E–166B). So arithmetic deals with the natural numbers, and logistic concerns the relations among the numbers. In ancient works, logistic is usually understood as the theory of *calculation.* Most writers take it to be a practical discipline, concerning measurement, business dealings, and so forth (e.g., Proclus [485, 39]; see Heath [1921, chapter 1]). For Plato, however, logistic is every bit as theoretical as arithmetic. As Jacob Klein [1968, 23] puts it, theoretical logistic "raises to an explicit science that knowledge of relations among numbers which . . . precedes, and indeed must precede, all calculation."

The structuralist rejects this distinction between Plato's arithmetic and theoretical logistic. There is no more to the individual numbers "in themselves" than the relations they bear to each other. Klein [1968, 20] wonders what is to be studied in arithmetic, as opposed to logistic. Presumably, the art of counting—reciting the numerals—is arithmetic par excellence. Yet "addition and also subtraction are only an extension of counting." Moreover, "counting itself already presupposes a continual relating and distinguishing of the numbered things as well as of the numbers." In the *Republic* (525C–D), Plato said that guardians should pursue *logistic* for the sake of knowing. It is through this study of the *relations* among numbers that their soul is able to grasp the nature of numbers as they are in themselves. We structuralists agree.[1]

The natural-number structure is exemplified by the strings on a finite alphabet in lexical order, an infinite sequence of strokes, an infinite sequence of distinct moments of time, and so on. Similarly, group theory studies not a single structure but a type of structure, the pattern common to collections of objects with a binary operation, an identity element thereon, and inverses for each element. Euclidean geometry studies Euclidean-space structure; topology studies topological structures, and so forth.[2]

One lesson we have learned from Plato is that one cannot delineate a philosophical notion just by giving a list of examples. Nevertheless, the examples point in a certain direction. To continue the dialectic, I define a *system* to be a collection of objects with certain relations. An extended family is a system of people with blood and marital

1. Klein [1968, 24] tentatively concludes that logistic concerns ratios among pure units, whereas arithmetic concerns counting, addition, and subtraction. In line with the later dialogues, it might be better to think of logistic as what we would call "arithmetic," with Plato's "arithmetic" being a part of higher philosophy. I am indebted to Peter King for useful conversations on this historical material.

2. Sometimes mathematicians use phrases like "the group structure" and "the ring structure" when speaking loosely about groups and rings. Taken literally, these locutions presuppose that there is a single structure common to all groups and a single structure common to all rings. Here, I prefer to use "structure" to indicate the subject of nonalgebraic theories, those that mathematicians call "concrete." To say that two systems have the same structure is to say that they share something like an isomorphism type. This is what allows us to speak of numbers as individual objects. So, in the present sense, group theory is not about a single structure, but rather a class of similar structures. My fellow structuralist, Michael Resnik (e.g. [1996]) denies the importance of this distinction. For Resnik, it seems, all mathematical theories are algebraic, none are "concrete" (or to be precise, there is no "fact of the matter" whether a given structure is "concrete").

relationships, a chess configuration is a system of pieces under spatial and "possible-move" relationships, a symphony is a system of tones under temporal and harmonic relationships, and a baseball defense is a collection of people with on-field spatial and "defensive-role" relations. A *structure* is the abstract form of a system, highlighting the interrelationships among the objects, and ignoring any features of them that do not affect how they relate to other objects in the system.

Although epistemology is treated in the next chapter, it will help here to mention a few ways that structures are grasped. One way to apprehend a particular structure is through a process of pattern recognition, or abstraction. One observes a system, or several systems with the same structure, and focuses attention on the relations among the objects—ignoring those features of the objects that are not relevant to these relations. For example, one can understand a baseball defense by going to a game (or several games) and noticing the spatial relations among the players who wear gloves, ignoring things like height, hair color, and batting average, because these have nothing to do with the defense system. It is similar to how one comes to grasp the type of a letter, such as an "E," by observing several tokens of the letter and focusing on the typographical pattern, while ignoring the color of the tokens, their height, and the like.

I do not offer much to illuminate the psychological mechanisms involved in pattern recognition. They are interesting and difficult problems in psychology and in the young discipline of cognitive science. Nevertheless, it is reasonably clear that humans do have an ability to recognize patterns.[3] Sometimes, ostension is at work. One points to the system and somehow indicates that it is the pattern being ostended, and not the particular people or objects. That is, one points to a system that exemplifies the structure in order to ostend the structure itself. Similarly, one can point to a capital "E," not to ostend that particular token but to ostend the type, the abstract pattern. Ordinary discourse clearly has the resources to distinguish between pattern and patterned, the psychological problems with pattern recognition and the philosophical problems with *abstracta* notwithstanding.

A second way to understand a structure is through a direct description of it. Thus, one might say that a baseball defense consists of four infielders, arranged thus and so, three outfielders, and so on. One can also describe a structure as a variation of a previously understood structure. A "lefty shift defense" occurs when the shortstop plays to the right of second base and the third baseman moves near the shortstop position. Or a "softball defense" is like a baseball defense, except that there is one more outfielder. In either case, most competent speakers of the language will understand what is meant and can then go on to discuss the structure itself, independent of any particular exemplification of it.[4]

---

3. Dieterle [1994, chapter 3] contains a brief survey of some of the psychological literature on pattern recognition, relating the process to structuralism. She argues that pattern recognition is the central component of a reliabilist epistemology of abstract patterns. I return to this in chapter 4.

4. An anecdote: Several years ago, I was called on to observe a remedial mathematics class. The students were among the worst prepared in mathematics. While waiting for the class to begin, I over-

For our first (or second) approximation, then, pure mathematics is the study of structures, independently of whether they are exemplified in the physical realm, or in any realm for that matter. The mathematician is interested in the internal relations of the places of these structures, and the methodology of mathematics is, for the most part, deductive. As Resnik puts it:

> In mathematics, I claim, we do not have objects with an 'internal' composition arranged in structures, we have only structures. The objects of mathematics, that is, the entities which our mathematical constants and quantifiers denote, are structureless points or positions in structures. As positions in structures, they have no identity or features outside a structure. ([1981, 530])

> Take the case of linguistics. Let us imagine that by using the abstractive process . . . a grammarian arrives at a complex structure which he calls *English.* Now suppose that it later turns out that the English corpus fails in significant ways to instantiate this pattern, so that many of the claims which our linguist made concerning his structure will be falsified. Derisively, linguists rename the structure *Tenglish.* Nonetheless, much of our linguist's knowledge about *Tenglish qua* pattern stands; for he has managed to describe *some* pattern and to discuss some of its properties. Similarly, I claim that we know much about Euclidean space despite its failure to be instantiated physically. ([1982, 101])

Of course, some of the examples mentioned above are too simple to be worthy of the mathematician's attention. What can we prove about an infield structure, or about the type of the letter "E"? There are, however, nontrivial theorems about chess games. For example, it is not possible to force a checkmate with a king and two knights against a lone king. This holds no matter what the pieces are made of, and even whether or not chess has ever been played. This fact about chess is a more or less typical mathematical theorem about a certain structure. Here, it is the structure of a certain game.

Most of the structures studied in mathematics have an infinite, indeed uncountable, number of positions. The set-theoretic hierarchy has a proper class of positions. It is contentious to suggest that we can come to understand structures like this by abstraction, or pattern recognition, from perceptual experience. That would require a person to view (or hear) a system that consists of infinitely many objects. There is thus an interesting epistemological problem for structuralism, which will be dealt with in due course (chapter 4). My present purpose is to point to the notion of structure and to characterize mathematics as the science of structure.

There is a revealing error in Hartry Field's *Science without numbers* [1980]. The purpose of that book is to articulate a view, now called "nominalism," that there are no abstract objects. According to Field, everything is concrete (in the philosophers' sense of that word). Because, presumably, numbers are *abstracta* par excellence, the

---

heard a conversation between two of them concerning the merits of a certain basketball defense. The discussion was at a rather high level of abstraction and complexity, at least as great as that of the subject matter of the class that day (the addition of fractions). It seems to me that there is not much difference in kind between abstract discussions of basketball defenses and the addition of fractions (but see section 5).

nominalist rejects the existence of numbers. A central item on Field's agenda is to show how science can proceed, at least in principle, without presupposing the existence of numbers and other abstract objects. He develops one example, Newtonian gravitational theory, in brilliant detail. The ontology of Field's nominalistic theory includes points and regions of space-time, but he argues that points and regions are concrete, not abstract, entities. There is no need to dispute the last claim here (but see Resnik [1985]). Whether abstract or concrete, Field's Newtonian space-time is Euclidean, consisting of continuum-many points and even more regions. Space-time exemplifies most (but not all) of the structure of $\mathbb{R}^4$, the system of quadruples of real numbers. Field himself insightfully exploits the fact that any model of space-time can be extended to a model of $\mathbb{R}^4$ by adding a reference frame and units for the metrics. Each line of space-time is then isomorphic to $\mathbb{R}$, and so addition and multiplication can be defined on a line. So something like addition and multiplication, as well as the calculus of real-valued functions, can be carried out in this nominalistic theory. All of this is supposed to be consistent with nominalistic rejection of *abstracta*.

Field considers the natural objection that "there doesn't seem to be a very significant difference between postulating such a rich physical space and postulating the real numbers." He replies, "[T]he nominalistic objection to using real numbers was not on the grounds of their uncountability or of the structural assumptions (e.g., Cauchy completeness) typically made about them. Rather, the objection was to their abstractness: even postulating *one* real number would have been a violation of nominalism. . . . Conversely, postulating uncountably many *physical* entities . . . is not an objection to nominalism; nor does it become any more objectionable when one postulates that these physical entities obey structural assumptions analogous to the ones that platonists postulate for the real numbers" (p. 31). The structuralist balks at this point. For us, a real number *is* a place in the real-number structure. It makes no sense to "postulate one real number," because each number is part of a large structure. It would be like trying to imagine a shortstop independent of an infield, or a piece that plays the role of the black queen's bishop independent of a chess game. Where would it stand? What would its moves be? One can, of course, ask whether the real-number structure is exemplified by a given system (like a collection of points). Then one could locate objects that have the *roles* of individual numbers, just as on game day one can identify the people who have the roles of shortstop on each team, or in a game of chess one can identify the pieces that are the bishops. But it is nonsense to contemplate *numbers* independent of the structure they are part of.

It is common for mathematicians to claim that mathematics has not really been eliminated from Field's system. Even if the title, *Science without numbers*, is an accurate description of the enterprise, it is not science without mathematics.[5] Some philosophers might be inclined to let the response of the mathematicians settle the

5. This response was made by the mathematicians who attended an interdisciplinary seminar I once gave on *Science without numbers*. In correspondence, Field himself reported similar observations from colleagues in mathematics departments. Of course, mathematicians are not the only ones to balk at the claim that Field's system does not significantly reduce the mathematical presuppositions of Newtonian gravitational theory. A number of philosophers and prominent logicians have joined the chorus.

matter. After all, mathematicians should be able to recognize their subject when they see it. In response, Field could point out that these mathematicians are simply not interested in questions of ontology, or that they do not understand or care about the distinction between abstract and concrete. This, of course, may be true. But observations about the typical interests of mathematicians miss the point. Field concedes that nominalistic physics makes substantial "structural assumptions" about space-time, and he articulates these assumptions with admirable rigor. Although Field would not put it this way, the "structural assumptions" characterize a structure, an uncountable one. This is a consequence of the fact that (the second-order version of) Field's theory of space-time is categorical—all of its models are isomorphic. Field's nominalistic physicist would study this structure *as such*, at least sometimes. Field himself proves theorems about this structure. As I see it, he thereby engages in mathematics. The activity of proving things about space-time is the same kind of activity as proving theorems about real numbers. Both are the deductive study of a structure, no more and certainly no less.

Field might reply that he is interested in one particular (concrete) exemplification of the structure, not the structure itself. This is fair, but it misses the point. As far as mathematics goes, it does not matter where, how, or even if the relevant structure is exemplified. The substructure of $\mathbb{R}^4$ is in the purview of mathematics, and both Field and his nominalistic physicist use typical mathematical methods to illuminate this structure, along with the concrete system that exemplifies it. I suggest that these observations underlie the mathematicians' response to Field. They are correct.

2 Ontology: Object

On the ontological front, there are two groups of issues. One is the status of whole structures, such as the natural-number structure, the real-number structure, and the set-theoretic hierarchy, as well as more mundane structures like a symphony, a chess configuration, and a baseball defense. The other issue concerns the status of mathematical objects, the places within structures: natural numbers, real numbers, points, sets, and so on.

We begin with the issue concerning mathematical objects. The existence of structures will be addressed directly later, but because of the interconnections, we will go back and forth between the issues. Once again, a natural number is a place in the natural-number structure, a particular infinite pattern. The pattern may be exemplified by many different systems, but it is the same pattern in each case. The number 2 is the second place in that pattern. Individual numbers are analogous to particular offices within an organization. We distinguish the office of vice president, for example, from the person who happens to hold that office in a particular year, and we distinguish the white king's bishop from the piece of marble that happens to play that role on a given chess board. In a different game, the very same piece of marble might play another role, such as that of white queen's bishop or, conceivably, black king's rook. Similarly, we can distinguish an object that plays the role of 2 in an exemplification of the natural-number structure from the number itself. The number is the office, the place in the structure. The same goes for real numbers, points of

Euclidean geometry, members of the set-theoretic hierarchy, and just about every object of a nonalgebraic field of mathematics. Each mathematical object is a place in a particular structure. There is thus a certain priority in the status of mathematical objects. The structure is prior to the mathematical objects it contains, just as any organization is prior to the offices that constitute it. The natural-number structure is prior to 2, just as "baseball defense" is prior to "shortstop" and "U.S. Government" is prior to "vice president."

Structuralism resolves one problem taken seriously by at least some Platonists—or realists in ontology—and which has been invoked by its opponents as an argument against realism. Frege [1884], who has been called an "arch-Platonist," argued that numbers are objects. This conclusion was based in part on the grammar of number words. Numerals, for example, exhibit the trappings of singular terms. Frege went on to give an insightful and eminently plausible account of the use of number terms in certain contexts, typically forms like "the number of F is y," where F stands for a predicate like "moons of Jupiter" or "cards on this table." But then Frege noted that this preliminary account does not sustain the conclusion that numbers are objects. For this, we need a criterion to decide whether any given number, like 2, is the same or different from any other object, say Julius Caesar. That is, Frege's preliminary account does not have anything to say about the truth-value of the identity "Julius Caesar = 2." This quandary has come to be called the *Caesar problem*. A solution to it should determine how and why each number is the same or different from *any object whatsoever*. The Caesar problem is related to the Quinean dictum that we need criteria to individuate the items in our ontology. If we do not have an identity relation, then we do not have bona fide objects. The slogan is "no entity without identity." Frege attempted to solve this problem with the use of extensions. He proposed that the number 2 is a certain extension, the collection of all pairs. Thus, 2 is not Julius Caesar because, presumably, persons are not extensions. This turned out to be a tragic maneuver, because Frege's account of extensions (in [1903]) is inconsistent. With the wisdom that hindsight brings, Frege should have quit while he was ahead.[6]

Paul Benacerraf's celebrated [1965] and Philip Kitcher [1983, chapter 6] raise a variation of this problem. After the discovery that virtually every field of mathematics can be reduced to (or modeled in) set theory, the foundationally minded came to think of the set-theoretic hierarchy as the ontology for all of mathematics. An economy in regimentation suggests that there should be a single type of object. Why have sets and numbers when sets alone will do? But there are several reductions of arithmetic to set theory, an embarrassment of riches. If numbers are mathematical objects and all mathematical objects are sets, then we need to know *which* sets the natural num-

6. A number of writers have shown that the essence of Frege's account of *arithmetic* is consistent (e.g., Boolos [1987]). The idea is to speak of numbers directly, not mediated by extensions. But then the Caesar problem remains unsolved. One can also consistently identify numbers with extensions, as long as one does not maintain that every open formula determines an extension and that two formulas determine the same extension if and only if they are coextensive (cf. Frege's infamous Principle V). I return to Frege's notion of "object" in chapter 5.

bers are. According to one account, due to von Neumann, it is correct to say that 1 is a member of 4. According to Zermelo's account, 1 is not a member of 4. Moreover, there seems to be no principled way to decide between the reductions. Each serves whatever purpose a reduction is supposed to serve. So we are left without an answer to the question, "Is 1 really a member of 4, or not?" What, after all, are the natural numbers? Are they finite von Neumann ordinals, Zermelo numerals, or some other sets altogether? From these observations and questions, Benacerraf and Kitcher conclude, against Frege, that numbers are not objects. This conclusion, I believe, is not warranted. It all depends on what it is to *be an object*, a matter that is presently under discussion.[7]

I would think that a good philosophy of mathematics need not answer questions like "Is Julius Caesar = 2?" and "Is $1 \in 4$?" Rather, a philosophy of mathematics should show why these questions need no answers, even if the questions are intelligible. It is not that we just do not care about the answers; we want to see why there is no answer to be discovered—even for a realist in ontology. Again, a number is a place in the natural-number structure. The latter is the pattern common to all of the models of arithmetic, whether they be in the set-theoretic hierarchy or anywhere else. One can form coherent and determinate statements about the identity of two numbers: $1 = 1$ and $1 \neq 4$. And one can look into the identity between numbers denoted by different descriptions *in the language of arithmetic*. For example, 7 is the largest prime that is less than 10. And one can apply arithmetic in the Fregean manner and assert, for example, that the number of cards in a deck is 52. But it makes no sense to pursue the identity between a place in the natural-number structure and some other object, expecting there to be a fact of the matter. Identity between natural numbers is determinate; identity between numbers and other sorts of objects is not, and neither is identity between numbers and the positions of other structures.

Along similar lines, one can ask about *numerical* relations between numbers, relations definable in the language of arithmetic, and one can expect determinate answers to these questions. Thus, $1 < 4$ and 1 evenly divides 4. These are questions internal to the natural-number structure. But if one inquires whether 1 is an element of 4, there is no answer waiting to be discovered. It is similar to asking whether 1 is braver than 4, or funnier.

Similar considerations hold for our more mundane structures. It is determinate that the shortstop position is not the catcher position and that a queen's bishop cannot capture the opposing queen's bishop, but there is something odd about asking whether positions in patterns are identical to other objects. It is nonsense to ask whether *the shortstop* is identical to Ozzie Smith—whether the person is identical to the po-

7. In chapter 2, we encountered an argument of Putnam [1981, 72–74] against metaphysical realism: "[C]ouldn't there be some kind of abstract isomorphism, or . . . mapping of concepts onto things in the (mind-independent) world? . . . The trouble with this suggestion is . . . that *too many* correspondences exist. To pick out just *one* correspondence between words or mental signs and mind-independent things we would have to already have referential access to the mind-independent things." Again, it all depends on what it is to be a "thing," and it depends on what "reference" is.

sition. Ozzie Smith is, of course, *a* shortstop and, arguably, he is (or was) the quintessential shortstop, but is he the position? There is also something odd about asking whether the shortstop position is taller or faster or a better hitter than the catcher position. Shortness, tallness, and batting average do not apply to positions.

Similar, less philosophical questions are asked on game day, about a particular lineup, but those questions concern the people who occupy the positions of shortstop and catcher that day, not the positions themselves. When a fan asks whether Ozzie Smith is the shortstop or whether the shortstop is a better hitter than the catcher, she is referring to the people in a particular lineup.[8] Virtually any person prepared to play ball *can* be a shortstop—anybody can occupy that role in an infield system (some better than others). Any small, moveable object can play the role of (i.e., can be) black queen's bishop. Similarly, and more generally, anything at all can "be" 2—anything can occupy that place in a system exemplifying the natural-number structure. The Zermelo 2 ({{φ}}), the von Neumann 2 ({φ, {φ}}), and even Julius Caesar can each play that role. The Frege–Benacerraf questions do not have determinate answers, and they do not need them.

One can surely *ask* the Frege–Benacerraf questions. Are Julius Caesar or {φ, {φ}} places in the natural-number structure? Do the monarch and the ordinal have essential properties relating them to other places in the natural-number structure? If the question is taken seriously, the answer will surely be "of course not." The retort is "How do you know?" or, to paraphrase Frege, "Of course these items are not places in the natural-number structure, but this is no thanks to structuralism." A structuralist could reply that Julius Caesar and {φ, {φ}} have essential properties other than those relating to other places in the natural-number structure, but that would miss the point.

We point toward a relativity of ontology, at least in mathematics. Roughly, mathematical objects are tied to the structures that constitute them. Benacerraf [1965, §III.A] himself espoused a related view, at least temporarily. In order to set up his dilemma, he "treated expressions of the form $n = s$, where n is a number expression and s a set expression as if . . . they made perfectly good sense, and . . . it was our job to sort out the true from the false. . . . I did this to dramatize the kind of answer that a Fregean might give to the request for an analysis of number. . . . To speak from Frege's standpoint, there is a world of objects . . . in which the identity relation [has] free reign." Benacerraf's suggestion is to hold that at least some identity statements are meaningless: "Identity statements make sense only in contexts where there exist possible individuating conditions. . . . [Q]uestions of identity contain the presupposition that the 'entities' inquired about both belong to some general category." We need not go this far, but notice that items from the same structure are certainly in the same "general category," and there are "individuating conditions" among them. Whether Benacerraf has given the only ways to construe identity statements remains

8. In professional baseball, shortstops are generally faster than catchers, and they are better hitters (even though there are notable exceptions). Statements like this have to do with the particular skills needed to play each position well.

to be seen. He concluded, "What constitutes an entity is category or theory-dependent. . . . There are . . . two correlative ways of looking at the problem. One might conclude that identity is systematically ambiguous, or else one might agree with Frege, that identity is unambiguous, always meaning sameness of object, but that (contra-Frege now) the notion of *object* varies from theory to theory, category to category." In mathematics, at least, the notions of "object" and "identity" are unequivocal but thoroughly relative. Objects are tied to the structures that contain them. It is thus strange that Benacerraf should eventually conclude that natural numbers are not objects. Arithmetic is surely a coherent theory, "natural number" is surely a legitimate category, and numbers are its objects.[9]

Suppose that mathematicians develop a new field. Call its objects "hypernumbers." Analogues in (reconstructed) history are the study of negative, irrational, and complex numbers, and quaternions. It would surely be pompous of the philosopher to suggest that the field of hyperarithmetic is somehow illegitimate and is destined to remain so until we know how to individuate hypernumbers. The mathematicians do not have to tell us, once and for all, how to figure out whether, say, the additive identity of the hypernumbers is the same thing as the zero of arithmetic or the zero of analysis or the empty set. It is enough for them to differentiate hypernumbers from each other.

As hinted earlier, there is an important caveat to this relativity. I do not wish to go as far as Benacerraf in holding that identifying positions in different structures (or positions in a structure with other objects) is always meaningless. On the contrary, mathematicians sometimes find it convenient, and even compelling, to identify the positions of different structures. This occurs, for example, when set theorists settle on the finite von Neumann ordinals as the natural numbers. They stipulate that 2 is $\{\phi, \{\phi\}\}$, and so it follows that $2 \neq \{\{\phi\}\}$. For a more straightforward example, it is surely wise to identify the positions in the natural-number structure with their counterparts in the integer-, rational-, real-, and complex-number structures. Accordingly, the natural number 2 is identical to the integer 2, the rational number 2, the real number 2, and the complex number 2 (i.e., $2 + 0i$). Hardly anything could be more straightforward. For an intermediate case, mathematicians occasionally look for the "natural settings" in which a structure is best studied. An example is the embedding of the complex numbers in the Euclidean plane, which illuminates both structures. It is not an exaggeration to state that some structures grow and thrive in certain environments. This phenomenon will occupy us several times, in chapters 4 and 5. The point here is that cross-identifications like these are matters of *decision*, based on convenience, not matters of discovery.

Parsons [1990, 334] puts the relativity into perspective:

> [O]ne should be cautious in making such assertions as that identity statements involving objects of different structures are meaningless or indeterminate. There is an obvi-

---

9. I will return to this relativity throughout the book, notably in section 6 of chapter 4 on epistemology, section 3 of chapter 5 on Frege, and chapter 8. I am indebted to Crispin Wright and Bob Hale for pressing the Caesar issue.

> ous sense in which identity of natural numbers and sets is indeterminate, in that different interpretations of number theory and set theory are possible which give different answers about the truth of identities of numbers and sets. In a lot of ordinary, mathematical discourse, where different structures are involved, the question of identity or non-identity of elements of one with elements of another just does not arise (even to be rejected). But of course some discourse about numbers and sets makes identity statements between them meaningful, and some of that . . . makes commitments as to the truth value of such identities. Thus it would be quite out of order to say (without reference to context) that identities of numbers and sets are meaningless or that they lack truth-values.

Even with this caveat, the ontological relativity threatens the semantic uniformity between mathematical discourse and ordinary or scientific discourse. Of course, this depends on what is required for uniformity, of which more later (section 9 of chapter 4, on reference). The threat also depends on the extent to which ordinary objects are not relative. I briefly return to this in chapter 8.

On a related matter, Azzouni [1994, 7–8, 146–147] accuses structuralists of being "ontologically radical" in the sense that we "replace the traditional metaphysically inert mathematical object with something else." It depends on what one thinks was there to be accepted or replaced. Structuralism is a view about what the objects of, say, arithmetic *are*, not what they should be, and we claim to make sense of what goes on in mathematics. Mathematicians do not usually use phrases like "metaphysically inert." Perhaps Azzouni's view is that we structuralists are being radical with respect to traditional *philosophies* of mathematics, Platonism in particular. I leave it to the reader to determine the extent to which I am proposing a replacement or further articulation of prior realist philosophies of mathematics.

One slogan of structuralism is that mathematical objects are places in structures. We must be careful here, however, because there is an intuitive difference between an object and a place in a structure, between an officeholder and an office. We can accommodate this intuition and yet maintain that numbers and sets are objects by invoking a distinction in linguistic practice. There are, in effect, two different orientations involved in discussing structures and their places (although the border between them is not sharp). Sometimes the places of a structure are discussed in the context of one or more systems that exemplify the structure. We might say, for example, that the shortstop today was the second baseman yesterday, or that the current vice president is more intelligent than his predecessor. Similarly, we might say that the von Neumann 2 has one more element than the Zermelo 2. Call this the *places-are-offices* perspective. This office orientation presupposes a background ontology that supplies objects that fill the places of the structures. In the case of baseball defense and that of government, the background ontology is people; in the case of chess games, the background ontology is small, movable objects—pieces with certain colors and shapes. In the case of arithmetic, sets—or anything else—will do for the background ontology. With mathematics, the background ontology can even consist of places from other structures, when we say, for example, that the negative, whole real numbers exemplify the natural-number structure, or that a Euclidean line exemplifies the real-number structure. Indeed, the background ontology for the places-are-offices perspective can even consist of the places of *the very structure under discussion*, when it is noted, for example, that the even natural numbers exemplify the natural-number structure. We will have occasion later to consider structures whose places are occupied by other structures. One consequence of this is that, in mathematics at least, the distinction between office and officeholder is a relative one. What is an object from one perspective is a place in a structure from another.

In contrast to this office orientation, there are contexts in which the places of a given structure are treated as objects in their own right, at least grammatically. That is, sometimes items that denote places are bona fide singular terms. We say that the vice president is president of the Senate, that the chess bishop moves on a diagonal, or that the bishop that is on a black square cannot move to a white square. Call this the *places-are-objects* perspective. Here, the statements are about the respective structure as such, independent of any exemplifications it may have. Arithmetic, then, is about the natural-number structure, and its domain of discourse consists of the places in this structure, treated from the places-are-objects perspective. The same goes for the other nonalgebraic fields, such as real and complex analysis, Euclidean geometry, and perhaps set theory.

It is common to distinguish the "is" of identity from the "is" of predication. The sentence "Cicero is Tully" does not have the same form as "Cicero is Roman." When, in the places-are-objects perspective, we say that 7 is the largest prime less than 10, and that the number of outfielders is 3, we use the "is" of identity. We could just as well write "=" or "is identical to." In contrast, when we invoke the places-are-offices perspective and say that $\{\{\phi\}\}$ is 2 and that $\{\phi, \{\phi\}\}$ is 2, we use something like the "is" of predication, but here it is predication *relative to a system* that exemplifies a structure. Let us call this the "is" of office occupancy. We are saying that $\{\{\phi\}\}$ plays the role of 2 in the system of Zermelo numerals and that $\{\phi, \{\phi\}\}$ plays the role of 2 in the system of finite von Neumann ordinals. When we say that Ozzie Smith is the shortstop, or that Al Gore is the vice president, we also invoke the "is" of office occupancy.

This does not exhaust the uses of the copula in mathematics. I noted earlier that for convenience, mathematicians sometimes identify places from different structures. For example, when set theorists settle on the von Neumann account of arithmetic, and thereby declare that 2 is $\{\phi, \{\phi\}\}$, they invoke what may be called the "is of identity by fiat."

The point here is that sometimes we use the "is" of identity when referring to offices, or places in a structure. This is to treat the positions *as objects*, at least when it comes to surface grammar. When the structuralist asserts that numbers are objects, this is what is meant. The places-are-objects perspective is thus the background for the present realism in ontology toward mathematics. Places in structures are bona fide objects.

My perspective thus presupposes that statements in the places-are-objects perspective are to be taken literally, at face value. Bona fide singular terms, like "vice president," "shortstop," and "2" denote bona fide objects. This reading might be questioned. Notice, for example, that places-are-objects statements entail generalizations over all systems that exemplify the structure in question. Everyone who is vice president—whether it be Gore, Quayle, Bush, or Mondale—is president of the Senate in that government. Every chess bishop moves on a diagonal, and none of those on black squares ever move to white squares (in the same game). No person can be shortstop and catcher simultaneously; and anything playing the role of 3 in a natural-number system is the successor of whatever plays the role of 2 in that system. In short, places-are-objects statements apply to the particular objects or people that happen to occupy the positions with respect to any system exemplifying the structure. Someone might hold, then, that places-are-objects statements are no more than a convenient rephrasing of corresponding generalizations over systems that exemplify the structure in question. If successful, a maneuver like this would eliminate the places-are-objects perspective altogether. The apparent singular terms mask implicit bound variables. This rephrasing plan, however, depends on being able to generalize over all systems that exemplify the structure in question. To assess this idea, we turn to our other main ontological question, the status of structures themselves.

3 Ontology: Structure

Because the same structure can be exemplified by more than one system, a structure is a one-over-many. Entities like this have received their share of philosophical attention throughout the ages. The traditional exemplar of one-over-many is a universal, a property, or a Form. In more recent philosophy, there is the type/token dichotomy. In philosophical jargon, one says that several tokens *have* a particular type, or *share* a particular type; and we say that an object *has* a universal or, as Plato put it, an object *has a share of*, or *participates in* a Form. As defined above, a structure is a pattern, the form of a system. A system, in turn, is a collection of related objects. Thus, structure is to structured as pattern is to patterned, as universal is to subsumed particular, as type is to token.

The nature and status of types and universals is a deep and controversial matter in philosophy. There is no shortage of views on such matters. Two of the traditional views stand out. One, due to Plato, is that universals exist prior to and independent of any items that may instantiate them. Even if there were no red objects, the Form of Redness would still exist. This view is sometimes called "*ante rem* realism," and universals so construed are "*ante rem* universals." The main alternative, attributed to Aristotle, is that universals are ontologically dependent on their instances. There is no more to redness than what all red things have in common. Get rid of all red things, and redness goes with them. Destroy all good beings, all good things, and all good actions, and you destroy goodness itself. A sobering thought. Forms so construed are called "in re universals," and the view is sometimes called "in re realism." Advocates of this view may admit that universals exist, after a fashion, but they deny that universals have any existence independent of their instances.

Of course, there are other views on universals. Conceptualism entails that universals are mental constructions, and nominalism entails that they are linguistic constructions or that they do not exist at all. For present purposes, I lump these alternate views with in re realism. The important distinction is between *ante rem* realism and the others. Our question is whether, and in what sense, structures exist independently of the systems that exemplify them. Is it reasonable to speak of the natural-number structure, the real-number structure, or the set-theoretic hierarchy on the off chance that there are no systems that exemplify these structures?

One who thinks that there is no more to structures than the systems that exemplify them—an advocate of an in re view of structures—might be attracted to the program suggested at the very end of the previous section. Recall that from the structuralist perspective, it is the places-are-objects perspective that sanctions the thesis that numbers are objects. On the program in question, however, places-are-objects statements are not taken at face value but are understood as generalizations in the places-are-offices perspective. So "3 + 9 = 12" would come to something like "in any natural-number system *S*, the object in the 3 place of *S* *S*-added to the object in the 9 place of *S* results in the object in the 12 place of *S*." When paraphrased like this, seemingly bold ontological statements become harmless—analytic truths if you will. For example, "3 exists" comes to "every natural-number system has an object in its 3 place," and "numbers exist" comes to "every natural-number system has objects in its places."

In sum, the program of rephrasing mathematical statements as generalizations is a manifestation of structuralism, but it is one that does not countenance mathematical objects, or structures for that matter, as bona fide objects. Talk of numbers is convenient shorthand for talk about all systems that exemplify the structure. Talk of structures generally is convenient shorthand for talk about systems. A slogan for the program might be "structuralism without structures."[10]

Dummett [1991, chapter 23] makes the same distinction concerning the nature of structures. According to "mystical" structuralism, "mathematics relates to *abstract structures*, distinguished by the fact that their elements have no non-structural properties" (p. 295). Thus, for example, the zero place of the natural-number structure "has no other properties than those which follow from its being the zero" of that structure. It is not a set, or anything else whose nature is extrinsic to the structure. Dummett's mystical structuralist is thus an *ante rem* realist about structures. The other version of structuralism takes a "hardheaded" orientation: "According to it, a mathematical theory, even if it be number theory or analysis which we ordinarily take as intended to characterize *one* particular mathematical system, can never properly be so understood: it always concerns all systems with a given structure" (Dummett [1991, 296]). If the hardheaded structuralist countenances structures at all, it is only in an in re sense.

Parsons [1990, §§. 2–7] presents, but eventually rejects, a hardheaded view like this, which he dubs *eliminative structuralism*: "It . . . avoids singling out any one . . . system as the natural numbers. . . . [Eliminative structuralism] exemplifies a very natural response to the considerations on which a structuralist view is based, to see statements about a kind of mathematical objects as general statements about struc-

---

10. This slogan was adopted by Hellman [1996], after he read a draft of this chapter.

tures of a certain type and to look for a way of eliminating reference to mathematical objects of the kind in question by means of this idea" (p. 307). Benacerraf [1965, 291] settles on a hardheaded, eliminative, in re version of structuralism, when he writes, "Number theory is the elaboration of the properties of all structures of the order type of the numbers." This, of course, is of a piece with his rejection, noted earlier, of the thesis that numbers are objects.[11]

Thus, the eliminative structuralist program paraphrases places-are-objects statements in terms of the places-are-offices perspective. Recall that the places-are-offices orientation requires a background ontology, a domain of discourse. This domain contains objects that fill the places in the requisite (in re) structures. In the case of baseball defenses, the background ontology consists of people ready to play ball; in the case of chess configurations, the ontology consists of pieces of marble, wood, plastic, metal, and so on, manufactured in a certain way. In the case of mathematics, any old objects will do—so long as there are enough of them.

The main stumbling block of the eliminative program is that to make sense of a substantial part of mathematics, the background ontology must be quite robust. The *nature* of the objects in the final ontology does not matter, but there must be a lot of objects there. To see this, let Φ be a sentence in the language of arithmetic. According to eliminative structuralism, Φ amounts to something in the form:

(Φ') for any system S, if S exemplifies the natural-number structure, then $\Phi[S]$,

where $\Phi[S]$ is obtained from Φ by interpreting the nonlogical terminology and restricting the variables to the objects in S. If the background ontology is finite, then there are no systems that exemplify the natural-number structure, and so Φ' and $(\neg\Phi)'$ are both true. Because mathematics is not vacuous, this is unacceptable. We do not end up with a rendering of arithmetic if the background ontology is finite. Similarly, an eliminative-structuralist account of real analysis and Euclidean geometry requires a background ontology whose cardinality is at least that of the continuum, and set theory requires a background ontology that has the size of a proper class (or at least an inaccessible cardinal).

I suppose that one can maintain that there are infinitely many *physical* objects, in which case an eliminative account of arithmetic may get off the ground with a physical ontology. As we have seen (section 1), Field [1980] holds that each space-time point is a physical object. If this claim is plausible, then an eliminative structuralist might follow this lead with an account of analysis and geometry. Nevertheless, it seems reasonable to insist that there is *some* limit to the size of the physical universe. If so, then any branch of mathematics that requires an ontology larger than that of the physical universe must leave the realm of physical objects if these branches are not to be doomed to vacuity. Even with arithmetic, it is counterintuitive for an account of mathematics to be held hostage to the size of the physical universe.[12]

11. Here, Benacerraf uses "structure" as I use "system."

12. See Parsons [1990]. This is why Field [1980] himself does not attempt to reduce analysis and geometry to a theory of space-time points and regions. Incidentally, according to the view developed

There are three structuralist responses to this threat of vacuity. One is to maintain an eliminative program but postulate that enough abstract objects exist for all of the structures under study to be exemplified. That is, for each field of arithmetic, we assume that there are enough objects to keep that field from being vacuous. I call this the *ontological option*.

On this program, if one wants a single account for all (or almost all) of mathematics, then the background ontology of abstract objects must be quite big. As noted earlier, several logicians and philosophers think of the set-theoretic hierarchy as the ontology for all of mathematics. The universe is V. If one assumes that every set in the hierarchy exists, then there will surely be enough objects to exemplify just about any structure one might consider. Because, historically, one purpose of set theory was to provide as many isomorphism types as possible, set theory is rich fodder for eliminative structuralism. A structure, on this account, is an order type of sets, no more and no less.

The crucial feature of this version of eliminative structuralism is that the background ontology is not understood in structuralist terms. If the iterative hierarchy is the background, then set theory is not, after all, the theory of a particular structure. Rather, it is about a particular class of objects, the background ontology V. Perhaps from a different point of view, set theory can be thought of as the study of a particular structure *U*, but this would require another background ontology to fill the places of *U*. This new background ontology is not to be understood as the places of another structure or, if it is, we need yet another background ontology for *its* places. On the ontological option, we have to stop the regress of system and structure somewhere. The final ontology is not understood in terms of structures, even if everything else in mathematics is.

To be sure, there is nothing sacrosanct about Zermelo-Fraenkel set theory. Foundationalists have shown that mathematics can be rendered in theories other than that of the iterative hierarchy (e.g., Quine [1937]; Lewis [1991], [1993]). Among these are a dedicated contingent of mathematicians and philosophers who hold that the category of categories is the proper foundation for mathematics (see, for example, Lawvere [1966]).[13] The ultimate background ontology for eliminative structuralism can thus be the domain of any of several set theories or category theories.

A structuralist might be tempted to step back from this competition of background theories and wonder if there is a structure common to all of them. However, on the ontological option, this temptation needs to be resisted. The structures studied in two theories can be compared only in terms of a more inclusive theory.

Of course, eliminative structuralists need not consider their most powerful theory (or theories) to be about the background ontology. They may regard, say, ZFC and

in Maddy [1990, chapter 5], if the transitive closure of a set *s* contains only physical objects, then *s* itself is a physical object. It follows that there is a proper class of physical objects, and there are systems of physical objects that exemplify the set-theoretic hierarchy. Such systems are located where the original objects are. The thesis that sets of physical objects are themselves physical objects is criticized by Chihara [1990, chapter 10] and Balaguer [1994].

13. McLarty [1993] is a lucid and insightful start of a structuralist program in terms of category theory—topos theory in particular.

topos theory both as theories of specific structures. On the present version of eliminative structuralism, they do need to acknowledge the existence of the ultimate background ontology, but they need not develop a formal theory of this system.

On all versions of structuralism, the nature of the objects in the places of a structure does not matter—only the relations among the objects are significant. On the ontological option, then, the only relevant feature of the background ontology is its *size*. Are there enough objects to exemplify every structure the mathematician might consider? If a theory of this ontology is developed, the only relevant factor is the size of the ontology.

In correspondence and conversation, some nominalists express sympathy with a structuralist account of mathematics, but they quickly add that one should speak of "possible structures" rather than just structures. Our second option pursues this suggestion by modalizing eliminative structuralism. Instead of saying that arithmetic is about all systems of a certain type, one says that arithmetic is about all *possible* systems of a certain type. Again, let Φ be a sentence in the language of arithmetic. Earlier, on behalf of eliminative structuralism, I rendered Φ as "for any system S, if S exemplifies the natural-number structure, then $\Phi[S]$." With the present option, Φ is understood as

> for any *possible* system S, if S exemplifies the natural-number structure, then $\Phi[S]$,

or

> *necessarily*, for any system S, if S exemplifies
> the natural-number structure, then $\Phi[S]$.

The problem, of course, is to keep arithmetic from being vacuous without assuming that there is a system that exemplifies the structure. The solution here is to merely assume that such a system is possible. The same goes for real analysis and even set theory. Unlike the ontological option, here we do not require an actual, rich background ontology. Instead, we need a rich background ontology to be *possible*. I call this the *modal option*.

Hellman [1989] carries out a program like this in meticulous detail. The title of the book, *Mathematics without numbers*, sums things up nicely. It is a structuralist account of mathematics that does not countenance the existence of structures—or any other mathematical objects for that matter. Statements in a nonalgebraic branch of mathematics are understood as generalizations inside the scope of a modal operator. Instead of assertions that various systems exist, Hellman has assertions that the systems might exist.

Probably the central issue with the modal option is the nature of the invoked modality. What are we to make of the "possibilities" and "necessities" used to render mathematical statements? I presume that thinking of the possibility as *physical* possibility is a nonstarter, for reasons already given. Perhaps it is physically possible for there to be a system that exemplifies the natural-number structure, the real-number structure, or Euclidean space, but it is stretching this modal notion beyond recognition to claim that a system that exemplifies the set-theoretic hierarchy is physically possible (Maddy [1990, chapter 5] notwithstanding; see note 12 above). The relevant modal operator is not to be understood as *metaphysical* possibility either. Intuitively, if mathematical objects–like numbers, points, and sets–exist at all, then their existence is metaphysically necessarily. According to this intuition, "the natural numbers exist" is equivalent to both "possibly, the natural numbers exist" and "necessarily, the natural numbers exist" (assuming that the modal logic system S5 is sound for metaphysical necessity). Now, recall that on the first, ontological option there must be a sufficiently large realm of objects. Presumably, the items in the ontology are not metaphysically different from natural numbers. Thus, the existence and the possible existence of the items in the background ontology are equivalent. Thus, the use of metaphysical modality does not really weaken the ontological burden of eliminative structuralism (for an elaboration of a similar point, see Resnik [1992]).

For this reason, Hellman mobilizes the *logical* modalities for his eliminative structuralism. Our arithmetic sentence Φ becomes

for any *logically possible* system S, if S exemplifies
the natural-number structure, then $\Phi[S]$.

This maneuver gives the modal option its best shot. The modal structuralist needs to assume only that it is logically possible that there is a system that exemplifies the natural-number structure, the real-number structure, and so on.

Recall that in contemporary logic textbooks and classes, the logical modalities are understood in terms of sets. To say that a sentence is logically possible is to say that *there is* a certain set that satisfies it. According to the modal option of eliminative structuralism, however, to say that there is a certain set is to say something about every logically possible system that exemplifies the structure of the set-theoretic hierarchy. This is an unacceptable circularity. It does no good to render mathematical "existence" in terms of logical possibility if the latter is to be rendered in terms of existence in the set-theoretic hierarchy. Putting the views together, the statement that a sentence is logically possible is really a statement about all set-theoretic models of set theory. Who says there are such models? Once again, we have a menacing threat of vacuity. Hellman accepts this straightforward point, and so he demurs from the standard, model-theoretic accounts of the logical modalities. Instead, he takes the logical notions as *primitive*, not to be reduced to set theory. I return to this exchange of ontology for modality in chapter 7.

The third option avoids the eliminative program altogether and adopts an *ante rem* realism toward structures. Structures exist whether they are exemplified in a nonstructural realm or not. On this option, statements in the places-are-objects mode are taken literally, at face value. In mathematics, anyway, the places of mathematical structures are as bona fide as any objects are. So, in a sense, each structure exemplifies itself. Its places, construed as objects, exemplify the structure.

First, a disclaimer: In the history of philosophy, *ante rem* universals are sometimes given an explanatory primacy. It might be said, for example, that the reason the White House is white is that it participates in the Form of Whiteness. Or what *makes* a basketball round is that it participates in the Form of Roundness. No such explanatory claim is contemplated here on behalf of *ante rem* structures. I do not hold, for example, that a given system is a model of the natural numbers because it exemplifies the natural-number structure. If anything, it is the other way around. What makes the system exemplify the natural-number structure is that it has a one-to-one successor function with an initial object, and the system satisfies the induction principle. That is, what makes a system exemplify the natural-number structure is that it is a model of arithmetic. In much of the current literature, types do not carry this sort of explanatory burden, either. Thus, in this respect, *ante rem* structures are like types.

Michael Hand [1993, 188] states that an *ante rem* structuralist is indeed committed to structures that bear the explanatory burden: "Admittedly, . . . [t]he motivation behind structuralism has nothing to do with the possible explanatory function of abstract patterns. . . . Nonetheless, the structuralist is committed to more than [the] limited motivation suggests. After all, abstract patterns, structures, are not entities newly posited by the structuralist. . . . Instead, the structuralist is making use of things we already know something about, and that we already put to use metaphysically in various ways. Since this is so, she is responsible . . . to the metaphysical uses to which we already put them." The idea behind this pronouncement seems to be that since we *ante rem* structuralists are using a notion like the traditional one-over-many, we are committed to all of the features *and uses* of *ante rem* universals *as traditionally conceived*. Hand goes on to argue, quite insightfully, that nothing can bear the explanatory burden, and he concludes that structures do not exist. According to his pronouncement, it seems, once a (philosophical) notion has been debunked, no one is allowed to use a variation on that notion—even a variation that survives the debunking. Readers sympathetic with this pronouncement are invited to construe structures as a *new* sort of notion, one that is similar in some ways to traditional *ante rem* universals but does not bear their explanatory burden. To paraphrase Kripke, call structures "shmuniversals." Hand suggests that this maneuver leaves *ante rem* structuralism unmotivated. I take this book to provide some motivation, and I leave it to the reader to judge the matter.

To sum up, the three options are ontological eliminative structuralism, modal eliminative structuralism, and *ante rem* realism. I believe that the *ante rem* option is the most perspicuous and least artificial of the three. It comes closest to capturing how mathematical theories are conceived. Nevertheless, I do not mean to rule out the other options. Indeed, it follows from the thesis of structuralism that, in a sense, all three options are equivalent. As will be shown, each delivers the same "structure of structures." The next section provides a brief account of each option and a defense of their equivalence (see also chapter 7).

4 Theories of Structure

No matter how it is to be articulated, structuralism depends on a notion of two systems that exemplify the "same" structure. That is its point. Even if one eschews structures themselves, we still need to articulate a relation among systems that amounts to "have the same structure."

There are several relations that will do for this. I mention two, both of which are equivalence relations. The first is *isomorphism*, a common (and respectable) mathematical notion. Two systems are isomorphic if there is a one-to-one correspondence from the objects and relations of one to the objects and relations of the other that preserves the relations. Suppose, for example, that the first system has a binary relation R. If f is the correspondence, then $f(R)$ is a binary relation of the second system and, for any objects m, n, of the first system, R holds between m and n in the first system if and only if $f(R)$ holds between $f(m)$ and $f(n)$ in the second system. Informally, it is sometimes said that isomorphism "preserves structure."

Isomorphism is too fine-grained for present purposes. Intuitively, one would like to say that the natural numbers with addition and multiplication exemplify the same structure as the natural numbers with addition, multiplication, and less-than. However, the systems are not isomorphic, for the trivial reason that they have different sets of relations. The first has no binary relation to correspond with the less-than relation, even though that relation is definable in terms of addition: $x < y$ if and only if $\exists z(z \neq 0 \ \& \ x + z = y)$. There is, in a sense, nothing new in the "richer" system. Similarly, we would like to say that the various formulations of Euclidean plane geometry with different primitives all exemplify the same structure.

Resnik [1981] has formulated a more coarse-grained equivalence relation among systems (and structures) for this purpose. First, let R be a system and P a subsystem. Define P to be a *full subsystem* of R if they have the same objects (i.e., every object of R is an object of P) and if every relation of R can be defined in terms of the relations of P. The idea is that the only difference between P and R is that some definable relations are omitted in P. So the natural numbers with addition and multiplication are a full subsystem of the natural numbers under addition, multiplication, and less-than. Let M and N be systems. Define M and N to be *structure-equivalent*, or simply *equivalent*, if there is a system R such that M and N are each isomorphic to full subsystems of R. Equivalence is a good candidate for "sameness of structure" among systems.[14]

Notice that structure equivalence is characterized in terms of *definability*, a blatant linguistic notion. One consequence is that equivalence is dependent on the resources available in the background metalanguage (or the U-language). For example, in a standard first-order background, the natural numbers with successor alone are not equivalent to the natural numbers with addition and multiplication, because addition cannot be defined from successor in the first-order theory. However, the theories are equivalent in a second-order background (see Shapiro [1991, chapter 5]). The dependence on the background theory and, in particular, on its language should not be surprising. A recurring theme in this book is that a number of ontological matters are tied to linguistic resources.

Let us briefly consider what would be involved in rigorously developing each of the three options for structuralism: the ontological in re route, the modal in re route, and the *ante rem* route. Recall that the ontological option presupposes an ultimate (nonstructural) background ontology for all of mathematics. The first item on the agenda would thus be a detailed account of this background ontology. As above, the set-theoretic hierarchy V is a natural choice for the background, in which case there

---

14. Structure equivalence is analogous to definitional equivalence among theories (see chapter 7).

already is a developed theory—Zermelo-Fraenkel set theory. Next, we need an account of *systems* in this background ontology. This, too, has been done already, via standard model theory. An n-ary "relation" is a set of n-tuples and an n-place function is a many-one set of $(n + 1)$-tuples. A system is an ordered pair that consists of a domain and a set of relations and functions on it. Model theorists sometimes use words like "structure," "model," and "interpretation" for what I call a "system." In set theory, isomorphism and structure equivalence are also easily defined, thus completing the requisite eliminative theory of structuralism. In other words, using common model-theoretic techniques, set theorists can speak of systems that share a common structure. Notice that we do not find what I am calling "structures" in the ontology. All we have is isomorphism and structure-equivalence *among systems*. Recall that the slogan of eliminative structuralism is "structuralism without structures."

Because isomorphism and structure-equivalence are equivalence relations, one can informally take a structure to be an isomorphism type or a structure-equivalence type. So construed, a structure is an equivalence class in the set-theoretic hierarchy. Notice, however, that each nonempty "structure" is a proper class, and so it is not in the set-theoretic hierarchy. The relevant notions could be expanded to include proper class systems, but then we could not take a structure to be an equivalence class of systems unless we moved to a third-order background.

With admirable rigor and attention to detail, Hellman [1989] develops the modal option. Modal operators are added to a standard formal language, and the aforementioned notions of "system" and isomorphism are invoked. A sentence of arithmetic, say, is rendered as a statement about all *possible* systems that satisfy the (second-order) Peano axioms.[15] Although the program is correctly characterized as "structuralist," there is no notion of *structure* in the official modal language.

Finally, the *ante rem* option requires a *theory* of structures. The plan is to stop the regress of system and structure at a universe of structures. Because structures themselves are in the ontology, we need an identity relation on structures. Resnik [1981] seems to hold that there is no such identity relation, arguing that there is no "fact of the matter" as to whether two structures are the same or different, or even whether two systems exemplify the same structure (but see Resnik [1988, 411 note 16]). Notice that this goes against the Quinean dictum "no entity without identity." Quine's thesis is that within a given theory, language, or framework, there should be definite criteria for identity among its objects. There is no reason for structuralism to be the single exception to this. If we are to have a theory of structures, we need an identity relation on them. Perhaps Resnik demurs at the development of such a theory (see Resnik [1996]). It seems to me, however, that if one is to speak coherently about structures and avoid the ontological and modal options, then such a theory is needed, at least at some stage of analysis. In Quinean terms, the need to regiment one's infor-

15. Hellman's account avoids the use of the notion of "possible system," because he does not countenance an ontology of possibilia. The program also does not directly use semantic notions like "satisfaction."

mal language applies to its philosophical parts as well as the more respectable scientific neighborhoods.

When Resnik states that there is no "fact of the matter" concerning the identity of structures, he may just mean that the ordinary use of the relevant terms does not determine a unique identity relation. This much is quite correct. To regiment our language, we would need to *define* the requisite identity relations, but there is no uniquely best candidate for this. Like the identification of places from different structures (see section 2), the identity relation we need is more a matter of decision or invention, based on convenience, rather than a matter of discovery. But we do need to decide.

We take identity among structures to be primitive, and isomorphism is a congruence among structures. That is, we stipulate that two structures are identical if they are isomorphic. There is little need to keep multiple isomorphic copies of the same structure in our structure ontology, even if we have lots of systems that exemplify each one.[16] We could also "identify" structures that are structure-equivalent, but it is technically inconvenient to do so.

With the ontological option just delimited, systems are constructed from sets in the fashion of model theory, and structures are certain equivalence types on systems. For the *ante rem* option, we axiomatize the notion of structure directly. The envisioned theory has variables that range over structures, and thus a quantifier "all structures." Each structure has a collection of "places" and relations on those places. Once again, the places-are-objects perspective is taken seriously. The theory thus has a second sort of variable that ranges over places in structures.

The category theorist characterizes a structure or a type of structure in terms of the structure-preserving functions, called "morphisms," between systems that exemplify the structures. For many purposes, this is a perspicuous approach (see McLarty [1993]), but here I provide an outline of a more traditional axiomatic treatment. In effect, structure theory is an axiomatization of the central framework of model theory.

Because it appears to be necessary to speak of relations and functions on places, I adopt a second-order background language (see Shapiro [1991]). An alternative to this would be to include a rudimentary theory of collections as part of the theory.

First, a *structure* has a collection of *places* and a finite collection of functions and relations on those places. The isomorphism relation among structures and the satisfaction relation between structures and formulas of an appropriate formal language are defined in the standard way. We could stipulate that the places of different structures are disjoint, but there is no reason to do so. Our first axiom, concerning the existence of structures is simpleminded but ontologically nontrivial:

Infinity: There is at least one structure that has an infinite number of places.

Because structures, places, relations, and functions are the only items in the ontology, everything else must be constructed from those items. Thus, a *system* is defined to be a collection of places from one or more structures, together with some

16. The sequence of natural numbers contains many isomorphic copies of itself, but there is only one natural-number structure. In structure theory, the copies are systems.

relations and functions on those places. For example, the even-number places of the natural-number structure constitute a system, and on this system, a "successor" function could be defined that would make the system exemplify the natural-number structure. The "successor" of n would be $n + 2$. Similarly, the finite von Neumann ordinals are a system that consists of places in the set-theoretic hierarchy structure, and this system also exemplifies the natural-number structure, once the requisite relations and functions are added. Other systems consist of the places of several structures, with relations defined on their "objects." For example, a nonstandard model of simple first-order arithmetic (with successor alone) consists of the natural numbers "followed by" the integers.

The places of a given structure—considered from the places-are-objects perspective—are objects. As characterized here, then, each structure is also a system.

Our next axioms concern what may be called "substructures":

Subtraction: If S is a structure and R is a relation of S, then there is a structure S' isomorphic to the system that consists of the places, functions, and relations of S except R. If S is a structure and f is a function of S, then there is a structure S'' isomorphic to the system consisting of the places, functions, and relations of S except f.

Subclass: If S is a structure and c is a subclass of the places of S, then there is a structure isomorphic to the system that consists of c but with no relations and functions.

Addition: If S is a structure and R is any relation on the places of S, then there is a structure S' isomorphic to the system that consists of the places, functions, and relations of S together with R. If S is a structure and f is any function from the places of S to places of S, then there is a structure S'' isomorphic to the system that consists of the places, functions, and relations of S together with f.

That is, one can remove places, functions, and relations at will; and one can add functions and relations.

The remaining objective for my theory is to assure the existence of large structures. The next axiom is an analogue of the powerset axiom of set theory:

Powerstructure: Let S be a structure and s its collection of places. Then there is a structure T and a binary relation R such that for each subset $s' \subseteq s$ there is a place x of T such that $\forall z(z \in s' \equiv Rxz)$.

Each subset of the places of S is related to a place of T, and so there are at least as many places in T as there are subsets of the places of S. Thus, the collection of places of T is at least as large as the powerset of the places of S. The powerstructure axiom can be formulated in the second-order background language.

So far, structure theory resembles what is called Zermelo set theory. We have the existence of the natural-number structure, the real-number structure, a structure whose size is the powerset of that, and so on. The smallest standard model of the theory has the size of $V_{2\omega}$, the smallest standard model of Zermelo set theory.

To get beyond the analogue of Zermelo set theory, my next item is the analogue of the replacement principle:

Replacement: Let S be a structure and f a function such that for each place x of S, fx is a place of a structure, which we may call S_x. Then there is a structure T that is (at least) the size of the union of the places in the structures S_x. That is, there is a function g such that for every place z in each S_x there is a place y in T such that $gy = z$.

The idea is the same as in set theory. There is a structure at least as large as the result of "replacing" each place x of S with the collection of places of a structure S_x. With this axiom, every standard model of structure theory is the size of an inaccessible cardinal. In effect, structure theory is a reworking of second-order Zermelo-Fraenkel set theory.

The main principle behind structuralism is that any coherent theory characterizes a structure, or a class of structures. For what it is worth, I state this much:

Coherence: If Φ is a coherent formula in a second-order language, then there is a structure that satisfies Φ.

The problem, of course, is that it is far from clear what "coherent" comes to here. The question of when a theory is coherent, and thus describes a structure (or class of structures), will occupy us later several times (e.g., section 5 of this chapter, and chapter 4).[17] Notice, for now, that because we are using a second-order language, simple (proof-theoretic) consistency is not sufficient to guarantee that a theory describes a structure or class of structures. Because the completeness theorem fails, there are consistent second-order theories that are not satisfiable (see Shapiro [1991, chapter 4]). Consider, for example, the conjunction P of the axioms of Peano arithmetic together with the statement that P is not consistent. Contra Hilbert, consistency does not imply existence even for a structuralist. We need something more like satisfiability, but the latter is usually formulated in terms of the set-theoretic hierarchy (or some other ontology): a theory is satisfiable if *there is a model* for it. There is no getting away from this problem, but perhaps the circle is not vicious.

We can, of course, add an axiom that, say, second-order ZFC is coherent, and thus conclude that there is a structure the size of an inaccessible cardinal. Another, less ad hoc route to large structures is to assume that structure theory itself is coherent, and so is any theory consisting of structure theory plus any truth of structure theory. This suggests a reflection scheme. Let Φ be any (first- or second-order) sentence in the language of structure theory. Then the following is an axiom:

Reflection: If Φ, then there is a structure S that satisfies the (other) axioms of structure theory and Φ.

Letting Φ be a tautology, the principle entails the existence of a structure the size of an inaccessible cardinal. Letting Φ be the conjunction of the other axioms of structure theory (or ZFC) plus the existence of a structure the size of an inaccessible cardinal, the reflection principle entails the existence of a structure the size of the second inaccessible cardinal, and it goes on from there.[18]

One might think that I am inviting a version of Russell's paradox. Is there a structure of all structures? The answer is that there is not, just as there is no set of all sets. Because a "system" is a collection of places in structures (together with relations),

---

17. A more general principle is that every coherent *collection* Γ of formulas is satisfied by a structure, but to be picky, one should add a proviso that Γ is not the size of a proper class.

18. A variation of the reflection principle, along the lines of Bernays [1961], entails the existence of structures the size of a Mahlo cardinal, a hyper-Mahlo cardinal, up to an indescribable cardinal. A suitable reflection also entails the powerstructure and replacement axioms. See Shapiro [1987] and Shapiro [1991, chapter 6] for a study of higher-order reflection principles. See also Levy [1960].

some systems are "too big" to exemplify a structure. This defect could be avoided, as it is in set theory, by stipulating that there are no systems the size of a proper class. The relevant axioms would be that for every system *S*, there is no function from the places of *S* onto the class of all places in all structures.

The point is that structuralism is no more (and no less) susceptible to paradox than set theory, modal structuralism, or category theory. Some care is required in regimenting the informal discourse, but it is a familiar sort of care. One can ascend to another level and interpret the objects of the domain of the structure language as the places in a superstructure. But, as with set theory, we cannot take this structure to be *in* the range of the (structure) variables of the original theory. The ontology changes as we go to a metalanguage. This, however, is rarified analysis. Normally, there is little need to ascend beyond the original structuralist language, at least not in this way, just as there is little need to ascend to some sort of superset theory.

Enough on the details of structure theory. For someone familiar with axiomatic set theory, everything is straightforward. The reason the development goes smoothly is that structure theory, as I conceive it, is about as rich as set theory. It has to be if set theory itself is to be accommodated as a branch of mathematics. In a sense, set theory and the envisioned structure theory are notational variants of each other. In particular, structure theory without the reflection principle is a variant of second-order ZFC, and structure theory with the reflection principle is a notational variant of set theory with a corresponding reflection principle.

Nevertheless, for present purposes, structure theory is a more perspicuous and less artificial framework than set theory. If nothing else, structure theory regards set theory (and perhaps even structure theory itself) as one branch of mathematics among many, whereas the ontological option makes set theory (or another designated theory) the special foundation. However, even this is not a major advantage, because the equivalence and mutual interpretability of the frameworks are straightforward. Anything that can be said in either framework can be rendered in the other. Talk of structures, as primitive, is easily "translated" as talk of isomorphism or equivalence types over a universe of (primitive) sets. In the final analysis, it does not really matter where we start.

The same goes for the modal option, but the articulation and details of that equivalence will be postponed (chapter 7). The upshot is the same as with set theory and structure theory. Anything that can be said in the modal structural system (of Hellman [1989]) can be rendered in either the set language or the structure language.

In short, on any structuralist program, *some* background theory is needed. The present options are set theory, modal model theory, and *ante rem* structure theory. The fact that any of a number of background theories will do is a reason to adopt the program of *ante rem* structuralism. *Ante rem* structuralism is more perspicuous in that the background is, in a sense, minimal. On this option, we need not assume any more about the background ontology of mathematics than is required by structuralism itself.[19] But when all is said and done, the different accounts are equivalent.

19. McLarty [1993] makes the same claim on behalf of a category-theoretic foundation of mathematics.

The smooth translations between the various theories also suggest that none of them can claim a major epistemological advantage over the others. The sticky epistemic problems get "translated" as well. Probably the deepest epistemic problem with standard set theory is how we can know anything about the abstract, acausal universe of sets. Which sets exist? How do we know? Formidable problems indeed. In the case of structure theory, the corresponding problems concern how we can know anything about the realm of structures. Which structures exist? How can we tell? The very same problem, in the case of modal structuralism, is how we know anything about the various possibilities. Which structures are possible? How can we tell?

The upshot of this section, then, is that there are several ways to render structuralism in a rigorous, carefully developed background theory, and there is very little to choose among the options. In a sense, they all say the same thing, using different primitives. The situation with structuralism is analogous to that of geometry. Points can be primitive, or lines can be primitive. It does not matter because, in either case, the same structure is delivered. The same goes for structuralism itself. Set theory and structure theory are equivalent, in the sense defined above. To speak loosely, the same "structure of structures" is delivered. Modal structuralism also fits, once the notion of "equivalence" is modified for the modal language.

5 Mathematics: Structures, All the Way Down

I articulate the picture of *ante rem* structuralism here, to demonstrate why this account is more perspicuous than the others, and to continue the dialectic of articulating the notions of structure, theory, and object.

Thus far, I have spoken freely of ordinary, nonmathematical structures, such as baseball defenses, governments, and chess configurations, along with mathematical structures like the natural numbers and the set-theoretic hierarchy. One might wonder whether the word "structure" is univocal across these uses. What if anything distinguishes *mathematical* structures from the others?

One possible answer is that in principle, there is no difference in kind between mathematical and nonmathematical structures. This has a clean, holistic ring to it—at least on the ontological front. A cocky holist might go on to claim that the only difference between the "mathematical" structures and the others is that the former are the ones studied by mathematicians qua mathematicians. If enough mathematicians took a professional interest in baseball defenses, then baseball defenses would be mathematical structures. If mathematicians took a professional interest in chess, then chess configurations would be mathematical structures. Typically, the structures studied by mathematicians are complex and interesting, but this does not mark a difference in kind.[20]

A slightly more cautious claim would be that the difference between mathematical and ordinary structures is not so much in the structures themselves but in the way

20. Mea culpa. In the past, when responding to questions, I would usually take this cocky holistic line. This would be greeted with frowns and incredulous stares from my audiences—with the possible exception of ontic holists.

they are studied. Mathematics is the *deductive* study of structures. The mathematician gives a description of the structure in question, independently of any systems this structure may be the structure of. Anything the mathematician, qua mathematician, goes on to say about the structure must follow from this description. Ordinary structures are not usually studied this way, or not studied this way exclusively. Recall the passage from Resnik [1982] in section 1. When the imaginary linguists discover that Tenglish is not the structure of spoken English, presumably by comparing the structure as defined to the spoken language, they lose interest in the structure. Their methodology is focused on what Tenglish is supposed to be a structure of. In contrast, if Tenglish is internally coherent, a mathematician can go on to study the structure, independently of whether it is exemplified in any real or even possible linguistic community. On this orientation, ontic holism is maintained, but mathematics is distinguished by its deductive epistemology.

This account is not cautious enough. Although there are interesting borderline cases between mathematical and ordinary structures, which will further occupy us when we get to applications (chapter 8), there are important differences between the two types of structures. A vague border is still a border.

One difference between the types of structures concerns the nature of the *relations* between the officeholders of exemplifying systems. Consider our standby, the baseball-defense structure. Imagine a system that consists of nine people placed in the configuration of a baseball defense but hundreds of miles apart—the "right fielder" in New York, the "center fielder" in Detroit, and so on. This system does not exemplify the structure of baseball defense, although one might say that it simulates or models the structure. There is an implicit requirement that the player at first base be within a certain distance of first base, the pitcher, and so forth. If not, then it is no baseball defense. In mathematical structures, on the other hand, the relations are all *formal*, or *structural*. The only requirements on the successor relation, for example, are that it be a one-to-one function, that the item in the zero place not be in its range, and that the induction principle hold. No spatiotemporal, mental, personal, or spiritual properties of any exemplification of the successor function are relevant to its being the successor function.

Although these examples may point in a certain direction, there is a problem of precisely formulating this notion of a "formal" relation. There are clear cases of formal relations and there are clear cases of nonformal relations. Surely, if a relation involves a *physical* magnitude like distance or a personal property like intelligence or age, then it is not formal. Being thirty-five years of age or older is not a formal property. One can leave things at this intuitive level, letting borderline cases take care of themselves. Accordingly, the border between mathematical and nonmathematical structures may not be sharp. Perhaps we can do better. If each relation of a structure can be completely defined using only logical terminology and the other objects and relations of the system, then they are all formal in the requisite sense. A slogan might be that formal languages capture formal relations. This is still not an adequate definition of "formal" or "structural" relation, however, because it is not clear how to formulate the logical/nonlogical boundary without begging any questions (see Shapiro [1997]).

Tarski [1986] proposed a criterion for the logical/nonlogical boundary that seems particularly apt here—whatever its fate in the philosophy of logic (see Sher [1991] for an insightful elaboration). His idea is that a notion is logical if its extension is unchanged under every permutation of the domain. Thus, for example, the property among sets of being nonempty is logical, because any permutation of the domain takes nonempty sets (of objects in the domain) to nonempty sets, and any such permutation takes the empty set to itself. The property of being thirty-five years of age or older is not logical, because there are permutations of the domain (of people) that take someone older than thirty-five to someone younger.

Because, in a permutation, any object can be replaced by any other, a notion that is invariant under all permutations ignores any nonstructural or intrinsic features of the individual objects. In these terms, the present proposal is that a relation is *formal* if it can be completely defined in a higher-order language, using only terminology that denotes Tarski-logical notions and the other objects and relations of the system, with the other objects and relations completely defined at the same time. All relations in a mathematical structure are formal in this sense.

If this definition of "formal" is adopted, then it is immediate that any relation that is logical in Tarski's sense is formal. However, it does not follow that all formal relations are logical. For example, neither 0 nor the successor function is Tarski-logical, because there are permutations of the natural numbers that take 0 to something else and there are permutations that do not preserve the successor function. Suppose, however, that we go up a level. Notice that any permutation of the natural numbers takes the successor function to the successor function of a (possibly different) natural-number system on the same domain. Likewise for 0. That is, if f is a permutation of the natural numbers, then $f(0)$ occupies the 0 place in a new system S, and m is the successor of n in S if $f^{-1}(m)$ is the successor of $f^{-1}(n)$ in the original natural numbers. The new system S exemplifies the natural-number structure. Thus, the notion of *natural-number system* $<N, 0, s>$ is logical in Tarski's sense: any permutation of the domain takes a natural-number system to a natural-number system. In general, for any mathematical structure S, the notion of "exemplifies structure S" is logical in Tarski's sense. This is a pleasing feature of the combination of structuralism and the given account of logical notions. It manifests the two slogans that mathematics is the science of structure, and that logic is topic-neutral.

Another important difference between mathematical and ordinary structures concerns the sorts of items that can occupy the places in the structures. Imagine a system that consists of a ballpark with nine piles of rocks, or nine infants, placed where the fielders usually stand. Imagine also a system of chalk marks on a diagram of a field, on which a baseball manager makes assignments and discusses strategy. Intuitively, neither of these systems exemplifies the defense structure. A system is not a baseball defense unless its positions are filled by people prepared to play ball. Piles of rocks, infants, and chalk marks are excluded. Prima facie, these requirements on the officeholders in potential defense systems are not "structural." For example, the requirement that the officeholders be people prepared to play is not described solely in terms of relations among the offices and their occupants. The system of rock piles and the system of chalk marks can perhaps be said to *model* or *simulate* the baseball-defense structure, but they do not exemplify it. Similarly, there is no possible system that exemplifies the U.S. government (before the year 2017) in which my eldest daughter is president. The president must be thirty-five years of age, chosen by the electoral college, and be a native-born citizen. The age, birth, and election requirements are not structural, in that those requirements are not described in terms of relations of the officeholders to each other. There are, again, systems that model or simulate the government, and my daughter has the office of president in some of these, but simulating and exemplifying are not the same thing.

In contrast, *mathematical* structures are *freestanding*. Every office is characterized completely in terms of how its occupant relates to the occupants of the other offices of the structure, and any object can occupy any of its places. In the natural-number structure, for example, there is no more to holding the 6 office than being the successor of the item in the 5 office, which in turn is the successor of the item in the 4 office. Anything at all can play the role of 6 in a natural-number system. Any *thing*. There are no requirements on the individual items that occupy the places; the requirements are solely on the relations between the items. A consequence of this feature is that in mathematics there is no difference between simulating a structure and exemplifying it.[21]

The freestanding nature of mathematical structures and the "formal" or "structural" nature of their relations are connected to each other. Suppose that a structure *S* has a nonformal relation, say, one that involves a physical magnitude, such as distance. For example, let it be required that the occupants of two particular places be ninety feet apart. Then *S* cannot be free-standing. The places of *S* that bear the distance relations cannot be filled with abstract objects, for example, because such objects do not have distance relations with each other. Similarly, if some relations of *S* require the objects to be movable, then objects that cannot be easily moved, like stars, cannot fill those places. If, on the other hand, all of the relations in a structure are formal, then any objects at all can fill the places. Insofar as the relations are formal, the structure is freestanding.

As we have seen, the places in the natural-number structure can be occupied by places in other structures (like finite von Neumann ordinals). Even more, the places in the natural-number structure can be occupied by the same or other natural numbers. The even numbers and the natural numbers greater than 4 both exemplify the natural-number structure. In the former, 6 plays the 3 role, and in the latter 8 plays the 3 role. In the series of primes, 7 plays the 3 role. The *ante rem* account of structures easily accommodates this freestanding feature of mathematical structures. Places of structures, considered from the places-are-objects perspective, can occupy places in the same or in different structures.

As noted earlier, there is one trivial example. In the system of natural numbers, 3 itself plays the 3 role. That is, the number 3, in the places-are-objects perspective, occupies the 3 office. The natural-number structure itself exemplifies the natural-

21. I am indebted to Diana Raffman and Michael Tye for several insightful conversations on these matters.

number structure. Hand [1993] argues that the freestanding feature of structures, construed *ante rem*, invites a Third Man regress. It is a Third ω. Both the system of finite von Neumann ordinals and the system of Zermelo numerals exemplify the natural-number structure. So do the natural numbers themselves, qua places-are-objects. Thus, the argument goes, we need a new structure, a super natural-number structure, which the original natural-number structure shares with the finite von Neumann ordinals and the Zermelo numerals. Actually, we need no such thing. The best reading of "the natural-number structure itself exemplifies the natural-number structure" is something like "the places of the natural-number structure, considered from the places-are-objects perspective, can be organized into a system, and this system exemplifies the natural-number structure (whose places are now viewed from the places-are-offices perspective)." In each case, there is no need for a Third.[22] The natural-number structure, as a system of places, exemplifies itself. The Third ω is the first ω.

Eliminative in re structuralist programs do not fully accommodate the freestanding nature of mathematical structures. As we have seen, on both eliminative options, there is no places-are-objects perspective. On this view, numbers are not objects, and so cannot be organized into systems. Strictly speaking, on either eliminative program, neither the natural-number structure nor numbers exist (as objects), and so such items cannot fill the places of structures.

On the other side of the ledger, there is not even a prima facie Third Man concern with eliminative in re structuralism. In general, if a structure is not freestanding, then there is no problem with a Third. No one would say, for example, that the baseball-defense structure is itself a baseball defense. You cannot play ball with the places of a structure; people are needed. Thus, if one is still bothered by the possibility of a Third ω, it might be best to eschew freestanding structures and adopt an eliminative program.

Parsons [1990] delimits an important distinction between different levels of *abstracta*: "Pure mathematical objects are to be contrasted not only with concrete objects, but also with certain abstract objects, that I call quasi-concrete, because they are directly 'represented' or 'instantiated' in the concrete. Examples might be geometric figures (as traditionally conceived), symbols whose tokens are physical utterances or inscriptions, and perhaps sets or sequences of concrete objects" (p. 304). Parsons's contrast is aligned with the matters under discussion here. His quasi-concrete objects are naturally organized into systems; his point is that the structures of such systems are not freestanding. Prima facie, only inscriptions of some sort can exemplify linguistic types, and, at least traditionally, only points in space can exemplify geometric points.

Parsons argues that a "purely structuralist account does not seem appropriate for quasi-concrete objects, because the representation relation is something additional

22. See Dieterle [1994, chapter 1] for a further discussion of the Third Man argument in the context of structuralism and a more detailed reply to Hand [1993]. Dieterle relates the present issue to some contemporary treatments of the traditional Third Man problem.

to intra structural relations." Because quasi-concrete objects "have a claim to be the most elementary mathematical objects," structuralism is not the whole story about mathematics. There is more to mathematics than what is indicated by the slogan "the science of structure."

Several responses to Parsons's charge are available. First, the structuralist might argue that quasi-concrete objects are not really mathematical objects. This is surely counterintuitive, because sets, geometric figures, and strings seem to be mathematical if anything is. Second, one might argue that Parsons's distinction is ill founded. There are not really any levels of *abstracta* to accommodate. As indicated, I demur from this option. The distinction is well taken, if not precise. Third, and less radically, one might claim that structures of quasi-concrete *abstracta* lie on the border between mathematical and ordinary structures and that the structures of quasi-concrete *abstracta* can be replaced with freestanding, formal ones. As far as mathematics goes, this replacement is virtually without loss. In particular, we can concede Parsons's point and try to delimit the role of quasi-concrete objects, showing how they are perhaps restricted to motivation and epistemology. The latter strategy is consistent with Parsons's own conclusions.

A brief look at the history of mathematics shows that the structures of quasi-concrete objects have been gradually supplanted by freestanding structures whose relations are formal. Consider geometry. From antiquity through the eighteenth century, geometry was the study of physical space, perhaps idealized. The points and lines of Euclidean geometry are points and lines of space. Thus, they are concrete or quasi-concrete, and their structure is not freestanding. Moreover, relations like "betweenness" and "congruence" are not formal. For point B to be between A and C, it must lie on a line connecting them, with A (physically) on one side and C on the other. For two line segments to be congruent, they must be the same length. Because of various internal developments, however, geometry came to be construed more and more formally, and thus more and more structurally. Along the way, nonspatial systems were construed as exemplifying the structure of various geometries. In analytic geometry, for example, the structure of Euclidean geometry is exemplified with a system of triples of real numbers. There is, of course, a "betweenness" relation of real analysis, in which π is between 3.1 and 3.2. This relation is similar to the "betweenness" of geometry, but the similarity is just structural, or formal. Real numbers are not actually parts of locations in space—but, as we now know, the structures are the same.[23] The subsequent use of idealized "points" and the use of analogues of complex analysis in geometry provided the crucial motivation for the move to a formal, structural construal of geometry. It became ever more difficult to understand the techniques, "constructions," and even the ontology of geometry as connected essentially to physical space. In chapter 5, I take a further look at some of these developments.

Unlike geometry, string theory does not have a long and hallowed history, but one can see a similar, if abbreviated development. Intuitively, strings are linguistic

---

23. Dedekind [1872] effectively exploited the structural similarities between the points on a line and real numbers in his celebrated treatment of continuity.

types, the forms of *written marks*. Because only written marks can be tokens of these types, the structures are not freestanding. Moreover, the central operation of the theory is concatenation, which is not formal. Two strings are concatenated when they are placed (physically) next to each other. Thus, strings are quasi-concrete. Nevertheless, it is not much of a stretch to see a more formal look to contemporary string theory. For one thing, it is now common to consider more abstract models of string theory. With Gödel numbering, for example, logicians consider natural numbers or sets to be strings, in which case concatenation is given an interpretation as an arithmetic or set-theoretic operation. In other words, systems of numbers or sets exemplify the structure of strings. Moreover, logicians now regularly consider infinitely long strings in a variety of contexts. Surely those strings cannot be instantiated with physical or spoken inscriptions. I return to the nature of strings and their role in epistemology in chapter 4.

Set theory is a most interesting case study for *ante rem* structuralism, and Parsons himself treats it at some length (in [1990] and in much more detail in [1995]). Like geometry and string theory, the intuitive ideas that underlie and motivate current axiomatic set theory are not structural. Teachers and elementary textbooks usually define a set to be a *collection* of its elements. Although it is quickly added that a set is not to be thought of as the result of a physical or even a mental collecting, there still seems to be more to membership than a purely formal relation between officeholders. Parsons and others note that there are actually different conceptions of "set" that are invoked in the motivation of axiomatic set theory.[24] One of them "is the conception of a set as a totality 'constituted' by its elements, so that it stands in some kind of ontological dependence on its elements, but not vice versa. This would give to the membership relation some additional content, still very abstract but recognizably more than a pure structuralism would admit" (Parsons [1990, 332]). A second motivating notion is the idea of a set as the extension of a predicate, so that each set is somehow ontologically dependent on the predicate and not on its elements. Parsons argues that neither of these motivating notions quite matches the one delivered in Zermelo-Fraenkel set theory. The Zermelo-Fraenkel notion departs "from concrete intuition at least when it admits infinite sets," and it departs from the predicative notion when it "admits impredicatively defined sets" ([1990, 336]). The upshot is that it may be best to view the structure delivered in modern set theory as freestanding and formal: "The result of these extensions . . . is that the elements of the original [nonstructural] ideas that are preserved in the theory have a purely formal character. For example, the priority of the elements of a set to the set, which is usually motivated by appealing to the first of [the] two informal conceptions is reflected in the theory itself by the fact that membership is a well-founded relation" (p. 336). Well-foundedness can be characterized in a second-order language using no nonlogical terminology: a relation E is well-founded if and only if $\forall P[\exists x Px \rightarrow \exists x(Px \mathbin{\&} \forall y(Eyx \rightarrow \neg Py))]$ (see

24. Parsons is a major contributor to a substantial literature on the philosophical underpinnings of axiomatic set theory. See Benacerraf and Putnam [1983, part 4].

Shapiro [1991, chapter 5]). Thus, well-foundedness is a formal property in the Zermelo-Fraenkel structure, and it replaces the quasi-concrete notion of "priority."

Parsons's careful conclusion is that one can and should overcome some intuitive, prereflective reasons for taking the domain of Zermelo-Fraenkel set theory non-structurally. The "universe of sets" is a collection of places "related in a relation called 'membership' satisfying conditions that can be stated in the language of set theory" (Parsons [1990, 332]). The set-theoretic hierarchy is a freestanding structure.[25]

This conclusion generalizes to all of mathematics, or at least to all of pure mathematics: "A structuralist view of higher set theory will then oblige us to accept the idea of a system of objects that is really no more than a structure. But then there is no convincing reason not to adopt it in other domains of mathematics, in particular in the case of the natural numbers. It would be highly paradoxical to accept Benacerraf's conclusion that numbers are not objects and yet accept as such the sets of higher set theory" (Parsons [1990, 332]). Amen. The path urged here, via *ante rem* structuralism, is to accept both numbers and sets, on a par, as objects. They are places-as-objects. Parsons comes close to the same conclusion: "The absence of notions whose non-formal properties really matter . . . makes mathematical objects on the structuralist view continue to seem elusive, and encourages the belief that there is some scandal to human reason in the idea that there are such objects. My claim is that something close to the conception of objects of this kind, already encouraged by the modern developments of arithmetic, geometry, and algebra, is forced on us by higher set theory" (p. 335).

So far, so good; but where is the problem? We have spoken of the "transition . . . from dealing with domains of a more concrete nature to speaking of objects only in a purely structural way." The problem is that this transition "leaves a residue. The more concrete domains, often of quasi-concrete objects, still play an ineliminable role in the explanation and motivation of mathematical concepts and theories. . . . The explanatory and justificatory role of more concrete models implies . . . that [structuralism] is not the right legislation even for the interpretation of modern mathematics" (p. 338). So Parsons proposes a caveat to structuralism. If we kick away the ladder of the concrete or quasi-concrete objects, then we cannot motivate or even justify some mathematical theories. For example, teachers often refer to sequences of linguistic types in order to motivate the natural-number structure. Hilbert [1925] himself invoked a collection of sequences of strokes, a quasi-concrete structure, to define the objects of finitary mathematics. Parsons also notes that at least the lower portions of the set-theoretic hierarchy have quasi-concrete instantiations. The quasi-concrete seem to be a main exemplar of mathematical structures.

Maddy [1990, 174–175] makes a related point, claiming that there is an epistemological disanalogy between arithmetic and set theory. She agrees that a structuralist understanding of the natural numbers is "appealing partly because our understanding of arithmetic doesn't depend on which instantiation of the number structure

---

25. Hellman [1989, 53–73] also treats set theory structurally, but not as the theory of a freestanding structure.

we choose to study." Set theory is different: "Experience with any endless row might lead us to think that every number has a successor, but it is experience with sets themselves that produces the intuitive belief that any two things can be collected into a set. . . . [T]hough any instantiation of the natural number structure can give us access to information about that structure, our information about the set-theoretic hierarchy structure comes from our experience with one particular instantiation."

I can put the point in present terms. Recall the coherence principle in the development of structuralism in section 4:

> If Φ is a coherent formula in a second-order language,
> then there is a structure that satisfies Φ.

This is a central (albeit vague) principle of structuralism. There is no getting around the fact that systems of quasi-concrete objects play a central role in convincing us that some mathematical theories are coherent and thus characterize well-defined structures (especially because consistency is not sufficient for coherence). The best way to show that a structure exists is to find a system that exemplifies it. At some point, we have to appeal to items that are not completely structural (unless somehow everything—every thing—is completely structural; see chapter 8). And at some point, we have to appeal to items that are not completely concrete, given the size of most mathematical structures. So we appeal to the quasi-concrete. If we completely eschew quasi-concrete systems, we lose any motivation or intuitive justification that even arithmetic and geometry (and string theory) are well motivated or even coherent.

Another reason to think that the quasi-concrete cannot be eliminated is that I have appealed to quasi-concrete items in order to characterize the very notion of a *structure*. Recall that a structure is the form of a system, and a system is a collection of objects under various relations. The notion of "collection" is an intuitive one. There is something fishy about appealing to the set-theoretic hierarchy, as a freestanding *ante rem* structure, in order to explicate the notion of "collection" in the characterization of "system" and thus "structure." Where did we get on this merry-go-round, and how do we get off?

A related point concerns the practice of characterizing specific structures using a second-order language. Such languages make literal use of intuitive notions like "predication" or "collection." A crucial step in the defense of second-order languages is that we have a serviceable, intuitive grasp of notions like "all subsets" (see Shapiro [1991]). This notion is also quasi-concrete. Boolos [1984] (and [1985]) has proposed an alternate understanding of monadic, second-order logic, in terms of plural quantifiers, which many philosophers have found attractive. Parsons [1995] contains an insightful discussion of pluralities in the context of structuralism. He shows that we are dealing with yet another quasi-concrete notion.

In all cases, then, the conclusion is the same. We can try to hide the quasi-concrete, but there is no running away from it. Parsons's caveat is well taken. However, the caveat does not undermine the main ontological thesis of *ante rem* structuralism, the idea that the subject matter of a branch of pure mathematics is well construed as a class of freestanding structures with formal relations. The role of concrete and quasi-concrete systems is the motivation of structures and the justification that structures with certain properties exist. The history of mathematics shows a trend from concrete and quasi-concrete systems to more formal, freestanding structures. There is no contradiction in the idea of a system of quasi-concrete objects' exemplifying a freestanding *ante rem* structure. Nevertheless, Parsons's caveat is a reminder not to forget the roots of each theory. Without reference to the quasi-concrete, some mathematical theories are left unmotivated and unjustified. In the next chapter, I turn to the epistemology of structuralism. However, to recall the dialectical pattern of this book, I am not finished with ontological matters. First, a brief interlude to note a connection with the philosophy of mind.

6 Addendum: Function and Structure

In contemporary philosophy, several views go by the name of "functionalism." If we limit ourselves to philosophy of mind and philosophy of psychology, the framework of this chapter provides convenient terminology in which to recapitulate some common themes. Functionalism is an in re structuralism of sorts.

Ned Block [1980] describes three types of functionalism. First, *functional analysis* is a research strategy aimed at finding explanations of a certain type: "A functional explanation is one that relies on a decomposition of a system into its component parts; it explains the working of the system in terms of the capacities of the parts and the way that the parts are integrated with one another. For example, we can explain how a factory can produce refrigerators by appealing to the capacities of the various assembly lines, their workers and machines, and the organization of these components" (p. 171). Block uses the word "system" to refer to a collection of related objects or people, just as I do here. A functional explanation is an account of what a system is like and what it does. The explanation begins by noting that the system exemplifies a certain structure and then invokes features of the structure itself, ignoring properties of the system (and its constituents) that do not relate to the structure. In the sketch cited, the only relevant facts about the people on the assembly lines are their relationships to each other and to the items playing other roles in the structure. Their hair color and gender do not matter. I take it that an explanation of why a shift defense is effective against a left-handed pull hitter is also a functional analysis.

Second, Block defines *computation-representation functionalism* to be a special case of functional analysis in which "psychological explanation is seen as akin to providing a computer program for the mind. . . . [F]unctional analysis of mental processes [is taken to] the point where they are seen to be composed of [mechanical] computations. . . . The key notions . . . are representation and computation. Psychological states are seen as systematically representing the world via a language of thought, and psychological processes are seen as computations involving these representations" (p. 179). Again, the connections with structuralism are straightforward. According to computation-representation functionalism, the theorist is to find an equivalence between psychological processes and something like a natural language, a formal language, or a computer language. This equivalence is of a piece with isomorphism and structure equivalence. The plan is to establish a systematic correlation between microprocesses and something like grammatical transformation rules or machine-language instructions. The brain is an ensemble of microprocesses and is seen to be "equivalent" to either the functioning of language as a whole or to the function of a programmed computer. Much of the work in the emerging discipline of cognitive science can be seen as an attempt to fit this mold.

Block's third theme, *metaphysical functionalism*, has the most interesting connections with structuralism. This functionalism is not, or is not merely, a theory of psychological explanation but is rather a theory of the nature of the mind and mental states like pain, belief, and desire. The metaphysical functionalist is "concerned not with how mental states account for behavior, but with what they are" (p. 172). According to the metaphysical functionalist, mental states are functional states. In present terms, the metaphysical functionalist characterizes a structure, and *identifies* mental states with places in this structure. In other words, a functional state just is a place in a structure. As Block puts it, "Metaphysical functionalists characterize mental states in terms of their causal roles, particularly, in terms of their causal relations to sensory stimulations, behavioral outputs, and other mental states. Thus, for example, a metaphysical functionalist theory of pain might characterize pain in part in terms of its tendency to be caused by tissue damage, by its tendency to cause the desire to be rid of it, and by its tendency to produce action designed to separate the damaged part of the body from what is thought to cause the damage" (p. 172). According to metaphysical functionalism, then, pain is to be characterized in terms of its relation to other mental states and to certain inputs and outputs. This is not much different from characterizing a natural number in terms of its relations to other numbers. Of course, the characterization of the natural numbers is rigorous and precise, whereas the above characterization of pain is admittedly inadequate. The metaphysical functionalist envisions a program for filling it in, much as the Peano postulates fill in the details of the natural-number structure.

Block describes this functionalist program in terms much like those of the present chapter. He envisions that we start with a psychological theory T that describes the relations among pain, other mental states, sensory inputs, and behavioral outputs. Reformulate T as a single sentence, with mental-state terms all as singular terms. So T has the form

$$T(s_1, \ldots, s_n),$$

where $s_1, \ldots, s_n$ are the aforementioned singular terms for mental states. Now, if s_i is the term for "pain," then we can define an organism y to *be in pain* as follows, adapting the technique of Ramsey sentences:

$$y \text{ has pain if and only if } \exists x_1 \ldots \exists x_n(T(x_1, \ldots, x_n) \text{ and } y \text{ has } x_i).$$

In other words, y is in pain if and only if y has states that relate to each other in various ways and tend to produce such and such outputs when confronted with thus and so inputs. Block illustrates this with a nonmental example: "Consider the 'theory' that says: 'The carburetor mixes gasoline and air and sends the mixture to the ignition chamber, which, in turn . . . ' [Block's ellipsis] Let us consider 'gasoline' and 'air' to be input terms, and let x_1 replace 'carburetor', and x_2 replace 'ignition chamber'" (p. 175). Then, according to the metaphysical functionalist, we can say that y is a carburetor if and only if "$\exists x_1 \ldots \exists x_n$[(The x_1 mixes gasoline and air and sends the mixture to the x_2, which, in turn, . . .) and y is an x_1]" (p. 175).

Block thus uses the term "functional state" for something like the present "place in a structure." To continue the automotive examples, "valve-lifter" is a functional term, because anything "that lifts valves in an engine with a certain organizational structure is a valve-lifter." Similarly, "carburetor" is a functional term, as are mental-state terms like "pain," "belief," and "desire." They all denote places in structures. Block uses "structural term" to refer to something like an officeholder. For example, "camshaft" is said to be a structural term, relative to "valve-lifter," because a "camshaft is *one* kind of device for lifting valves" (p. 174). In contemporary philosophy, "C-fiber" would be a structural term.

Presumably, the theory of pain is more sophisticated than the theory of carburation, but the form of the metaphysical functionalist analysis is the same. Notice that the structuralist definition of the natural numbers also has this form. We say that a given object z plays the 2 role in a certain system S if and only if S satisfies the Peano axioms and z is the S-successor of the S-successor of the zero object of S.

The structures delimited by metaphysical functionalism are not freestanding, and most of their places are not formal. Carburetors must *mix gasoline and air*. One cannot locate a carburetor in anything but an internal-combustion device. Computers and humans do not have systems that mix gasoline and air in preparation for combustion, and so computers and humans do not have carburetors. In the case of (physical) pain, the indicated inputs and outputs must also be held fixed. If an organism does not have something like the capacity for tissue damage, then it is not capable of pain. We can locate the exemplification of the pain structure in humans and animals, and perhaps we can locate a pain system in extraterrestrials and in future machines, but certainly not in abstract objects or planets.

If the concepts given functional definitions are made a little more formal and freestanding—along the lines of the development of geometry—then borderline cases of the concepts are produced. Eventually, the boundary with mathematics is crossed. Suppose there were a device that mixed two things other than gasoline and air, and sent the mixture to an ignition chamber. The functional definition would be something like this:

$$\exists w_1 \exists w_2 \exists x_1 \ldots \exists x_n[(\text{The } x_1 \text{ mixes } w_1 \text{ and } w_2 \text{ and sends the mixture to the } x_2, \text{ which, in turn, } \ldots) \text{ and } y \text{ is an } x_1].$$

Would the y be a carburetor? Perhaps. Suppose it did not mix the two things but did something else to them, and rather than sending the result somewhere, did something else with it:

$$\exists X \exists w_1 \exists w_2 \exists z_1 \ldots \exists z_m \exists x_1 \ldots \exists x_n[(X x_1 w_1 w_2 \;\&\; \ldots) \;\&\; y = x_1].$$

Clearly, this does not define "carburetor" in any sense of the word. At the limit, we would produce a purely formal definition, which characterizes a freestanding structure. In theory, any object could play the x_1 role, including the number 2 and Julius Caesar. There would be systems of sets and numbers that exemplify the resulting structure. "Carburetor" would be an object of pure mathematics, and carburetor theory would have gone the route of geometry, dealing with an *ante rem* structure.

4

Epistemology and Reference

1 Epistemic Preamble

For a philosopher who takes the full range of contemporary mathematics seriously, the most troublesome issues lie in epistemology. The situation is especially acute for traditional realism in ontology. Almost every realist agrees that mathematical objects are *abstract.* Although there is surprisingly little discussion of the abstract/concrete dichotomy in the literature,[1] the idea seems to be that *abstracta* are not located in space-time and are (thus) outside the causal nexus. We do not bump up against abstract objects, nor do we see them or hear them. If mathematical objects are like this, then how can we know anything about them? How can we formulate warranted beliefs about mathematical objects and have any confidence that our beliefs are true? Most of us believe that every natural number has a successor, and I would hope that at least some of us are fully justified in this belief. But how?

Benacerraf's celebrated [1973] develops this difficulty into an objection to realism in ontology by invoking the so-called causal theory of knowledge. According to this epistemology, there is no knowledge of a type of object unless there is some sort of causal connection between the knower and at least samples of the objects. On this account, it seems, knowledge of *abstracta* is impossible, because, by definition, there is no causal contact with such objects. In recent decades, the causal theory of knowledge has been roundly criticized from several quarters, and not just by friends of *abstracta.* There is no consensus on any epistemology, causal or otherwise. There is no leading contender.

1. One very notable exception to the lack of discussion on the abstract/concrete dichotomy is the fine study in Hale [1987]. See also Zalta [1983]. As noted in previous chapters, mathematicians use the "abstract/concrete" label for a different distinction. For them, arithmetic is a "concrete" study, because its subject is a single structure (up to isomorphism). Group theory is more "abstract." The mathematicians' "abstract/concrete" is my "algebraic/nonalgebraic."

In the present climate, then, one cannot claim (without further ado) that realism in ontology has been refuted simply because there is no causal contact with abstract objects. Nevertheless, the problem remains. Whatever the fate of the causal theory, Benacerraf is quite correct that there is something troublesome about ontological realism on the epistemic front. The realist cannot note the lack of consensus and cheerfully declare, "What, me worry?" nor can he make a simple announcement that he will produce an epistemology for mathematics as soon as epistemologists are finished.

To sharpen the critique of realism in ontology, note that the causal theory of knowledge is an instance of a widely held genre called "naturalized epistemology," whose thesis is that the human subject is a thoroughly natural being situated in the physical universe. Any faculty that the knower has and can invoke in pursuit of knowledge must involve only natural processes amenable to ordinary scientific scrutiny. The realist thus owes some account of how a physical being located in a physical universe can come to know about *abstracta* like mathematical objects. There may be no refutation of realism in ontology, but there is a deep challenge to it.[2] The burden is on the realist to show how realism in ontology is compatible with naturalized epistemology.

One option is to follow Gödel [1964] and postulate an epistemic faculty that allows humans to grasp or otherwise understand how things are in the realm of mathematical objects. This faculty, sometimes called "intuition," is said to be analogous to sense perception. Prima facie, a special faculty of mathematical intuition is at odds with naturalized epistemology. What natural process can illuminate a causally isolated realm of objects? Antirealists are fond of pointing out how ad hoc and otherwise unsupported the special-faculty hypothesis is, often poking fun at Gödel's philosophical ideas. Maddy [1990] proposes another tactic, arguing that at least some mathematical objects are concrete and are apprehended by ordinary sensory perception. Here, at least, there is no conflict with naturalism. A third strategy is an indirect approach to epistemology. Mathematical objects are taken to be posits, something like theoretical entities of science (e.g., Putnam [1971]; Resnik [1990]; and, again, Maddy [1990]). Current scientific theory tells us that some physical entities, like electrons, quarks, and black holes, cannot be perceived either—and so are in the same class as mathematical objects as far as epistemology goes. According to this third strategy, the existence of mathematical objects is justified by the same sorts of criteria that apply to ordinary scientific posits, whatever those criteria might be.

Advocates of some contemporary strategies thus propose to square mathematics with naturalized epistemology by showing that the boundary of the natural extends to include the mathematical. Numbers, points, Hilbert spaces, and maybe even the set-theoretic hierarchy are natural objects within the bounds of ordinary scientific scrutiny. Mathematics is part of science and cannot be exorcized from it. Notice, incidentally, that if there is a significant overlap between the mathematical and the

2. Field [1989, essay 7] develops a related argument against realism in ontology.

natural, then even the Gödelian strategy may be a live option. Mathematical intuition might be found sufficiently similar to ordinary perception or to other epistemic mechanisms to remove its stigma as a philosophically suspect faculty. We may not know much about how the mechanisms function, but perhaps they do not violate reasonable naturalistic scruples. But we should not forget the cliché about pudding and proof. There is work to be done.

One upshot of most epistemic proposals brought by contemporary realists in ontology (concerning mathematics) is a blurring of the abstract/concrete boundary (see Resnik [1985]; Maddy [1990]). In a sense, this blurring is endorsed, articulated, and defended both in the discussion of epistemology and reference to follow and in the treatment of applications in chapter 8. It is not that there is no difference—a fuzzy border is still a border—but the difference is not a sharp one and does not allow for crisp philosophical pronouncements, or easy answers to deep questions. Perhaps it is reasonable to shift the burden to defenders of the abstract/concrete boundary. Why is the distinction important? What role does it play?

One motivation for the dichotomy may be the long-standing view that mathematical knowledge is, or can be, a priori. Because most knowledge about individual concrete objects is a posteriori, mathematical objects are not individual concrete objects. Thus, they are abstract. To be sure, the notion of a priori knowledge is under such severe attack today (especially in North America) that many philosophers do not take the notion seriously. The assault dates to Quine [1951] (see Kitcher [1983] for an instance directed at mathematical knowledge). A priori knowledge is lumped with the analytic/synthetic distinction and put on the philosophical scrap heap. Notice, incidentally, that it follows from the causal theory that all knowledge of objects is a posteriori, because on that theory such knowledge depends on matters of causality. We cannot be as definite about naturalized epistemology generally, but it is at least prima facie difficult to square a priori knowledge with it.

Nevertheless, an important and influential group of ontological realists maintain that mathematical knowledge is a priori. One prominent project is an attempt to sustain the thesis that mathematical knowledge is of the same general kind as logical knowledge, so that mathematical knowledge is based on pure thought. This view, called "logicism," dates to Frege [1884] and is pursued in Wright [1983] and Hale [1987].

The notion of a priori knowledge is thus alive, if not well. Whatever the fate of the logicist project, one burden on the philosopher of mathematics is to explain the appeal of the idea that mathematical objects are abstract and mathematical knowledge is a priori. Even if these theses are mistaken, what is it about mathematics that led so many to accept them?

To get down to the business of this daunting task, this chapter contains three ways that structures are apprehended and knowledge of them obtained. The first is simple *abstraction* or *pattern recognition*. With processes much like—or even identical to—ordinary sensory perception, a subject comes to recognize and learn about patterns. The subject does not see or hear patterns themselves, of course. We do not literally see or hear abstract entities. Our subject comes to recognize a pattern by observing patterned systems. The mechanisms involved pose deep and interesting problems for cognitive psychology and the budding field of cognitive science, but pattern recognition is, I believe, philosophically unproblematic. The interesting and speculative part of this approach is the thesis that we come to understand some patterns to be freestanding, and ultimately *ante rem.*

However speculative and tentative this may be, it is a scant beginning. Most of the structures studied in mathematics are infinite, and all but a few of those are uncountable. It is contentious, to say the least, to claim that infinite structures are apprehended by pattern recognition. To grasp a structure this way, the subject would have to perceive an infinite system and then abstract a pattern from it. I suggest variations of pattern recognition that can lead to knowledge of small infinite structures, such as the natural-number structure and perhaps the continuum. This, however, is about as far as anything resembling pattern recognition can take us. We do not get anywhere near the set-theoretic hierarchy via that route. Any grasp of those patterns must come from another source. I propose two further techniques for arriving at structures. One is a linguistic abstraction, of sorts, and the other is implicit definition. Note that both of these are linguistic. This tie to language is one source of the longstanding belief that mathematical knowledge is a priori—to the extent that the grasp of language and the knowledge of logical consequence are a priori. The epistemological development leads to the discussion of reference and semantics in the closing section of the chapter. I finally make good on the suggestion, from chapter 2, that model-theoretic semantics is the hallmark or central framework of realism in ontology and realism in truth-value.

The strategies sketched here are not tied to any particular global epistemology.[3] To be sure, mathematical objects—places in structures—are abstract and causally inert. Thus, the present program is not compatible with crude versions of the causal theory of knowledge. In line with chapter 1, I take the existence of mathematical knowledge to be something close to a philosophical datum, just about incorrigible. If an epistemology entails that mathematical knowledge is impossible, I would be inclined to reject the epistemology. Of course, the confidence in mathematical knowledge does not guarantee that a successful epistemology will be consistent with *structuralism*, or with the particular epistemic tactics invoked below.

2 Small Finite Structure: Abstraction and Pattern Recognition

In contemporary philosophical jargon, the phrase "abstract object" means something like "object that is not part of space-time" (but see Hale [1987, chapter 3]). Numbers are the paradigms of abstract objects. However, the etymology of the word "abstract" indicates the results of a process of *abstraction.* According to the *Oxford English Dictionary*, the adjective "abstract" is derived from "abstracted." In this sense, col-

3. See Dieterle [1994, chapter 2] for a defense of structuralism in terms of a reliabilist epistemology, the view that a belief P is knowledge if P is true and the belief was produced by a reliable process (see, for example, Goldman [1986]). Dieterle argues that if the ontological themes of structuralism are true, then certain psychological processes are reliable and thus do produce knowledge of structures.

ors, shapes, and types are paradigm cases of abstract objects. The plan here begins by bringing this traditional usage of "abstract" in line with the contemporary, at least in part. Some nonspatiotemporal objects are in fact apprehended (but not perceived) by abstraction. In particular, one process for apprehending small structures is *pattern recognition*. I do not claim to understand the psychological mechanisms involved, but pattern recognition is a faculty that humans clearly do have.[4] My modest purpose is to illustrate a few instances of the procedure at work, showing how it can lead to an apprehension of freestanding, *ante rem* structures.

Let us start with the recognition of letters, numerals, and short strings. This is about as mundane as the type/token dichotomy gets. Some realists argue that simple types can be apprehended through their tokens, via abstraction. We see (or hear) the tokens and somehow obtain knowledge of the types. If this idea is sustained, then we have at least one case of knowledge of (abstract) objects with no causal, or any other, contact with them. We do not see or hear types; the contact is only with the tokens. Antirealists might regard this maneuver as invoking magic, much as they criticize Gödelian intuition. Here, however, realism does not seem quite so ad hoc. In the first place, pattern recognition is not philosophically occult. To sustain the realists' intuition, however, we must see what pattern recognition delivers and we must examine how types can be construed as objects.

The primary mechanism for introducing characters to the uninitiated is ostensive definition. A parent points to several instances of, say, a capital "E" and pronounces "eee." Eventually, the child comes to understand that it is the letter—the type—that is ostended, and not the particular tokens. Wittgenstein [1953] is noted for his reminder that the practice of ostension presupposes abilities on the part of both teacher and learner. They must already be able to recognize the sorts of things being ostended—whatever those sorts of things might be. Again, I do not propose to shed much light on the relevant psycholinguistic faculties involved. The more modest task is to show how these faculties might lead to knowledge of *abstracta*.

At this point, of course, some antirealists object that the child has merely learned to apply the *predicate* "capital E." Although it is hard to quibble with this assessment just yet, the maneuver does not help in the end. The difference between an object and a predicate extension (or a property) is a relative one, depending on context. It is similar to the relativity of system and structure, as presented in chapter 3. For now, however, this is just a promissory note. Stay tuned (section 5).

If our antirealists concede that there are character types and string types, they will claim that such types are in re. There is no more to the type "E" than the sum of written

4. Dieterle [1994, chapter 2] surveys some of the relevant psychological literature, relating the results to the apprehension of small structures. Resnik [1990] is more speculative, delimiting a process by which our "ancestors" (and us) may have become "committed" to (small) abstract structures. Although the "genetic" part of Resnik's account seems to involve the recognition of patterns, he does not invoke abstraction. Resnik follows Quine in holding that both mathematical and physical entities are *postulated*, and their existence is justified on holistic grounds. A detailed contrast between abstraction and postulation is beyond the scope of this book.

capital "E" tokens. Destroy all the tokens and the type goes with them. In terms of the previous chapter, however, there is a tendency toward thinking of character and string types as freestanding. Throughout the learning process, each character type is seen to be exemplified by more and more kinds of objects. At first, of course, the child associates the type "E" with tokens that have roughly the same shape: a straight vertical line with three smaller horizontal lines protruding to the right. Soon, however, the child learns to identify tokens with different shapes, such as "$\mathscr{E}$," as capital "E"s. The child then learns that there is a type whose tokens include both capital and lowercase "E"s.

At this point, there is nothing like a common shape to focus on, and so we have moved beyond simple abstraction. An opponent might insist that we are no longer dealing with a single type but a loose conglomeration of types (the "E" type, the "$\mathscr{E}$" type, etc.). There is no need to adjudicate the proper use of the word "type" here. The reader is free to use a different word. The important point is that we must leave the simple property/object dichotomy and think in terms of *places in a pattern* or structure. What the various "E"s have in common is that they all have the same role in an alphabet and in various strings. That is, our child has learned to recognize an alphabet structure and "E" as a place in it—the fifth place.

Thus far, all of the various "E" tokens are physical inscriptions, consisting of hunks of ink, graphite, chalk, burned toner, pixels, moisture on glass, and so on. But the child also learns that there are tokens among certain *sounds*. The sound "eee" is also an "E." There is a short "E," a long "E," and so forth. When it comes to phonemes, we also supplant the simple property/object dichotomy with places in a pattern. What makes a phoneme recognizable as a short "E" has something to do with its place in a larger system, because the very same wave pattern can sound different when embedded in different words. The relevant psycholinguistic studies would take us too far afield.

For the ontologist, the important point is that our child eventually comes to see all kinds of objects as tokens of "E." In addition to those listed above, there is sign language, flag semaphores, smoke signals, and Morse code. At play, our child might make up an alphabet that consists of balls of different sizes and colors. In coding, a character might even be tokened by (tokens of) *other characters*. "Look Watson, the 'H' here is an 'A,' the 'C' is a 'B' . . ." In structuralist terms, I would say that a code is a case in which certain types are used as tokens of other types. Holmes discovered a system that exemplifies the alphabet structure in which "H" plays the "A" role and "C" plays the "B" role. We have a direct analogue of the fact that the even numbers themselves exemplify the natural-number structure.

Returning to our eager child, if she is fortunate enough to take a course in advanced logic and learns about Gödel numbering, she will see how characters and strings can have *natural numbers* as tokens. By then, it should be clear that strings and natural numbers share a common structure.

The process outlined here may not go all the way to characters and strings as completely freestanding abstract objects, but the development goes pretty far in that direction. Presumably, nothing philosophically occult or scientifically disrespectable has been invoked along the way. In the end, we either demystify numbers or we mystify more mundane items.

Let us consider another simple sort of pattern, small cardinal numbers. For each natural number n, there is a structure exemplified by all systems that consist of exactly n objects. For example, the *4 pattern* is the structure common to all collections of four objects. The 4 pattern is exemplified by the starting infielders on a baseball team (not counting the battery), the corners of my desk, and two pairs of shoes. We define the 2 pattern, 3 pattern, and so on, similarly. Let us call these "cardinal structures," or "finite cardinal structures." The finite cardinal structures have no relations and so are as simple as structures get. We include the 1 pattern as a degenerate case. It is exemplified by a "system" that consists of a single object under no relations. The 0 pattern is an even more degenerate case. It is exemplified by a degenerate system that consists of no objects.

In part, our child starts to learn about cardinal structures by ostensive definition. The parent points to a group of four objects, says "four," then points to a different group of four objects and repeats the exercise. Eventually, the child learns to recognize the pattern itself. Virtually everything said here about characters and strings applies mutatis mutandis to (small) cardinal structures.

The freestanding nature of cardinal structures is even more prominent than that of strings. At first, perhaps, our child may believe that the 4 pattern applies only to systems of physical objects that happen to be located near each other, but she soon learns to count all kinds of systems and sees that the 4 pattern applies universally. We count the planets in the solar system, the letters in a given word, the chimes of a clock, the colors in a painting, and even properties: "Justice and mercy are two cardinal virtues." Because anything can be counted, systems of all sorts exemplify the cardinal patterns. Cardinal structures are paradigms of freestanding structures. We even count *numbers* when we note that there are four primes less than 10. That is, systems of numbers like {2, 3, 5, 7} exemplify finite cardinal structures. As noted, this is a motivation for thinking of cardinal structures as *ante rem*.

The freestanding nature of cardinal structures seems to underlie Frege's [1884] contention that numbers are *logical* objects. Although he does not speak of patterns directly, the idea is that cardinal structures are topic-neutral and universally applicable. If one can speak of objects at all, then one can count them. Consider, for example, Frege's argument against the view (which he attributes to Mill) that arithmetic truths are empirical generalizations. According to the view under attack, the equation "$1 + 2 = 3$" is a statement that any object and any pair of objects can always be rearranged as the vertices of a triangle. This generalization is verified empirically. Frege noted that this simple-minded view does not account for larger numbers, and it does not touch the application of number to things that cannot be moved. We cannot arrange one clock chime and a pair of chimes to make a triangle. Moreover, if Frege is right that properties are not objects, then even some nonobjects can be counted. In structuralist terms, cardinal structures are exemplified beyond the realm of objects (see section 5).

We should distinguish finite cardinal structures from what may be called finite *ordinal* structures. The ordinal 4 pattern, for example, is the structure of any system of four objects *considered in a particular order*—a first, a second, a third, and a fourth. This pattern is exemplified by my sister-in-law's children, in birth order;

by those same children arranged according to the alphabetical order of their middle names; by the numbers 6, 7, 12, 15 in numerical order; and by the first four infinite cardinal numbers, in reverse cardinal order (i.e., $<\aleph_3, \aleph_2, \aleph_1, \aleph_0>$). In mathematical terminology, we are speaking here of (the structures of) finite sequences, as opposed to finite unordered sets.

One might argue that the ordinal is the primary number concept. The usual way to determine the cardinality of a collection of objects is to count it, which imposes an *order* on the collection and thus makes it an instance of the corresponding ordinal pattern. Dummett [1991, 293] chides Frege for focusing on cardinal numbers rather than ordinals:

> [Frege's] definition of the natural numbers did not achieve the generality for which he aimed. He assumed . . . that the most general application of the natural numbers is to give the cardinality of finite sets. The procedure of counting does not merely establish the cardinality of the set counted: it imposes a particular ordering on it. It is natural to think this ordering irrelevant, since any two orderings of a finite set will have the same order type; but, if Frege had paid more attention to Cantor's work, he would have understood what it revealed, that the notion of ordinal number is more fundamental than that of cardinal number. . . . [A]fter all, when we count the strokes of a clock, we are assigning an ordinal number rather than a cardinal. . . . [Frege] was well aware that Cantor was concerned with ordinal rather than cardinal numbers; but . . . he dismissed the difference as a mere divergence of interest, and never perceived its significance.

This priority issue does not concern me here. Both number notions are taught to children, and everything said about cardinal numbers and strings applies to ordinals as well. In a sense, the system of finite cardinal patterns, the system of finite ordinal patterns, and the system of strings have the same structure, namely, the natural-number structure. I will get to this infinite structure soon, but first I briefly deal with larger finite structures.

Notice, incidentally, that by itself pattern recognition does not deliver anything resembling a priori knowledge. To obtain knowledge of strings and number structures via pattern recognition, our subject must encounter tokens and collections of objects. To be sure, no particular specimens are necessary—any token of the relevant type and any collection of the right size will do. Someone might argue that we have a priori knowledge of certain facts about finite structures: just as we can know a priori that all green objects are colored, we can know a priori that any system exemplifying the 4 pattern is larger than any system exemplifying the 3 pattern. Still, I will look elsewhere for the sources of the idea that mathematical knowledge is a priori.

3 Long Strings and Large Natural Numbers

It is widely believed that the serious epistemic problems with mathematical objects concern infinite systems. I argued in section 2 that because we can see or hear instances of types and finite patterns, these structures are not all that removed from us, and so their epistemology is perhaps tractable. But this applies only to *small* finite structures, such as individual characters, short strings, and the first few finite cardinal and ordinal structures. What of the larger ones?

At some point, still early in our child's education, she develops an ability to understand cardinal and ordinal structures beyond those that she can recognize all at once via pattern recognition and beyond those that she has actually counted, or even could count. What of the 9,422 pattern, and the quadrillion pattern? No one has ever seen a system that exemplifies the quadrillion pattern, nor has anyone counted such a system. Life is too short, and the eyes are too dim. Surely, we do not learn about and teach such patterns by simple abstraction and ostensive definition. The parent does not say, "Look over there, that is 9,422" or "Look at the national debt, counted in mills. That is the quadrillion pattern a few times over." Yet we speak of large numbers with ease. We learn about the number of molecules in physical objects and the distance to other galaxies. The rules for determining draws entail that there are only finitely many possible chess games, but it is a mind-boggling number. We calculate how much memory is required to look ahead, say, twenty moves in a brute-force strategy—more bits than there are particles in the known universe. We do not come to know about such finite cardinal structures through the workings of a simple abstractionist epistemology. Nevertheless, humans clearly are able to recognize, discuss, and manipulate large, finite cardinal structures and large, finite ordinal structures.

The same goes for strings. Long sequences of characters are not apprehended by simple abstraction. During linguistic development, our child learns to parse tokens of strings she has never seen. Indeed, the ability to parse extends to strings that never had any tokens, and may never have any. Moreover, some strings and some grammatical sentences are so long that there is not enough ink and paper in the universe to make a token of them. The Gödel sentence for a standard first-order Peano arithmetic may not be quite that long, but it falls in the same category. Yet that string can be comprehended and coherently discussed. Logic treatises are full of information about this particular string.

Thus, for anyone who invokes only simple abstraction in the epistemology of mathematics, many of the features that make the infinite problematic are shared by the large finite. Even at this stage, the structuralist needs epistemic strategies other than simple pattern recognition. Here we get more speculative.

Returning to our learning child, perhaps she reflects on the sequence of *numerals*, eventually noting that the sequence goes beyond the collections she has actually counted.[5] She then sees that any finite collection can be counted and thus has a cardinality.

---

5. An anecdote: My own children seemed to learn about numbers through numerals. They first learned to count by rote, reciting numerals in order—to the delight of us and their grandparents. They began to get the point of the exercise when they later learned how to count *objects*, by reciting numerals while pointing to different objects. That is, the children became competent in reciting the numerals well before they could count even small collections of objects. Of course, during this process, they learned more and more numerals. Eventually, they will understand that the numerals do not run out, and thereby begin to get the idea of the natural-number structure. Even if my children's development is typical, I do not pretend that this is even a hint of a psychological account of learning the numbers. Surely, other crucial skills are being acquired and refined during this process.

A related possibility is that humans have a faculty that resembles pattern recognition but goes beyond simple abstraction. The small finite structures, once abstracted, are seen to display a pattern themselves. For example, the finite cardinal structures come in a natural order: the 1 pattern, followed by the 2 pattern, followed by the 3 pattern, and so on. We then *project* this pattern of patterns beyond the structures obtained by simple abstraction. Consider our child learning the patterns represented by the following:

$$|, ||, |||, ||||.$$

For present purposes, we can think of these patterns either as strings, as finite cardinal structures, or as finite ordinal structures. Reflecting on these finite patterns, the subject realizes that the sequence of patterns goes well beyond those she has seen instances of. Perhaps this is her first hint of an *ante rem* structure. Our subject thus gets the idea of a sequence of 9,422 strokes, and she gets the idea of the 9,422 pattern. Soon she grasps the quadrillion pattern. I assume that the reader, realist or otherwise, understands these strings and sequences, and I will not bother giving tokens and instantiations of them. The cost in ink would drive up the cost of this book.

Somewhere along the line, antirealists might concede that pattern recognition and the other psycholinguistic mechanisms lead to *belief* in (perhaps *ante rem*) structures, and they may concede that we have an ability to coherently discuss these structures. But antirealists will maintain that these mechanisms do not yield *knowledge* unless the structures (or at least their places) exist. Have we established this last, ontological claim? Can this be done without begging the question?

I do not mean to take the philosophical high ground and simply scoff at the antirealist's serious and earnest charges, but I do request the reader's indulgence. In this book, I present an account of the existence of structures, according to which an ability to coherently discuss a structure is evidence that the structure exists (see, for example, section 8). This account is perspicuous and accounts for much of the "data"—mathematical practice and common intuitions about mathematical and ordinary objects. The argument for realism is an inference to the best explanation. The nature of structures guarantees that certain experiences count as evidence for their existence.

4 To the Infinite: The Natural-number Structure

Given the aforementioned routes to large finite structures (construed *ante rem*), the simplest infinite structure is near at hand. Our subject, no longer a child, continues to reflect on the sequence of larger and larger finite structures and grasps the notion of a *finite sequence* per se. The finite sequences are ordered as follows:

$$|, ||, |||, ||||, \ldots$$

Our subject learns that the sequence of sequences goes on indefinitely. She sees that the *system* of finite ordinal structures has a pattern. For each sequence, there is a unique next-longest sequence, and so there is no longest sequence. The system of finite sequences is potentially infinite. Eventually, the subject can coherently discuss the *struc-*

ture of these finite patterns, perhaps formulating a version of the Peano axioms for this structure. We have now reached the structure of the natural numbers.

Of course, the same point can be reached by reflection on the finite sequences of strokes or the finite cardinal structures. The system of finite ordinal structures, the system of finite cardinal structures, and the system of strokes all exemplify the same structure—the natural-number structure. Even strings on a finite alphabet will work. From a structuralist point of view, there is not much difference between strings on a finite alphabet and natural numbers. In terms of the previous chapter, the system of strings on a finite (or countable) alphabet is structure-equivalent to the natural numbers.

To briefly reiterate, then, we first contemplate the finite structures as *objects* in their own right. Then we form a *system* that consists of the collection of these finite structures with an appropriate order. Finally, we discuss the *structure* of this system. Notice that this strategy depends on construing the various finite structures, and not just their members, as *objects* that can be organized into systems. It is *structures* that exhibit the requisite pattern. *Ante rem* structuralism provides the most straightforward line on this. If one takes a more in re (or eliminative) approach, the move to the natural-number structure might not work, because there might not be enough finite structures.

We have a new wrinkle on the structure/system dichotomy here. What is structure from one point of view—the perspective of finite cardinal structures—is *object* from another. The finite structures are themselves organized into a system, and the structure of that system is contemplated. The 4 pattern itself plays the role of 4 in the natural-number structure.

Another route to the natural-number structure is for our subject to reflect on ever-increasing sequences of strokes and formulate the notion of a sequence of strokes that does not end (in one direction). This is an infinite string, and so I cannot give a token of it in this finite book. The practice is to write something like this instead:

| | | | | . . .

The point is that students eventually come to understand what is meant by the ellipses ". . ." The students can coherently discuss the infinite pattern and can teach it to others. When they do, they have grasped (an instance of) the natural-number structure. Perhaps the same point can be reached by reflecting on the passage of time. If the time line is thought of as divided into discrete moments, one second apart, then the moments from now on exemplify the natural-number structure. This is not too far removed from the approach of Kant and the traditional intuitionists.

After a given structure is understood, other structures may be characterized and understood in terms of it. To return to our favorite example from chapter 3, one can describe a "lefty shift defense" as one in which the shortstop plays to the right of second base, and the other infielders move to the right, with the player at first base guarding the line. Those who know about baseball will understand what is meant even if they have never seen this defense. Similarly, once the natural-number structure is understood, then other infinite structures can be described in those terms. The integer structure, for example, is like the natural-number structure, but unending in both directions:

. . . ||||||| . . .

Again, students eventually understand what is meant and can discuss the structure coherently. The rational-number structure is the structure of pairs of natural numbers, with the appropriate relations.

Notice that this technique invokes the structure/system relativity described in the previous chapter. The *places* of a given structure—here the integer structure—are treated as objects in their own right. Such objects are arranged into a system of pairs and collections of pairs. The structure of the constructed system is then apprehended.

So far, we have only denumerable structures. To obtain larger ones, our subject can contemplate certain *sets* of rationals, as in Dedekind cuts, or she can contemplate certain infinite sequences of rationals, as in Cauchy sequences (assuming that such talk of sets or sequences is coherent). These two techniques differ, of course, but the *same structure* results, the structure of the real numbers. The presentation is often a pedagogical challenge, but once the student acquires some facility working within the structures, in the appropriate language, no problems arise. We have at least the appearance of communication, and on the present account, it is communication of facts about structures. We are on our way to type theory.

The technical development here is well known, of course. Perhaps the philosophy has gone a bit too fast, however. The antirealist will surely balk here, arguing that at best I have only pointed toward *belief* in the respective structures. And again, I refer to our theory of sections 8 and 9, according to which the ability to coherently discuss a structure is evidence that it exists.

Notice, as an aside, that in the foregoing development, there is an ambiguity in the reference of numerals. The term "4" has been used to denote a certain finite cardinal *structure* (the pattern common to any collection of four things), and it has been used for the corresponding ordinal structure. The same word is also used to denote the appropriate *place* in the natural-number structure. The very same numeral is also used to denote a place in the structure of the integers, a place in the real-number structure, a place in the complex-number structure, and even a place in the set-theoretic hierarchy,[6] where it denotes a finite ordinal $\{0, 1, 2, 3\}$, or to be precise $\{\phi, \{\phi\}, \{\phi, \{\phi\}\}, \{\phi, \{\phi\}, \{\phi, \{\phi\}\}\}\}$.

5 Indiscernability, Identity, and Object

Although Robert Kraut's "Indiscernability and ontology" [1980] is not aimed at mathematics and does not explicitly deal with structures, the work provides another epistemic route to structures. Along the way, Kraut lends insight into the notions of object and identity, at least as the notions are construed in structuralism. His starting point is the Leibniz principle of the identity of indiscernibles: if two items cannot be

6. Are all of these 4s one and the same? Chapter 3 contains a discussion of the sense in which one can and should identify the places in different structures. There is no problem with identifying these places, nor is there a problem with keeping them separate or, in some contexts, in refusing to consider the question. It is a matter of decision, based on convenience.

distinguished—if anything true of one is true of the other—then the objects can be or should be identified. The objects can be taken as (or they are) one rather than two. Kraut combines this thesis with the observation that what is "discernible" depends on the conceptual resources available. The result is a theory relativity of objects, quite consonant with the present relativity of objects and the relativity of system and structure.[7]

The following pieces of philosophical fiction are not entirely coherent, but bear with me. Consider an imaginary economist who, while at work, speaks an impoverished version of (technical) English, a language that does not have the resources to distinguish between two people with the same income. Anything she says about a person P applies equally well to anyone else Q who has the same income as P. If she notes that P cannot afford the tuition at Harvard and that P is likely to get audited, then the same goes for Q. Someone who interprets the economist's language might apply the Leibniz principle and conclude that for those stuck with the impoverished resources, $P = Q$. That is, from the standpoint of the economist's scheme, people with the same income are identified and treated as a single object. To be fanciful, if the interpreter sees a certain woman, he might say (on behalf of the economist), "There is the $35,000." If a man with the same income walks by, the interpreter might remark, "There it is again." The economist herself might make the identification if she knows that the two are indiscernible and does not envision a framework for distinguishing them.

Of course, in the full background language, English, the two people can be distinguished in lots of ways: by gender, age, spatial location, and so on. By hypothesis, however, these resources are not available to our economist (while at work). In that context, people in the same income level are indiscernible. Nothing is lost by interpreting her language as being about income levels and not people (assuming sharp boundaries between levels, of course). A singular term, like "the Jones woman," denotes an income level.

Similarly, consider someone who speaks an impoverished version of English in which equinumerous collections of objects are not discernible. Call him a number person. In his language, anything true of one collection of objects is also true of any other collection with the same cardinality. So, following the Leibniz principle, equinumerous collections get identified, and numbers become objects. If our interpreter sees a clump of three trees, he might say, "For the number person, there is three." If three people walk by, he might remark, "There is three again."

A standard technique in algebra and number theory can be understood in these terms. Imagine a mathematician who decides to speak an impoverished language that cannot distinguish two integers if their difference is divisible by 7 (or, equivalently, if the numbers produce the same remainder when divided by 7). On her behalf, we make the indicated identifications: 2 is identified with 9, 16, -5, and so forth, whereas 3 is identified with 10. We interpret our mathematician as saying that $5 + 4 = 2$. Of

7. Kraut [1980] contains an insightful and compelling account of what is right and what is wrong with Geach's ([1967], [1968]) "relativity of identity."

course, 5 + 4 = 9 as well because, in her system, 2 and 9 are indiscernible and thus identical. Under this interpretation, "identity" is what we call "congruence modulo 7." The locution "2 = 9" can be rendered as "$2 \equiv 9 \pmod 7$" in the background language.

To be sure, Kraut's economist and our number person and mathematician are too far-fetched. In practice, no one forgets or suspends the background language to speak one of these impoverished languages. No one could. If someone did manage to forget the background framework and stick to the impoverished one, the thought experiment would probably fail, if it made sense at all. An aspect of Frege's [1884] aforementioned attack on simple-minded empiricism is relevant here (see section 2 of this chapter). Suppose that our number person looks at two decks of cards and we interpret him as saying "There is two." Then we assume that, at some level, the number person knows that it is the *decks* and not the cards or the colors that are being counted. Nothing in the hunk of mass itself determines that it is 2, 104, or any other number for that matter. If the subject loses the use of sortals like "deck," he will not see the stuff as 2. In other words, to see the decks as 2, the subject must be aware of the decks and must distinguish the two decks from each other.[8] Similarly, if he looks at two people and says "There is two again," it is the people and not the arms or the molecules that are relevant.

I submit that a better way to view the situation is that income levels and numbers are places in structures. If our economist's theory—including the background—is correct, it shows that an income-level structure is exemplified (more or less) in our economic system. The level roles are played by individual people or groups of people. In terms of chapter 3, this is called the "places-are-offices" perspective. The framework of our number person is understood similarly. If there are enough objects in the background ontology, equinumerous collections exemplify the natural-number structure. The role of 3 is played by either the higher-order property of being three-membered or the class of all three-membered classes. When we focus on the impoverished sublanguages and interpret them with the Leibniz principle, we take the places of the structure—income levels and numbers—as *objects* in their own right.

The "impoverishment" indicated here is a two-staged affair. We start with an interpreted base language or theory. This framework already has an ontology, in that there are objects in the range of its variables. Next, we focus on an equivalence relation over the ontology of the base language.[9] For our economist, the relationship is "sameness of income" between people. If the borders between levels are sharp, "same income" is an equivalence. The relation of equinumerosity, between collections of objects (or properties), is also an equivalence, as is the relation of congruence modulo

8. Frege's answer to "What is it that we number?" is *concepts*. The concept "decks before me" has number 2, whereas the concept "cards before me" has number 104. For Frege, the notion of "object" is bound up with that of "concept" (of which more later).

9. A relation R is *reflexive* if, for every x in its domain, Rxx holds; R is *symmetric* if, for every x, y, if Rxy then Ryx; and R is *transitive* if, for every x, y, z, if Rxy and Ryz, then Rxz. A relation is an *equivalence* if it is reflexive, symmetric, and transitive.

7. Any equivalence relation divides its domain into mutually exclusive collections, called "equivalence classes." Two items are in the same equivalence class if and only if the relation in question holds between them. The idea here is to see the equivalence classes as exemplifying a structure and to treat the places of this structure as objects.

Our next step is to formulate a sublanguage of the base language for which the equivalence is a *congruence*. Two conditions are necessary: if $\Phi(x)$ is a predicate of the sublanguage and p, q are in the same equivalence class (i.e., the equivalence relation holds between them), then $\Phi(p)$ iff $\Phi(q)$. That is, members in the same equivalence class cannot be distinguished in the sublanguage. If two items are equivalent, they are indiscernible. Second, if two items m, n are not equivalent, then there should be a predicate $\Psi(x)$ of the sublanguage that holds of one and fails of the other:[10] $\Psi(m)$ but not $\Psi(n)$.

In such cases, I suggest that in the sublanguage, the equivalence relation *is* the identity relation. The idea here is that the language and sublanguage together characterize a structure, the structure exemplified by the equivalence classes and the relations between them formulable in the sublanguage. It is thus possible to invoke the places-are-objects orientation, in which case the places in this structure are rightly taken to be its objects.

One further example of the sublanguage procedure closes a circle. In the previous chapter, a *system* is defined to be a collection of objects related in various ways. In attempting to characterize the relation of "having the same structure," we encountered two equivalence relations on systems, namely, isomorphism and structure equivalence. Suppose that we begin with one of these relations and apply the foregoing procedure to it. Is there an interesting sublanguage in which either isomorphism or structure equivalence is a congruence? Perhaps the framework of pure mathematics (properly regimented) might be such a sublanguage, as would the envisioned structure theory of chapter 3. According to this plan, structures themselves are seen to be objects in their own right.

---

10. I am being loose with some of the terminology. First, the envisioned sublanguage is called a "sublanguage" because it is constructed with a subset of the vocabulary from the base language. Although both languages are interpreted, they do not have the same ontology. The base language may not have the resources to refer to equivalence classes, and the sublanguage may not have the resources to refer to the objects of the base language. Second, in a formal language, a "predicate" is a formula $\Phi(x)$ with a free variable x. If we restrict ourselves to formulas with *only* one free variable, then no structure larger than the continuum is produced by the foregoing procedure (unless the language has uncountably many terms). Here we do allow the formula $\Phi(x)$ to contain other free variables. Suppose, for example, that Φ also contains y free. Then we say that Φ does not distinguish p and q only if $\forall y[\Phi(p, y) \equiv \Phi(q, y)]$. That is, no matter what value for the parameter y is chosen, Φ holds of p if and only if Φ holds of q. Notice, incidentally, that if the sublanguage contains the identity symbol (from the base language), then any two different objects in the original ontology can be distinguished in the sublanguage. If $m \neq n$, then the predicate "$x = m$" distinguishes them or, to be precise, the predicate "$x = y$" distinguishes them when the parameter y has the value m. Thus, if the sublanguage contains the identity symbol, then the only congruence is the identity relation, and the sublanguage procedure does not yield anything new.

The Kraut sublanguage procedure is remarkably similar to one elaborated in Wright [1983] and Hale [1987], on behalf of Fregean logicism. Their thesis is that we can introduce abstract objects by abstraction over an equivalence relation on a base class of entities. Frege himself suggested that *directions* might be obtained from lines in this manner:

> The direction of *l* is identical to the direction of *m* if and only if *l* is parallel to *m*.

Closer to home, numbers are obtained from concepts via a thesis sometimes called *Hume's principle*:

> The number of *F* is the number of *G* if and only if *F* is equinumerous with *G*.

Wright proposes that one requirement on the procedure is the formulation of a sublanguage in which the designated equivalence is a congruence.

The Frege–Wright–Hale project is a defense of *logicism* and thus of the long-standing view that arithmetic can be known a priori. Because the basic principles of arithmetic follow from Hume's principle (via second-order logic), arithmetic is a priori if Hume's principle and second-order logic are. Hume's principle, they argue, is grounded in our grasp of the relevant concepts. I return to Frege's development of arithmetic in chapter 5.[11]

Although the Kraut sublanguage procedure sketched above begins with the *objects* in a background framework, virtually the same process can start with items that are not objects. Because properties are notoriously difficult to individuate, many authors hold that properties are not objects. However, the sublanguage procedure does not require a clear identity relation on the items in the field of the equivalence relation. The above number language, for example, can start with properties, just as the Frege–Wright–Hale formulation does. For the purposes of the procedure, it does not matter how properties are individuated or even whether there is an explicit identity relation on them. All we need is a determinate *equinumerosity* relation on the properties. This relation is an equivalence, and a sublanguage in which it is a congruence provides all the resources we need to speak of indiscernability and identity among numbers.

Some writers take numbers themselves to be properties and, thus, I suppose, not objects. Harold Hodes [1984], for example, argues that Frege was too quick to conclude that numbers are objects (and thus to consider the question whether 2 is Julius Caesar). Rather, numbers are higher-order properties, that is, properties of properties (see also Maddy [1981] and Luce [1988]). Accordingly, 5 is the property of (ordinary) properties of having five items in its extension. The number 5 thus applies to

11. The Kraut sublanguage procedure and the Frege–Wright–Hale procedure are also similar to one that Tait [1986, 369 n. 12] calls "Dedekind abstraction" (see chapter 5 in this volume). Notice, incidentally, that the procedures in question invoke an inference from an equivalence relation to the existence of certain objects. Quine [1951] applies the converse of this inference in his arguments against meaning: if there were such a thing as the meaning of a sentence, then the relation of "same meaning" or "synonymy" would be an equivalence relation on sentences. Quine then argues that there is no such equivalence relation.

the property of being a starting player on a particular basketball team, and 9 applies to the property of Supreme Court justice at the beginning of 1989. Individual numbers are second-order properties, and the property of being a number is a third-order property, as are arithmetic operations and relations like addition, multiplication, less-than, and prime. Of course, the ordinary language of arithmetic does not read like this. In arithmetic, *first-order* variables range over numbers, and so numbers are treated as objects. Hodes says that this language/theory is a convenient *fiction* that simulates the relations between the aforementioned second-order number properties. In other words, the relations expressed in ordinary arithmetic are a convenient code for third-order arithmetic relations.

As a structuralist, I can accept second-order number properties, exactly as developed by Hodes on behalf of Frege. The resulting framework is useful and insightful, at least as far as it goes. If there are enough objects in the domain of the base language (or if one allows possible objects), Hodes's higher-order number properties exemplify the natural-number structure. Against Hodes, however, there is nothing fictional about the ordinary language of arithmetic. Its objects are the places in this structure.

Hodes's account does not go far enough. We do not just count objects at the base level. We also count across types. We say that there are four primes less than 10. What does this numeral "4" refer to? Type distinctions entail that this "4" cannot refer to a second-order number property, because it is numbers, second-order properties, that are being counted. I suppose we can formulate high-number properties along Hodes's lines: high-6, for example, would be a fourth-order property, which applies to a third-order property *P* just in case *P* applies to exactly six (second-order) number properties. But high-6 cannot be the same as 6, because high-6 is fourth-order and 6 is second-order. Strictly speaking, we should say that the high-number of primes less than 10 is high-4. No doubt, high-arithmetic is coherent. It is obtained by applying the foregoing process of impoverishment to the language of arithmetic. We can also develop linguistic resources for properties of mixed type, like "being a corner of my desk or a prime less than 10," and we could have numbers to count these. There is no point to this exercise, however, because high-arithmetic and ordinary arithmetic (and mixed arithmetic) are equivalent. They all describe the same structure. The natural-number structure is the form common to second-order number properties, fourth-order high-number properties, and so on.

A central item from chapter 3 is relevant here. In the early stages, we can be neutral about whether the structure yielded by the sublanguage procedure is best construed in re or *ante rem*. At first, of course, a structure characterized by a sublanguage is not freestanding. Its places are filled only by items from the original background ontology. Income levels are instantiated only by people, partnerships, corporations, and the like. If someone or something does not have an income, then it cannot occupy an income level (other than the zero level). In Frege's first example, only lines have directions. And at first only objects in the original ontology can be counted and only properties of those objects have numbers. However, as emphasized earlier, once a structure is characterized, it is sometimes taken to be *freestanding* in the sense that it is exemplified by objects beyond those of the original ontology. This is a distinctive feature of mathematical structures. In the case at hand, the natural-number structure is soon seen to be exemplified by systems of objects that are not (explicitly) in the original background framework. We see that anything can be counted, including numbers themselves. That is, once the natural-number structure is identified, it can be applied to count items beyond those in the background language used to characterize the structure in the first place.

6 Ontological Interlude

Before presenting my last epistemic route to structures, I pause and reflect on ramifications for the basic ontological notions of object and identity, at least in mathematics. How mathematical structures are grasped tells us something about what they are. I confirm and reinforce some of the ontological theses proposed earlier. The more-philosophical conclusions are anticipated in Kraut [1980], which speaks of objects and identity generally.

Various relativisms recur throughout this book. What is structure from one perspective is system from another. What is office from one point of view is officeholder from another. The present considerations suggest a variation on this theme. "Mathematical object" is to be understood as relative *to a theory*, or, loosely, to a background framework. Natural numbers are objects *of arithmetic*, but "natural numbers" may not designate objects in another theory or framework. In particular, natural numbers may not be objects in the original background language from which we began. They may be offices.

To what extent are the objects "produced" by the Kraut sublanguage procedure (and the Frege–Wright–Hale procedure) new? In particular, are numbers already in the background ontology from which we begin? On the present view, the numbers so produced are the places of a structure. I have spoken of the identity between the places of different structures and of the identity between the places of a structure and other items. My proposal is that there is a determinate statement of identity, one with a truth-value to be discovered, only if the context is held fixed—only if the terms on both sides of the identity sign denote places in the same structure. Otherwise, the identity is a matter of invention or stipulation, based on convenience. Of course, mathematicians sometimes find it convenient to identify the objects in different structures. In some cases, like that of embedding the natural numbers in the reals, a single identification suggests itself. The identification of the "2" of the natural numbers with the "2" of the reals seems inevitable. Notoriously, however, there are cases in which there is no single preferred identification. One such case is the embedding of the natural-number structure in the set-theoretic hierarchy. We *discover* that $16^2 = 256$, whereas we either stipulate that $2 = \{\phi, \{\phi\}\}$, or we stipulate that $2 \neq \{\phi, \{\phi\}\}$, or we insist that the identity has no truth-value and speak of the "is of office occupancy" instead.

Applying this to the sublanguage procedure, there is nothing demanding that a theorist identify the "new" numbers with objects in the old framework, the background ontology. Nor is there anything to prevent such an identification, provided only that the domain is infinite.[12] For example, if the background includes set theory, one can proceed with Zermelo or von Neumann and identify numbers with certain sets. With proper care, one can also proceed along the lines of Frege and Russell. However, there are instances of the foregoing sublanguage procedure in which a similar maneuver is not available. One of them has considerable historical and philosophical interest in its own right.

Two properties P, Q are *coextensive* if they apply to exactly the same objects: $\forall x(Px \equiv Qx)$. Coextensiveness is clearly an equivalence relation among properties. Moreover, in a purely extensional language or sublanguage, like that of classical mathematics, coextensiveness is a congruence. By the foregoing sublanguage procedure, we can identify properties with the same extension.[13] Under this interpretation, extensions become bona fide objects, and coextensiveness serves as the identity on extensions. That is, the sublanguage procedure characterizes an extension structure, and extensions are its places.

In this case, however, we cannot invoke a convention and locate the "new" objects in the original ontology, on pain of contradiction. Cantor's theorem entails that (under certain assumptions) there are more property extensions than objects. Thus, there are not enough objects in any domain to serve as extensions for that domain. In short, Frege's view that numbers are objects (in the original ontology) led to the Caesar problem. Frege's assumption that extensions are objects (in the original ontology) led to contradiction (I return to Frege in chapter 5).

The view here is starkly un-Fregean. Numbers are objects in the language of our number person, but numbers *need not* be objects in the original framework. Extensions are objects in extension theory but *cannot* be objects in the original background framework (on pain of contradiction). Our conclusion is that in mathematics, at least, one should think of "object" as elliptical for "object of a theory" (see Resnik [1975]). This is of a piece with the tie between individual objects and structures. Formal theories describe structures. Within arithmetic, numbers are objects and any well-formed identity in the *language of arithmetic* has a truth-value. It refers to the places in the natural-number structure. But there is no unique, preferred superstructure that has numbers and sets, nor is there a unique superstructure that contains numbers and people. The idea of a single, fixed universe, divided into objects a priori, is rejected here.

---

12. If the background ontology is finite, then it does not contain enough objects to serve as numbers. Hume's principle is not satisfied on any finite domain, but it is satisfiable on any infinite domain. So if we insist, with Frege, that natural numbers are objects in the original domain, then Hume's principle is an axiom of infinity. See Wright [1983] and Boolos [1987].

13. It is constitutive of extensional contexts that coextensiveness be a congruence. As early as [1941], Quine attacked the use of properties (or attributes, or propositional functions) in *Principia mathematica*, arguing that they have no clear criteria of identity. Quine proposed that extensional classes be used instead. The procedure sketched here delivers that result.

The [1987] version of Putnam's internal realism is a "conceptual relativity" that has many elements in common with the foregoing theses.[14] Here is a passage that we encountered in chapter 2 of this volume: "The device of *reinterpretation* . . . recogniz[es] that one person's 'existence' claim might be another person's something else" (p. 34). When our interpreter took our economist to be accepting the existence of income groups, along with an identity relation on groups, speakers who use the full background framework correctly understand them to be talking about people, and the economist's "identity" is an equivalence relation on that "ontology." Putnam speaks loosely of relativity to "conceptual schemes." He notes that in going from framework to framework, we are reinterpreting the *logical* terminology, the identity sign and the existential quantifier in particular. It follows that "the notions of object and existence are not treated as sacrosanct, as having just one possible use. It is very important to recognize that the existential quantifier can be used in different ways—ways consonant with the rules of formal logic" (p. 35). Putnam seems to agree that identity and existence have the same *logic* in each context. In that sense, the terms are always used the same way *internally*. The point is that when we interpret one discourse in another framework, the "translation" of the logical terminology is not homophonic. One person's identity is another's equivalence. The equivalence relation has the logic of identity in the sublanguage, because the equivalence is a congruence there. Our economist's first-order variables are taken to range over income groups where the variables of the background language range over people.

Putnam concludes that it would be a mistake to go on to "single out *one* use of the existential quantifier . . . as the only metaphysically serious one." Again, if "use" here is understood as something like "extension," then Putnam is rejecting the idea of a single universe of discourse, fixed once and for all. So do I, at least for mathematics. Putnam concludes that in cases like ours, if you "take the position that one will be equally 'right' in either case [then] you have arrived at the position I have called 'internal realism'" (p. 35). If I have accurately interpreted Putnam's suggestive remarks, then in mathematics at least, *ante rem* structuralism is a version of the [1987] incarnation of internal realism.

It is tempting to follow Putnam and apply the theory-relative conception of object outside of mathematics—to science and even to (properly regimented) ordinary discourse. The thesis would be that the universe does not come, nor does it exist, divided into objects a priori, independent of our language, or framework, or, to use another Wittgensteinian phrase, form of life. The complex web of beliefs, concepts, and theories that determines how we perceive and understand the world also determines what its objects are and when two of them are the same or different. This temptation will be resisted, at least until we turn to applications of mathematics in chapter 8.

---

14. Putnam's earlier characterization of internal realism, in [1981, chapter 3], includes an epistemic account of truth (as rational acceptability under ideal conditions). See also Putnam [1980]. This theme is not relevant here.

7 Implicit Definition and Structure

There are limits to the sizes of structures that can be apprehended by any of these epistemic techniques. First, as noted, one cannot grasp a structure *S* by simple pattern recognition unless one can perceive a system that exemplifies *S*. Such a structure can have at most a small, finite number of places. In sections 3–4, I suggested some extensions of pattern recognition beyond simple abstraction. These faculties yield knowledge of large, finite structures, denumerably infinite structures, and perhaps structures with continuum-many places, but not much more. If the Kraut sublanguage technique starts with an equivalence relation on the *objects* in the background ontology, it yields only structures of the same or smaller cardinality. For example, if the background ontology is denumerable, then only denumerable and finite structures are delivered by the technique. If the sublanguage procedure begins with an equivalence relation on *properties*, it can yield a structure as large as the powerset of the original ontology. So if we start with a denumerably infinite ontology, the technique might yield a structure the size of the continuum. One can then perform the sublanguage procedure on the properties in a language of this larger structure and produce a structure whose size is the powerset of the continuum. Continuing, one can work one's way through the finite levels of simple type theory, which will deliver enough structure to simulate most of classical mathematics. At this point, the theorist might follow one of the aforementioned extensions of pattern recognition and formulate the idea of a level in the type hierarchy per se. That is, we focus on the pattern of producing ever-larger structures and combine the results into a superstructure that contains all of the finite types. The theorist can go on from there, to transfinite levels.

Clearly, however, these routes to large structures are artificial and ad hoc, because the structures do not look at all like the ones studied in mathematics. Moreover, we are left well short of the set-theoretic hierarchy. Our final technique for apprehending and grasping structures is the most powerful and the most speculative and problematic.

One way to understand and communicate a particular structure is through a direct description of it. To return to the standard example from chapter 3, someone might describe a baseball defense like this: "There are three outfielders, arranged as the vertices of a triangle; there are four infielders, arranged thus and so; there is a pitcher in the middle of the infield and there is a catcher behind the plate." Similarly, the structure of the U.S. government can be described by listing the various offices and the ways that the various officeholders relate to each other. In either case, of course, listeners may misunderstand and think that a particular *system* is being described. They may display this confusion with inappropriate questions, like "What is the name of the centerfielder's mother?" or "Is the senior senator from South Carolina a Republican?" Eventually, however, a properly prepared listener will understand that it is the structure itself, and not any particular instance of it, that is being described. Again, I do not claim to illuminate the psycholinguistic mechanisms that underlie this understanding. There is a whole host of presuppositions on the part of the listener. Nevertheless, it is clear that at least some listeners get it.

We have here an instance of *implicit definition*, a technique familiar in mathematical logic.[15] Notice that a direct description can succeed in communicating a structure even if no instance of the structure is displayed. One may describe a variation of a baseball defense or a government that has not been tried yet. Someone might wonder how it would go if there were two presidents, one of whom is commander in chief of the armed forces and the other vetoes legislation. Structures described this way need not be freestanding, of course, but they clearly can go beyond those exemplified in the actual physical world that we all know and love. Structures successfully described by implicit definition are naturally construed as *ante rem* (if they exist at all, of course).

Notice also that in characterizing a structure by implicit definition, one uses singular terms to denote the places of the structure. That is, "the centerfielder" and "president" are definite descriptions or proper names. However, the terms do not denote people; they denote places in the respective structures. They denote the offices, not the officeholders. In chapter 3, this orientation toward the structure is called "places-are-objects."

The mathematical cases are more pristine. In the opening pages of a textbook on number theory, we might read that each natural number has a unique successor, that 0 is not the successor of any number, and that the induction principle holds. Similarly, a treatise in real analysis might begin with an announcement that certain mathematical objects, called "real numbers," are to be studied. The only thing we are told about these objects is that certain relations hold among them. We may be informed, for example, that the numbers have a dense linear ordering, that there are associative and commutative operations of addition and multiplication, and so on. One easily gets the impression that the objects themselves do not matter; the relations and operations or, in a word, the structure is what is to be studied.[16] Implicit definition and structuralism go together like hand and glove.

---

15. There is an ambiguity in the phrase "implicit definition." In one sense, an implicit definition presupposes that all but one of the terms of a language already has fixed meaning. Let L be a language and T a theory in L. Let c be a singular term that does not occur in L. A purported implicit definition of c is a set S of sentences that contains the new constant. It succeeds if, in each model of T, there is exactly one way to assign a denotation to c to make every member of S true. So, for example, "c is a perfect number and $c < 10$" is an implicit definition of 6 in arithmetic. This is *not* the sense of "implicit definition" used here. In the present context, an implicit definition is a *simultaneous* characterization of a number of items in terms of their relations *to each other*. In contemporary philosophy, such definitions are sometimes called "functional definitions" (see chapter 3, section 6).

16. To be sure, most mathematics texts do not have such austere beginnings. A study may begin by defining a particular subclass of previously understood mathematical objects, such as the analytic complex-valued functions or the partial recursive functions. Alternately, the book may indicate that its objects are constructed out of familiar mathematical material. For example, it may deal with sequences of real-valued functionals. These works can also be given a straightforward structuralist interpretation, because the objects they start with (e.g., functions or functionals) are the places of a structure. In terms of chapter 3, systems whose objects are the places of other structures are defined, and the structures of those systems are studied.

A traditional Platonist might take the early remarks in these mathematics books as giving the reader some truths about a particular collection of abstract objects—the natural numbers or the real numbers. The opening sections have a practical, epistemic purpose. To do deductions and learn more about the particular objects under study, we need a list of "axioms" with which to begin. We cannot prove everything. Nevertheless, on the traditional Platonist view, there is an important autonomy between the axioms and the subject matter. Moreover, it is not the purpose of the opening sections of the books to uniquely characterize anything. It is enough that the axioms be *true of* the intended subject matter. The fact that other systems of objects satisfy the axioms is irrelevant. We structuralists reject this autonomy between the axioms and the subject matter.

Because on the traditional Platonist view, the axioms are statements about a particular realm of objects, it is possible that the axioms can be mistaken. Perhaps there are natural numbers other than zero that have no successors. Perhaps the successor function is not one-to-one. As a thought experiment, try to consider the skeptical possibility that *all* of the Peano axioms are false of the natural numbers. A traditional Platonist is faced with such a possibility.[17] Because the referent of "the natural numbers" is somehow independent of the characterization in the language of arithmetic, any given belief and, indeed, every (nonlogical) belief we have about numbers might be false. For the structuralist, on the other hand, this extreme skeptical possibility can be dismissed out of hand. It is conceivable, barely, that arithmetic is incoherent, in which case *no* structure is characterized. Perhaps the theory of arithmetic is not categorical, in which case more than one structure is characterized. But it is nonsense to claim that the theory of arithmetic does successfully refer to a single, fixed structure (or a fixed class of structures) but says hardly anything true about it (or them). On our view, the language characterizes or determines a structure (or class of structures) if it characterizes anything at all.[18]

Suppose that a reader wonders whether 2 is Julius Caesar or whether 2 is $\{\phi, \{\phi\}\}$. As we have seen, on the traditional Platonist view, these queries have determinate answers, even if the information is useless and perhaps unknowable, and even if there is no information available to help answer the questions. By contrast, for us structuralists, the queries represent a misunderstanding of what is being accomplished at the early stage. In an implicit definition, asking about Julius Caesar is similar to the aforementioned listener who is wondering about the name of the centerfielder's mother. The speaker was describing a structure, not a system of particular people. The mathe-

---

17. This is an instance of the general charge that realism allows the possibility of global error. A Platonist might wonder how it is possible to successfully *refer* to the natural numbers and yet manage to get just about everything wrong about them. Of course, the matter of reference is another sticking point for virtually any realist account (of mathematics or anything else). A structuralist account of reference is sketched in section 9.

18. Implicit definition has played some role in the development of mathematics. Some of this history is recounted in chapter 5.

matics book is not describing a system of sets or Platonic objects or people. It describes a structure or a class of structures.

Some readers might find it strange that deduction or proof is not treated at any length in this chapter on the epistemology of mathematics. After all, deduction is the main technique for advancing mathematical knowledge, and the focus on deduction cannot be attributed entirely to a brute concern for rigor. I do not have much to add to the many fine studies of deduction. My present purpose is to help understand what it is that formal and informal proofs prove. Clearly, to see how we use deduction to extend knowledge of the subject matter of mathematics, we need some account of what that subject matter is. Given my emphasis on implicit definition and the role of language use, the present account would "predict" a central role for deduction. A structure is like a universal, a one-over-many of sorts, and we conceive of mathematical structures as freestanding and *ante rem*. Thus, one would expect mathematical epistemology to use a topic-neutral technique. It does not matter what objects, if any, fill the places of mathematical structures. Because, in principle, a deduction is independent of its subject matter, deduction fits structuralism well. If the axioms are part of a successful implicit definition, then they characterize a structure (or a class of structures) and are true of it (or them). So the theorems are also true of the structure(s), and of every system that exemplifies it (or them). That is, we study the structure itself via the logical consequences of the axioms of an implicit definition.

Implicit definition, together with deduction, also supports the long-standing belief that mathematical knowledge is a priori. Again, an implicit definition characterizes a structure or class of structures if it characterizes anything. Thus, if sensory experience is not involved in the ability to understand an implicit definition, nor in the justification that an implicit definition is successful, nor in our grasp of logical consequence, then the knowledge about the defined structure(s) obtained by deduction from implicit definition is a priori. I return to what is involved in understanding implicit definitions in section 9 of this chapter.

8 Existence and Uniqueness: Coherence and Categoricity

At its root, then, an implicit definition is a collection of sentences, which we can call "axioms." Of course, not every set of sentences successfully characterizes a structure, even if someone intends to use it for that purpose. I have not said much yet about what it takes for an implicit definition to succeed. This is where the structuralist account is most speculative. There are two requirements on an implicit definition. The first is that *at least* one structure satisfies the axioms. Call this the "existence condition." The second requirement is that *at most* one structure (up to isomorphism) is described. This is the "uniqueness condition."

Uniqueness is less important, but let us start with that. Consider an implicit definition of the natural numbers. Some philosophers and logicians insist that the axioms do not characterize the natural-number structure, because *no* theory characterizes any infinite structure up to isomorphism. The Löwenheim-Skolem theorems show that any theory with an infinite model has a model of every infinite cardinality. So unintended or nonstandard models of any substantial implicit definition cannot be ruled out.

This, of course, is the "Skolem paradox" (see Shapiro [1995]). The structuralist is free to accept the conclusion. Nothing in the philosophy entails that there is but one natural-number structure. One can maintain that each model of first-order Peano arithmetic is *a* natural-number structure. Just as group theory applies to many nonisomorphic structures and systems, so does arithmetic. In the terminology of chapter 2, on this view every branch of mathematics (with at least one infinite model) is algebraic. None are "concrete."[19]

I reject this concession to the Skolem paradox. The Löwenheim-Skolem theorems apply only to *first-order* formal theories. Thus, the Skolemite presupposes that first-order model theory captures everything that is relevant about reference in ordinary, informal mathematical discourse. Elsewhere, I argue that the *informal* language of mathematics has the resources to distinguish standard from nonstandard models (Shapiro [1991, especially chapter 8]). Mathematicians themselves commonly make and exploit the distinction, and I presume that they are not deluding themselves. In the case of arithmetic, either informal resources go beyond those captured in formal logic, or we have a sufficient grasp of the *second-order* induction axiom. That is, we understand the second-order quantifier well enough to see that all models of arithmetic are categorical.

For present purposes, then, I maintain the conclusion of Shapiro [1991] that second-order model theory provides a good picture of the semantics of mathematical languages (of which more in section 9). Thus, categorical characterizations of the prominent infinite mathematical structures are available. Because isomorphism, among systems, is sufficient for "same structure," a categorical theory characterizes a single structure if it characterizes anything at all.

The tools for resolving the uniqueness requirement are thus found in mathematics itself. The same goes for the question of the *existence* of structures, but that matter is not handled as easily. Several times in the early sections of this chapter, I inferred the existence of a pattern from an ability to *coherently* discuss the pattern. The same goes for implicit definitions. A structure is characterized if the axioms are coherent. Recall the coherence axiom presented as part of structure theory, in section 4 of chapter 3: if Φ is a coherent sentence in a second-order language, then there is a structure that satisfies Φ. It is time to tighten up this admittedly vague usage. What is it to be "coherent" in this sense?

The coherence principle is an attempt to address the traditional problem concerning the existence of mathematical objects and, with that, the problem of reference. No small feat. Mathematical objects are tied to structures, and a structure exists if there is a coherent axiomatization of it. A seemingly helpful consequence is that if it

19. Some of Resnik's published work suggests this orientation to structuralism. His view that there is no determinate identity relation among structures is of a piece with his acceptance of first-order logic and his views on the Skolem paradox. In addition to his works already cited, see Resnik [1966]. On the other hand, see Resnik [1988] and [1996].

is possible for a structure to exist, then it does. Once we are satisfied that an implicit definition is coherent, there is no further question concerning whether it characterizes a structure. Thus, structure theory is allied with what Balaguer [1995] calls "full-blooded platonism" if we read his "consistency" as "coherence." It is misleading to put things this way, however, because the modality that we invoke here is nontrivial, about as problematic as the traditional matter of mathematical existence. Balaguer holds that we can be sanguine and wait for an antirealist to develop an account of coherence/consistency—because the antirealist presumably needs this notion anyway. It is not fair to leave it like this, however. To flirt with paradox, structuralism will falter if the notion of "coherence" is incoherent. Moreover, it is not obvious that a notion of "consistency" suitable for an antirealist program will work here. In any case, more can be said about coherence.

A first attempt would be to read "coherent" as "consistent" and to understand consistency in *deductive* terms. On this view, then, the analogue of "coherence" for formal languages is "deductive consistency." The thesis would then be that if one cannot derive contradictory consequences from a set of axioms, then those axioms describe at least one structure. The slogan is "consistency implies existence." In his famous correspondence with Frege, Hilbert adopted a version of this slogan (see chapter 5).

The "consistency-implies-existence" thesis does get support from Gödel's completeness theorem: if a set of sentences in a first-order language is deductively consistent, then it has a model. The slogan and the completeness theorem together seem to provide a clean solution to the "existence problem" for structuralism and mathematics generally—if one sticks to first-order languages. Even so, the matter is not completely straightforward. First, the completeness theorem is a result *in* mathematics, set theory in particular. The various models for consistent axiomatizations are found in the set-theoretic hierarchy, another structure. Perhaps this circle is tolerable, because we are not out to put mathematics on a firm, extramathematical foundation. We can find support for structuralism within mathematics, even if the support is corrigible.

There is a second circularity in this coherence-is-consistency maneuver. Consistency is usually defined as the nonexistence of a sort of deduction. Surely, the consistency of an axiomatization does not follow from the lack of concrete *tokens* for the relevant deduction. On the contrary, consistency is the nonexistence of a certain *type*. As noted earlier, the structure of strings is the same as that of the natural numbers. We cannot very well argue that the natural-number structure exists because arithmetic is consistent if this consistency is understood as a fact about the structure of the natural numbers. Or can we? An alternative would be to define consistency in terms of "possible deduction tokens" or perhaps one can take consistency as an unexplicated primitive. It is not clear what this move to modality buys us. We would have a problem about the "possible existence" of strings and of structures. Under what circumstances can we conclude that a structure is possible? This "possible-structure-existence" problem seems awfully close to the present "structure-existence" problem (see chapter 7 for an elaboration of this point).

The situation is worse on the present, higher-order orientation to implicit definitions. There is no completeness theorem for standard higher-order logic (see Shapiro [1991, chapter 4]). Let P be the conjunction of the second-order axioms for Peano arithmetic and let G be a standard Gödel sentence that states the consistency of P. By the *in*completeness theorem, $P \& \neg G$ is consistent, but it has no models. Indeed, because every model of P is isomorphic to the natural numbers, G is true in all models of P. Clearly, $P \& \neg G$ is not a coherent implicit definition of a structure, despite its deductive consistency.

The relevant formal rendering of "coherence," then, is not "deductive consistency." A better analogue for coherence is something like "satisfiability." It will not do, of course, to *define* coherence as satisfiability. Normally, to say that a sentence Φ is satisfiable is to say that *there exists* a model of Φ. The locution "exists" here is understood as "is a member of the set-theoretic hierarchy," which is just another structure. What makes us think that set theory itself is coherent/satisfiable?

Perhaps the problem could be resolved if we were to define "coherent" as the existence of a structure within the structure theory of chapter 3. This, however, is a mere cosmetic maneuver. One of the axioms of structure theory invokes the notion of coherence. Moreover, how do we know that structure theory is true, or even coherent? Do we find a model of it in structure theory or somewhere else? The problem is that structure theory will be as ontologically rich as set theory, especially if set theory itself is to be accommodated as the theory of a structure.

There is no getting around this situation. We cannot ground mathematics in any domain or theory that is more secure than mathematics itself. All attempts to do so have failed, and once again, foundationalism is dead (see Shapiro [1991, chapter 2]). The circle that we are stuck with, involving second-order logic and implicit definition, is not vicious and we can live with it. I take "coherence" to be a primitive, intuitive notion, not reduced to something formal, and so I do not venture a rigorous definition.

Of course, we are not exactly in the dark about coherence. The notion can be usefully *explicated*. The set-theoretic notion of satisfiability is a good mathematical *model* of coherence. That is, satisfiability is a rigorous mathematical notion that captures much of the structure of coherence. Moreover, the extension of satisfiability seems to be reasonably close to the extension of the intuitive notion of coherence. The idea is that satisfiability is to coherence pretty much as recursiveness is to computability. We have something like a "coherence thesis" analogous to Church's thesis.[20] Of

20. The analogy with Church's thesis is not perfect, because the extensions of coherence and satisfiability may not exactly match. As noted, "satisfiable" is understood to be "satisfiable by a member of the set-theoretic hierarchy." But there is no set that is isomorphic (or structure-equivalent) to the set-theoretic hierarchy itself. Thus, if there are no inaccessibles, then second-order Zermelo-Fraenkel is not satisfiable, but it is presumably coherent. A better model of "coherent" would be something like "satisfiable by a set or a proper class" (see Shapiro [1987]; and Shapiro [1991, chapter 6]). For most purposes, however, the ordinary notion of satisfiability will do. Recall that we are not out to *define* coherence as satisfiability. That would be an intolerable circle.

course, I cannot prove that coherence is (close to) satisfiability, nor can I give an argument that will convince a skeptic. But there is still progress. Coherence is the intuitive notion that serves as the criterion for structure existence. Satisfiability is a rigorous mathematical analogue.

In mathematics as practiced, set theory (or something equivalent) is taken to be the ultimate court of appeal for existence questions. Doubts over whether a certain type of mathematical object exists are resolved by showing that objects of this type can be found or modeled in the set-theoretic hierarchy. Examples include the "construction" of erstwhile problematic entities, like complex numbers.[21] This much is quite consonant with structuralism. To "model" a structure is to find a system that exemplifies it. If a structure is exemplified, then surely the axiomatization is coherent and the structure exists. Set theory is the appropriate court of appeal because it is comprehensive. The set-theoretic hierarchy is so big that just about any structure can be modeled or exemplified there. Set theorists often point out that the set-theoretic hierarchy contains as many isomorphism types as possible. That is the point of the theory.

Surely, however, we cannot justify the coherence of set theory itself by modeling it in the set-theoretic hierarchy. Rather, the coherence of set theory is *presupposed* by much of the foundational activity in contemporary mathematics. Rightly or wrongly (rightly), the thesis that satisfiability is sufficient for existence underlies mathematical practice. One instance of this is the use of set-theoretic hierarchy as the background for model theory and mathematical logic generally. Structuralists accept this presupposition and make use of it like everyone else, and we are in no better (and no worse) of a position to justify it. The presupposition is not vicious, even if it lacks external justification.

According to Hallett [1990], Hilbert's position during his correspondence with Frege was that if an axiomatization accords with mathematics *as developed thus far*, then its "objects" exist. A smooth fit of the "new" theory into existing practice is all that is needed to establish existence. According to Hallett, Hilbert took the proof-theoretic notion of "consistency" to be a good gloss on this notion of "smooth fit" or "accordance with practice." He never took consistency to be a wholesale replacement of the intuitive criterion for existence. Proof-theoretic consistency is mathematically tractable, and a fruitful research program emerged. However, with the hindsight that Gödel provided, we see that consistency is not a good model for the criterion of existence, especially in light of Hilbert's advocacy of higher-order logic. My example above, $P \,\&\, \neg G$, is consistent, but we see that the "theory" does not accord with practice. The existence of a model for $P \,\&\, \neg G$ is inconsistent with how the higher-order quantifiers are understood.[22]

21. See Wilson [1993] for other manifestations of this "existence question." I am indebted to Wilson for his insistence on this question. See also chapter 5 in this volume.

22. I return to Hilbert in chapter 5.

9 Conclusions: Language, Reference, and Deduction

Most of these epistemic techniques suggest a tight link between grasp of language and knowledge of structures. This is especially true for implicit definition. For the fields of pure mathematics at least, grasping a structure and understanding the language of its theory amount to the same thing. There is no more to understanding a structure and having the ability to refer to its places than having an ability to use the language correctly. Recall that a structure is not determined by the places in it, considered in isolation from each other, but rather by the *relations* among the places. In essence, these relations are embodied in the language. In fact, the correct use of the language *determines* what the relations are.

Many realists, such as the traditional Platonists, balk at this epistemic link between ontology and language. They hold that the contents and the nature of the universe exist independently of us and our linguistic lives. Our language, if successful, more or less accurately tracks the true ontology, but we do not create the ontology with this language. To be sure, I agree with the spirit of this reflection. Structuralism is not a general skepticism nor a conventionalism. Mathematics is objective if anything is. The natural-number structure has objective existence and facts about it are not of our making. The point is that the way humans apprehend structures and the way we "divide" the mathematical universe into structures, systems, and objects depends on our linguistic resources. Through successful language use, we structure the objective subject matter. Thus, language provides our epistemic access to mathematical structures.

The close link between linguistic resources and epistemic grasp attenuates at least some of the standard epistemological and semantic puzzles concerning mathematics. Consider the simplest infinite structure, the natural numbers. Arithmetic is the theory of this structure. The epistemology of the natural numbers is especially tractable, because a typical *language* of arithmetic contains a system of *numerals*. Presumably, on all accounts, we do learn and understand a language of arithmetic. In accomplishing this, we learn how to generate and recognize the numerals. And, of course, the numerals exemplify the structure being characterized and studied.

I do not claim that the natural-number structure is somehow grasped by abstraction from numerals. This simple-minded idea puts the cart well before the horse. The point is that understanding a language that contains a (full) system of numeration—understanding how to work with the numerals—*presupposes* everything needed for arithmetic. This is especially true of the standard Arabic notation. Working with the base system of numerals itself involves addition and multiplication. In short, understanding how to use the language of arithmetic is sufficient for understanding and referring to a system that exemplifies the natural numbers. To grasp the natural-number structure itself, there is little more that one has to learn or do.[23]

---

23. The introductory essay of Parsons [1983] makes much of this observation in an account of mathematical intuition. The transparency of the natural-number structure, via a system of numerals,

One might think that this sketch is undermined by the fact that it is possible to give a categorical theory of the natural numbers without using numerals. Quine makes much of the fact that formal theories need not have any singular terms at all. This is quite beside the point here, however. The natural-number structure is not usually introduced with such an austere theory, and we know by experience that arithmetic is not usually learned that way, even if it could be.

Another common observation is that even if a language of arithmetic has a system of numerals, it need not *refer* to numbers. The first-order quantifiers of arithmetic can be construed substitutionally. This detail is also irrelevant to the present thesis. I do not claim that one who can successfully use a language of arithmetic is thereby "committed" to the existence of the natural numbers. The point is that those who grasp the workings of the language in question have just about everything they need to grasp the natural-number structure. The epistemology of arithmetic is tractable, because each number has a (canonical) name. Understanding and working with the language involves an understanding of (an exemplification of) the natural-number structure. There is no more to the natural-number structure than the relations embodied in the language and exemplified by the numerals.[24]

The situation in more advanced branches of mathematics is not quite as straightforward. Simple cardinality considerations entail that there is no (standard) model of analysis among the terms of the language of analysis or in any language that can be grasped and used by humans. Because there are uncountably many real numbers, the structure is not exemplified in the language itself, unless uncountably many terms are somehow introduced.[25]

Nevertheless, the general claim of this section stands. There is no more to understanding the real-number structure than knowing how to use the language of analysis. In learning the language, one comes to acquire facility with quantifiers over real numbers—one learns how to use them. This is about all that is involved in understanding the statements of analysis. In working with language, one learns the axioms of the implicit definition. These axioms, or the other descriptions of the real numbers, determine the relationships between real numbers, such as the operations on them and the continuity and Archimedean properties. The theme of this book is that

was probably a factor in Hilbert's adoption of finitary arithmetic as the epistemic basis for his program. The idea is that the structure of strings is necessary for any thinking at all. There is no more privileged standpoint than that of finitary arithmetic/string theory. See Sieg [1990, section 2]; Tait [1981]; and Hallett [1994]. For a similar idea, see Benacerraf [1965, §III.C].

24. An anecdote: A few years ago, a friend of mine, who is a civil engineer, saw my copy of Field's *Science without numbers* [1980] and, intrigued by the title, asked me about the book. I did my best to explain the program and, in particular, the thesis that numbers do not exist. My friend, who is extremely intelligent and remarkably patient, asked me whether there are numbers printed at the top of each page. I then explained the difference between numbers and numerals, but I think he was unimpressed. On the view defended here, he was right to be unimpressed.

25. On the other hand, the Henkin proof for the completeness theorem yields a model for any consistent first-order theory constructed from linguistic items like constants and predicates. For most theories, however, this produces an unnatural, ad hoc (nonstandard) model, and once again, the technique is limited to first-order axiomatizations.

these relations characterize the structure. There is no more to "the real numbers" than these relations. As is well known, the axioms determine the cardinality of the structure, setting it at the continuum. Similar considerations apply to complex analysis, and even set theory.

I briefly turn from epistemology to the related matter of semantics. Recall the conclusion from chapter 2, that model-theoretic semantics is an appropriate tool for realism in ontology and realism in truth-value. The central notion of model theory, of course, is the relation of "satisfaction," sometimes called "truth in a model." For a realist, the relation between a model and a sentence in a formal language is a decent mathematical model of the relationship between the world and a language that describes it. Our premise here is that understanding ordinary language involves some sort of grasp of relations like those of model theory.

Probably the most baffling, and intriguing, semantic notion is that of *reference*. The underlying philosophical issue is sometimes called the "'Fido'—Fido problem." How does a term come to denote a particular object? What is the nature of the relationship between a singular term ("Fido") and the object that it denotes (Fido), if it denotes anything? Notice that model theory, by itself, has virtually nothing to say on this issue. In textbook developments of model theory, reference is taken as an unexplicated *primitive*. It is simply *stipulated* that an "interpretation" includes a function from the individual constants to the domain of discourse. This is a mere shell of the reference relation.

In effect, model theory determines only the *relations* between truth conditions, the reference of singular terms, the extensions of predicates, and the extensions of the logical terminology. Model theory is thus a functional (or structural) definition of these semantic terms. To see a given system as a model-theoretic interpretation of a (formal) language, one needs only to think of the terms of the language as (somehow) referring to the domain of discourse, and to see the first-order quantifiers as ranging over that domain. As far as the model-theoretic scheme goes, it does not matter how this "reference" is to be accomplished or whether it can be accomplished in accordance with some theory or other. There is nothing problematic in the abstract consideration of models whose domains are beyond all causal contact. As far as model theory goes, reference can be *any* function between the singular terms of the language and the ontology. In the terminology of chapter 3, the relations of model theory, notably reference, are freestanding.

It is fair to say that when it comes to mathematics and theories of other *abstracta*, realism in ontology often falters over reference (about as much as it falters over epistemology). If we assume that ordinary languages are understood and if we accept the premise that model theory captures the structure of ordinary interpreted languages, then we can do better. There is, of course, no consensus on how reference to ordinary physical objects is accomplished. The theories are legion. I do presume, however, that reference to proverbial medium-sized physical objects *is* accomplished. Most of the time, speakers of natural language do manage to use names of people, places, and things to refer to those very people, places, and things—common errors and failures of reference notwithstanding. Understanding how to use ordinary language involves an understanding, at some level, of reference (however it works). Thus, if a model-theoretic definition is supplemented with the correct account of reference, we get a decent approximation to the truth conditions of those natural-language sentences that come closest to the formulas in a formal language. Perhaps model theory is not earth-shattering, but it is no small feat either.

Because mathematics is the science of structure, the "schematic" or structural semantic notions of model theory are all that we need. The details of the correct account of reference to physical objects are irrelevant.[26] Let us return to real analysis. Let S be any system that has the array of operations and relations needed to be a model-theoretic interpretation of the language of analysis. That is, S has a pair of binary functions, two designated elements, a binary relation, and so on. In the jargon of model theory, S has the "signature" of analysis. It does not matter how S is "given" or to what extent S can be grasped by someone. I submit that a subject's grasp of how her native mathematical language works is enough for her to see what it would be for S to be a model of real analysis, that is, for the theory of analysis to be "true of" S. It is just for the axioms to be true when the quantifiers range over the domain of S and for the singular terms and predicates to refer as S indicates that they do. As noted, the student already understands reference and quantification, at least schematically. She may not know that there *are* any models of the theory, but she does grasp what it would be for a system to be such a model. Now, because the theory is categorical and coherent, all of its models share a common structure. The suggestion of this book is that we think of real analysis as being about that very structure. Its variables range over the places of that structure, and its singular terms refer to some of those places. Knowing what it would be for a system to be a model of the axioms *is* to know what the real analysis structure is. Schematic knowledge about how language works leads to knowledge about structures.

We end up with a model-theoretic interpretation of analysis. The variables range over the places of a structure, and the singular terms refer to individual places in that structure. The rest is familiar model-theoretic semantics. Once we realize what the ontology is, we have realism in ontology.

If we insist on categorical characterizations of nonalgebraic theories, then we also have realism in truth-value. A categorical theory T is semantically complete in that for every sentence Φ in the language, either $T \models \Phi$ or $T \models \neg\Phi$. Semantically complete theories yield a straightforward bivalent model-theoretic semantics. If we stick to categorical (or semantically complete) theories, realism in ontology, as construed here, thus leads to realism in truth-value.[27]

My conclusions here are consonant with the earlier remarks on implicit definition. The "axioms" characterize a class of structures if they characterize anything at all. If the theory is categorical, then only one structure is delivered. When we move

---

26. This underscores the observation from chapters 1 and 2 that many of the tools of contemporary philosophy and logic were developed with mathematics, and not ordinary discourse, in mind.

27. Of course, to insist on categorical theories of rich mathematical structures, we require resources beyond first-order logic. Let A be any first-order axiomatization of arithmetic. If Φ is deductively independent of A, then there is no "fact of the matter" whether Φ is true in A. Indeed, Φ holds in some models and fails in others. The sentence Φ is a legitimate assertion in the theory, and yet it lacks a truth-value. Thus, realism in truth-value is not delivered without semantic completeness.

to the "places-are-objects" perspective, the freestanding, schematic notions captured by model theory give the reference to places in that structure. All that is required is a general and commonplace, if sophisticated, grasp of the workings of language. In other words, in learning some languages, one learns structures, and can refer to their places. To learn the language—and shift to the places-are-objects perspective—is to grasp the structure. On my view, understanding mathematics is not philosophically occult—no more so than our ability to learn and use a language. To the extent that knowledge of model theory is a priori, so is mathematical knowledge.

To close this long epistemological chapter, it will prove instructive to contrast the present conclusions with Quine's thesis of the inscrutability of reference. There are many structuralist tendencies in Quine's understanding of discourse, but ironically, his thesis of inscrutability blocks the final ratification of structuralism.

On the *ante rem* view of structures advocated here, there is little, if any, inscrutability of reference. Once we accept the Peano axioms as an implicit definition of the natural-number structure, the numeral "27," for example, refers to a place in this structure. There is no room for doubt or inscrutability concerning just which place this is.[28] Start at the 0 place and begin counting. If you count correctly, you will arrive at the 27 place, and no other, at the proper moment. Of course, as noted, there is some indeterminacy or, if you like, inscrutability, when the places from *different* structures are identified. It is convenient for some purposes to locate the natural numbers in the set-theoretic hierarchy, but there is no unique best way to accomplish this. Any of several identifications will do. Nevertheless, within arithmetic—within the natural-number structure—reference is determinate and scrutable.

No doubt, it is "inscrutable" whether *ante rem* structuralism is the uniquely correct account of mathematics. Notice, however, that on both an eliminative and a modal construal of structuralism there is no inscrutability of reference either, because, strictly speaking, there is no reference at all. The numeral "27" does not denote an object. When it occurs in context, the numeral represents a variable that ranges over a particular place in any natural-number system (see chapter 3).

A full-blown inscrutability of reference occurs if there is a single ontology for all of science *and* if the natural numbers, for example, are located somewhere in that ontology. On this perspective, numbers are objects, but we do not know and cannot know *which* objects they are. Quine seems to embrace just this combination of views. His thesis that "existence" is univocal suggests a single ontology for the entire "web of belief," and his realism toward arithmetic entails that numbers are objects. But which objects? We cannot know. This has at least a family resemblance to Frege's Caesar problem.

In a note [1992], Quine turns his attention to structuralism. He opens with a few remarks about arithmetic: "We are familiar with three adequate but incompatible ways of modeling number theory in . . . the theory of classes. We bandy our numbers without caring which classes we are bandying from among this wealth of alternatives. We are just content that we are operating somewhere within the ontology of classes to which we have committed ourselves anyway for other purposes. . . . [The] struc-

28. I am indebted to Pierluigi Miraglia for pointing out how a structuralist ontology alleviates the inscrutability of reference.

turalist treatment of number . . . is just a way of eliminating an idle question—'What is number?'—and a gratuitous decision among indifferent alternatives" (p. 5). So far, so good. It looks like inscrutability crops up once we decide that the natural-number structure needs to be "located" in the set-theoretic hierarchy. There is no unique way to do the locating. Quine goes on to discuss a structuralistic account of classes, due to David Lewis [1991] (see also Lewis [1993]). He likes the idea: "Structuralism for classes, hence for all abstract objects, is undeniably congenial. They are things that are known anyway only by their structural role in cognitive discourse; never by ostension. . . . Your class of cats and mine can . . . be different things, and your membership relation and mine can be different, if it makes any sense to say so, though we see eye to eye on every cat. Such is the appeal of structuralism concerning abstract objects" (p. 6).

Quine thus endorses a major theme of this book. He then proposes a breathtaking extension, well beyond the scope of the present treatment: "My own line is a yet more sweeping structuralism, applying to concrete and abstract objects indiscriminately" (p. 6). This "sweeping" view is then illustrated and defended with familiar Quinean themes concerning observation sentences, reification, the evidence for scientific statements, and proxy functions: "[I]f we transform the range of objects of our science in any one-to-one fashion . . . the entire evidential support of our science will remain undisturbed. . . . The conclusion is that there can be no evidence for one ontology as over against another, so long as we can express a one-to-one correlation between them. Save the structure and you save all. . . . For abstract objects this is unsurprising, and quite in the spirit of Ramsey, Lewis, and Benacerraf" (p. 8). As noted earlier, this very observation is a motivation for *ante rem* structuralism. Quine, however, stops just short of this: "My global structuralism should not . . . be seen as a structuralist ontology. To see it thus would be to rise above naturalism and revert to the sin of transcendental metaphysics. My tentative ontology continues to consist of quarks and their compounds, also classes of such things, classes of such classes, and so on . . ." (p. 9). The thesis of naturalism is that we should look to our best science for a description of the contents of the universe (see chapter 1). This science speaks of quarks, classes, and numbers, but, supposedly, it does not speak of structures and their places. So we are left with the inscrutability of reference.

Putting physical objects aside for now, I certainly agree with Quine that numbers exist and that they are pretty much as the scientist/mathematician says they are. Quine takes it to follow that natural numbers should not be construed as places in the natural-number structure and that there is something wrong with an ontology of structures generally. This is a non sequitur. Quine's own considerations suggest that nothing mathematicians say can rule out interpreting them as talking about the places of a structure. That is the point. It is the burden of this book to show that the *ante rem*, structuralist interpretation of modern mathematics is consistent with the practice of mathematics. Moreover, it provides a compelling account of that very same practice. The places of the natural-number structure, from the places-are-objects perspective, are objects with just the properties and relations that the mathematician attributes to the natural numbers. If this is "transcendental metaphysics," then so be it, but it is not pernicious.

5

How We Got Here

1 When Does Structuralism Begin?

Sooner or later, the philosopher of mathematics needs to ponder the historical scope of his or her proposed account. Is the claim that mathematics is the science of structure meant to include the work of Euclid, Archimedes, Descartes, Leibniz, Gauss, Cauchy, and Lebesgue, not to mention Thales and Mac Lane? Were these mathematicians engaged exclusively in the study of structures, whether they thought so or not? Should the portion of their work that does not fit the mold of structuralism be dismissed as nonmathematics? Certainly not. Mathematics has a long history, and it is still evolving. One would be hard put to come up with a single philosophical account of mathematics that accommodates every time slice from its roots in antiquity until today, let alone the future.[1] Structuralism is (only) a perspicuous account of the bulk of contemporary mathematics, putting its ontology, epistemology, and applications in perspective. The purpose of this chapter is to recount some historical themes and precursors to the thesis that mathematics is the science of structure. Structuralism is the natural outcome of some developments in mathematics and in philosophy.

The nineteenth century was a watershed for mathematics. Stein [1988, 238] claims that during this period, mathematics underwent "a transformation so profound that it is not too much to call it a second birth of the subject," the first birth having been in ancient Greece. Contemporary mathematics has a reputation for being a cut-and-dried discipline, lacking any unclarity or controversy. Whether this is accurate or not, an

1. Of course, similar issues apply to philosophical accounts of virtually any ongoing human activity. Can anything enlightening be said about the entire span of artistic activity? In this regard, the situation with mathematics is a lot less severe than that of other disciplines. The contemporary mathematician does not have to undergo a major paradigm shift in order to read, say, Euclid or Archimedes. There is more than a mild family resemblance between contemporary and Greek mathematics. There are differences, of course, but they are more subtle.

impartial observer would not make similar claims about mathematical activity in the nineteenth century. The friendly beast that we now call "classical mathematics" is the result of battles that took place then. In the next section, I take on the modest task of recounting some themes in the development of Euclidean, projective, and non-Euclidean geometry in the nineteenth and early twentieth century. There was a gradual transformation from the study of absolute or perceived space—matter and extension—to the study of freestanding, *ante rem* structures.

The same period also saw important developments in philosophy, with mathematics as its case study. According to Coffa [1991, 7], for "better or worse, almost every philosophical development since 1800 has been a response to Kant." A main item on the agenda was to account for the prima facie necessity and a priori nature of mathematics without invoking Kantian intuition. Can we understand mathematics independently of the forms of spatial and temporal intuition? Coffa argues that the most successful approach to this problem was that of the "semantic tradition," running through the work of Bolzano, Frege, the early Wittgenstein, and Hilbert, culminating with the Vienna Circle, notably Schlick and Carnap. The plan was to understand necessity and a priority in *formal* terms. These philosophers located the source of necessity and a priority in the use of language. Necessary truth is truth by definition. A priori knowledge is knowledge of language use. In one way or another, this philosophical tradition was linked to the developments in mathematics. One legacy left by both the developments in mathematics and the semantic tradition in philosophy is mathematical logic, model-theoretic semantics in particular. As we saw in the previous chapters, the emergence of model theory and the emergence of structuralism are, in a sense, the same.

The second part of my narrative includes sketches of early-twentieth-century theorists who either developed structuralist insights, or opposed these moves, or both. The list includes Dedekind, Poincaré, Russell, Frege, and Hilbert. I conclude with a brief account of the Bourbaki group.

2 Geometry, Space, Structure

The transition from geometry as the study of physical or perceived space to geometry as the study of freestanding structures (see chapter 3) is a complex tapestry.[2] One early theme is the advent and success of analytic geometry, with projective geometry as a response. Another is the attempt to accommodate ideal and imaginary elements, such as points at infinity. A third thread is the assimilation of non-Euclidean geometry into mainstream mathematics and into physics. These themes contributed to a growing interest in rigor and the eventual detailed understanding of rigorous deduction as independent of content. Anything beyond a mere sketch of a scratch of this rich and wonderful history is beyond the scope of this book—and my own competence.

2. Much of this chapter draws from Nagel [1939]; Freudenthal [1962]; Coffa [1986], [1991, chapters 3 and 7]; and Wilson [1992]. Most of the translations are from Nagel [1939]. Readers interested in these episodes of mathematical history are urged to consult those excellent works.

The traditional view of geometry is that its topic is matter and extension. The truths of geometry seem to be necessary, and yet geometry has something to do with the relations between physical *bodies*. A philosopher may find it difficult to reconcile these two features. Kant's account of geometry as synthetic a priori, relating to the forms of perceptual intuition, was a heroic attempt to accommodate both the necessity and the empirical applicability of geometry.

The traditional view of arithmetic is that its topic is *quantity*. Arithmetic was thought of as the study of the discrete, whereas geometry was the study of the continuous. The fields are united under the rubric of mathematics, but one might wonder what they have in common other than this undescribed genus. The development of analytic geometry went some distance to bridge the gap between the two subjects and to loosen the distinction between them. Mathematicians learned that the study of quantity can shed light on matter and extension. Good for mathematics, but this compounds the philosophical problem. Of course, Kant held that arithmetic is also synthetic a priori, based on the forms of the perception of time.

One result of the development of analytic geometry was that synthetic geometry, with its reliance on diagrams, fell into neglect. Lagrange even boasted that his celebrated treatise on mechanics did not contain a single diagram—although one might wonder how many of his readers appreciated this. The dominance of analytic geometry left a void that affected important engineering projects. For example, problems with plane representations of three-dimensional figures were not tackled by mathematicians. As the slogan goes, necessity is the mother of invention. The engineering gap was filled by the emergence of projective geometry (see Nagel [1939, §§7–8]). Roughly, projective geometry concerns spatial relations that do not depend on fixed distances and magnitudes, or on congruence. In particular, projective geometry dispenses with quantitative elements, like a metric.

Although all geometers continued to identify their subject matter as intuitable, visualizable figures in space, the introduction of so-called *ideal elements*, such as imaginary points, into projective geometry constituted an important move away from visualization. Parallel lines were thought to intersect, at a "point at infinity." Desargues proposed that the conic sections—circle, ellipse, parabola, and hyperbola—form a single family of curves, because they are all projections of a common figure from a single "improper point," also located at infinity. Circles that do not intersect in the real plane were thought to have a pair of imaginary points of intersection. As Nagel put it, the "consequences for geometrical techniques were important, startling, and to some geometers rather disquieting" (p. 150). Clearly, mathematicians could not rely on the forms of perceptual intuition when dealing with the new imaginary elements. The elements are not in perceivable space; we do not see anything like them.

The introduction and use of imaginary elements in analytic and projective geometry was an outgrowth of the development of negative, transcendental, and imaginary numbers in arithmetic and analysis. With the clarity of hindsight, there are essentially three ways that "new" entities have been incorporated into mathematical theories (see Nagel [1979]). One is simply to *postulate* the existence of mathematical entities that obey certain laws, most of which are valid for other, accepted entities. Complex numbers are like real numbers but closed under the taking of roots, and ideal points are like real points but not located in the same places. Of course, postulation begs the question against anyone who has doubts about the entities. Recall Russell's quip that postulation has the advantages of theft over honest toil. In reply, an advocate might point out the usefulness of the new entities, especially for obtaining results about established mathematical objects. But this benefit can be obtained with any system that obeys the stipulated laws. Thus, the second method is *implicit definition*. The mathematician gives a description of the system of entities, usually by specifying its laws, and then asserts that the description applies to any collection that obeys the stipulated laws (see chapter 4). At this point, the skeptic might wonder whether there are any systems of entities that obey the stipulated laws. The third method is *construction*, in which the mathematician defines the new entities as combinations of already-established objects. Presumably, this is the safest method, because it settles the question of whether the entities exist (assuming the already-established objects do). Hamilton's definition of complex numbers as pairs of reals fits this mold, as does the logicist definition of natural numbers as collections of properties.

One can think of the proposed constructions as giving fixed denotations to the new terms. The imaginary number i just is the pair $<0, 1>$. This is unnatural, because, to echo Benacerraf [1965], the pair has properties one would be loath to attribute to the complex number. A fruitful outlook would be to take implicit definition and construction in tandem. A construction of a system of objects establishes that there are systems of objects so defined, and so the implicit definition is not empty. Moreover, the construction also shows how the new entities can be related to the more established ones and may suggest new directions for research. Wilson [1992] shows that in some cases, construction is a search for the "natural setting" in which the structure is best studied. That is, the construction might provide the mathematical environment in which both the new field and the old one grow and thrive.

Nagel notes that all three methods were employed in the development of ideal points and points at infinity in geometry. In the early decades of the nineteenth century, imaginary numbers were scandalous but known to be useful. The problem was to figure out what to make of these pesky items. Poncelet came close to the method of postulation. In trying to explain the usefulness of such numbers in obtaining results about the real numbers, he claimed that mathematical reasoning can be thought of as a mechanical operation with abstract signs. The results of such reasoning do not depend on any possible referents of the signs, so long as the rules are followed. Perhaps this is a version of formalism or mechanism. Having thus "justified" new sorts of numbers in analysis, Poncelet went on to argue that geometry is equally entitled to employ abstract signs—with the same freedom from interpretation. He held that synthetic geometry is crippled by the insistence that everything be cast in terms of drawn or visualizable diagrams.

With hindsight, we see steps away from the traditional conception of geometry as the study of space. However, Poncelet himself was steeped in that tradition. He held that the introduction of ideal elements was a powerful tool for shedding light on spatial extension—just as imaginary numbers shed light on real numbers. Thus, to continue our hindsight, we might add instrumentalism to the views he expressed. Poncelet was surely correct that the introduction of ideal elements unifies projective geometry and gives it direction, but he did not see that this fruitful development also moved geometry further from its connection with matter and extension. When he declared that a straight line always has two points of intersection with a circle, even though the "usual" sort of straight line does not always intersect the "usual" sort of circle, he unwittingly contributed to a transformation of both the subject matter of geometry and the character of mathematical inquiry (see Nagel [1939, §§13–20]).

Poncelet's contemporaries were aware of the shortcomings of such bare postulation. Nagel cites authors like Gergonne and Grassmann, who more or less prefigured the method of implicit definition. Their work furthered the concern with rigor and the abandonment of the traditional view of geometry as concerned with extension. We move closer to the view of geometry as the study of structures.

By way of contrast, Euclid defines a "point" to be "that which has no parts," a "line" as a "length without breadth," and a "straight line" as a line "which lies evenly with the points on itself." Because these definitions play no role in the subsequent mathematical development, the modern reader might wonder why Euclid included them. Of course, he did not write the *Elements* for the modern reader doing this wondering. These definitions are in the mold of the traditional conception of geometry as the study of space.

Grassmann's *Ausdehnungslehre of 1844* developed geometry as "the general science of pure forms," considered in abstraction of any interpretation the language may have. He characterized the terms of geometry only by stipulated relations they have *to each other*: "No meaning is assigned to an element other than that. It is completely irrelevant what sort of specialization an element really is. . . . [I]t is also irrelevant in what respect one element differs from another, for it is specified simply as being different, without assigning a real content to the difference" (Grassmann [1972, 47]).

In this work, Grassmann dubbed the new study "the general science of pure forms"—to be considered independently of any intuitive content the theory might have. He distinguished "formal" from "real" sciences, along the same lies as the present distinction between pure and applied mathematics (see chapter 8). Traditional geometry is a real science—an applied mathematics—and so *this* geometry

> is not to be regarded as a branch of mathematics in the sense that arithmetic [is]. For geometry refers to something given by nature (namely, space) and accordingly there must be a branch of mathematics which develops in an autonomous and abstract way laws which geometry predicates of space. [In this mathematics], all axioms expressing spatial intuitions would be entirely lacking. . . . [T]he restriction that it be limited to the study of a three-dimensional manifold would . . . be dropped. . . . Proofs in formal sciences do not go outside the domain of thought into some other domain. . . . [T]he formal sciences must not take their point of departure from axioms, as do the real sciences, but will take definitions instead as their foundation. (ibid., 10, 22)

Writing in 1877, almost thirty years later, Grassmann explained that his *Ausdehnungslehre* is "the abstract foundation of the doctrine of space. . . . [I]t is free from all spatial intuition, and is a purely mathematical discipline whose application to space yields the structure of space. This latter science, since it refers to something given in nature (i.e., space) is no branch of mathematics, but is an application of mathematics to nature. . . . For while geometry is limited to the three dimensions of space, [the] abstract science knows no such limitation" (ibid., 297). As Nagel [1939, §36] put it, Grassmann was one of the first mathematicians who "explicitly recognized that mathematics is concerned with formal structures."

Grassmann's work did not achieve widespread attention until the 1870s. Much of the response the *Ausdehnungslehre* did generate was negative. Nagel [1939, §37] cites a number of his contemporaries who expressed serious misgivings about the dismissal of spatial intuition as the foundation of geometry. Even Grassmann agreed that geometry should be *applicable* to the study of space. Ideal elements remain problematic for the "real science" that corresponds to pure geometry. What in *space* is to answer to the ideal elements? How does the new formal science of geometry relate to the traditional study?

To resolve this, we move to the third method of accommodating ideal elements, construction. This time, our hero is von Staudt, who showed that if we interpret the imaginary points as complex constructions of real points on real lines, then all of the theorems of the new projective geometry come out true. In modern terms, von Staudt discovered a model of the new theory. It is a particularly useful model, because the supposed intended interpretation of original geometry—space—is a part of it.

Notice that with this construction, von Staudt not only ignored Euclid's definition of a point as that which has no parts, he contradicted it. When projective geometry is embedded in von Staudt's system, points have a complex internal structure. Some points do have parts.

This development furthered the idea that in studying geometry as such, it is immaterial how its "elements" are regarded, as long as the stipulated "definitions" are satisfied. They can be things with no parts, or they can be things with parts, or they can be configurations of things. Nagel concludes that von Staudt's method of construction supplements and reinforces the method of implicit definition (but see Wilson [1992]).

The discovery of duality in projective geometry marks a closely related development. We are accustomed, even today, to think of a line as a locus of points. However, one can just as well think of a point as a locus of lines. In projective geometry, with the ideal elements added, the symmetry between points and lines is deep. If the terminology for "points" and "lines" is systematically interchanged, all theorems still hold. Notice that with this duality, we once again contradict Euclid's definition of a point as that which has no parts. Interpreted via duality, the points do have parts.

Impressed with this duality and extending it, Plücker [1846, 322] wrote that geometric relations have validity "irrespective of every interpretation." When we prove a theorem that concerns, say, straight lines, we have actually proved many theorems, one for each interpretation of the theory. The analogy with algebraic equations is apparent. Chasles conceded that in light of the duality of projective geometry, the field should not be construed as the science of magnitude but as the science of order (see Nagel [1939, §55, §59]). The logical insight is that, with sufficient rigor, the derivation of theorems should depend only on the stipulated relations between the elements and not on any features of this or that interpretation of them.

Whether or not primitive terms are to be understood in terms of spatial or temporal intuition, the emergence of rigor led to the idea that the *inferences* of a branch of mathematics should be independent of intuition. The long-standing idea that logic is a formal and topic-neutral canon of inference has concrete application in the present story. The topic neutrality of logic dovetails with, and reinforces, the emancipation of geometry from matter and extension. Even if topic neutrality had been expressed earlier in the history of mathematics and logic, it was carried out in some detail for geometry in Pasch [1926].[3] He made the aforementioned complaint that when Euclid declared that a point has no parts, and so on, he was not "explaining these concepts through properties of which any use can be made, and which in fact are not employed by him in the subsequent development" (pp. 15–16). Pasch thought it important that geometry be presented in a formal manner, without relying on intuition or observation when making inferences:

> If geometry is to be truly deductive, the process of inference must be independent in all its parts from the meaning of the geometrical concepts, just as it must be independent of the diagrams; only the relations specified in the propositions and definitions may legitimately be taken into account. During the deduction it is useful and legitimate, but in no way necessary, to think of the meanings of the terms; in fact, if it is necessary to do so, the inadequacy of the proof is made manifest. If, however, a theorem is rigorously derived from a set of propositions . . . the deduction has value which goes beyond its original purpose. For if, on replacing the geometric terms in the basic set of propositions by certain other terms true propositions are obtained, then corresponding replacements may be made in the theorem; in this way we obtain new theorems . . . without having to repeat the proof. (p. 91)

As suggested by this passage, this development was an important factor in the emergence of model theory as the main tool in mathematical logic. Pasch expresses the idea of geometry as a hypothetical-deductive endeavor, what he calls a "demonstrative science." Nagel [1939, §70] wrote that Pasch's work set the standard for (pure) geometry: "No work thereafter held the attention of students of the subject which did not begin with a careful enumeration of the undefined or primitive terms and unproved or primitive statements; and which did not satisfy the condition that all further terms be defined, and all further statements proved, solely by means of this primitive base." This sounds much like the modern axiomatic method.

Concerning the source of the axioms, however, Pasch held that geometry is a natural science—an applied mathematics. He was a straightforward, old-fashioned empiricist, holding that the axioms are verified by experience with bodies. Pasch's empiricism died out; the need for formality and rigor did not. Structuralism is the result.

Similar axiomatic programs came up in other branches of mathematics. In Italy, for example, Peano's influence led to a project to systematize all branches of pure

3. The first edition appeared in 1882, Leipzig, Tuebner.

mathematics. Pieri, a member of this school, explicitly characterized geometry as a hypothetical-deductive enterprise. The subsequent development of n-dimensional geometry was but one fruit of this labor and constituted another move away from intuition. Even if spatial intuition provides a little help in the heuristics of four–dimensional geometry, intuition is an outright hindrance for five-dimensional geometry and beyond.

My final topic in this section is the emergence of non-Euclidean geometry. Here we see other steps toward a model-theoretic understanding of theories and thus another precursor to structuralism. I must be even more sketchy.

Michael Scanlan [1988] shows that the early pioneers of non-Euclidean geometry did not see themselves as providing models of uninterpreted axiom systems—in the present mold of model theory. The geometers were still thinking of their subject matter as physical or perceived space. Beltrami, for example, took Euclidean geometry to be the true doctrine of space and, *in that framework*, he studied the plane geometry on various *surfaces*. "Lines" on surfaces are naturally interpreted as geodesic curves.[4] In a flat space, like a Euclidean plane, the geodesic curves are the straight lines, and the geometry of such a plane is Euclidean. Beltrami showed how the geometries of some curved surfaces differ from Euclidean, thus moving toward non-Euclidean geometry. To be sure, Beltrami's procedure did involve some reinterpreting of the terms of geometry, and so it is a move in the direction of model theory and structuralism, but Scanlan shows that it is a small move. Beltrami could easily maintain that the terms all have their original meanings, transposed to the new contexts. "Line" means "geodesic curve," which is what it always meant anyway; "between" still means "between," and so on. The meanings of the terms are not as fixed as they were in Euclid, but the structures are not freestanding either. It seems fair to think of the items in Beltrami's geometries as quasi-concrete in the sense of Parsons [1990] (see chapter 3). The "point places" are still filled by *points*. Eventually, of course, mathematicians began to consider more interpretations of the primitives of geometry, in less purely spatial contexts. The formal, or structural, outlook was the natural outcome of this process.

On most of the views available, one important scientific question was whether physical space is Euclidean. Because several theories of space were out in the open, the question of how one is to adjudicate them naturally arises. Which one describes the space we all inhabit? The surprising resolution of this question further aided the transition of geometry to the study of freestanding, *ante rem* structures.

At first blush, it looks like the question of whether physical space is Euclidean is an empirical matter. Simply draw a triangle and carefully measure the angles to see if they add up to two right angles, or carefully measure the sides and see if the Pythagorean theorem holds.

---

4. Roughly, a geodesic on a surface P is a curve l with the property that, for any two points a, b on l, the shortest distance on P from a to b lies along l. If the earth were a sphere, then longitudes would be among the geodesic curves for its surface. The shortest distance from the North Pole to London on the surface would be along the longitude. The real shortest distance would be a line through the earth.

In an article entitled "Über den Ursprung und die Bedeutung der geometrischen Axiome" (1870), Helmholtz presented a thought experiment that eventually cast doubt on this simpleminded resolution (see Helmholtz [1921]). Consider a large spherical mirror that reflects events in our world (presumed to be Euclidean), and imagine a world in which physical objects are shaped and behave exactly as they appear to do in the mirror. For every object in our world, there is one in the mirror world represented by its image. The mirror world does not appear (to us) to be Euclidean—lines are not "straight," the "angles" of each "triangle" do not seem to total two right angles, and so on.

How would we confirm that our world is Euclidean? A person E draws a triangle and measures its sides with a ruler. They come to six, eight, and ten meters respectively, and so E concludes that the Pythagorean theorem holds and his world is Euclidean. So far so good, but when we look in the mirror world during the measurement, we see a funny-looking person E' (resembling E, sort of) measuring what looks to us like a curved triangle with a curved ruler. The person E' and his ruler (appear to us to) change shape as the measurement is performed, but lo and behold, the sides come to six, eight, and ten meters respectively. The mirror geometer E' uses exactly the same reasoning as above and concludes that *his* world is the Euclidean one. If E could somehow shout into the mirror world and tell E' that this conclusion is in error, announcing that the mirror ruler is curved and changes size during the measurement, E' would shout back that he saw no such change in size. Moreover, E' sees that the ruler of E is curved and changes size as *it* moves.[5]

Helmholtz himself did not officially stray from the empiricist view of geometry that his thought experiment apparently refutes. The practical application of a metric depends on the observation of congruence, and this presupposes freely movable, *rigid* bodies. Helmholtz adopted an axiom of "free mobility," which states that geometric objects can be moved around and placed on top of each other without changing their size or shape. Notice that many demonstrations in Euclid depend on this free mobility. Helmholtz held that free mobility is confirmed by observation, perhaps not realizing that his own thought experiment undermines this claim. Both E and E' *see* that their respective rulers are straight and do not change size during movement, and each sees that the other's ruler is curved and changes size as it moves. Which one is right? Which one performs measurements and observations that conform to true free mobility? In a sense both do, and this has ramifications for the nature of space and its study.

Although Helmholtz was not consistent in his philosophy of geometry, he foreshadowed the move away from observation and intuition. For an empiricist, the axioms must be tested through observation. Helmholtz noted, however, that to "test" the axioms, we must first know which objects are rigid and which edges are straight, and "we only decide whether a body is rigid, its sides flat and its edges straight, by means of the very propositions whose factual correctness the examination is supposed to show." Consider the postulate that if two magnitudes are identical to a third,

---

5. Helmholtz's thought experiment is described in Coffa [1991, chapter 3].

they are identical with each other. According to Helmholtz, this is "not a law having objective significance; it only determines which physical relations we are allowed to recognize as alikeness." This is an embryonic form of a structuralist insight. In applying geometry, we *impose* a structure on physical reality. Straightness, magnitude, and congruence are relations in such a structure. Physical space does not come with a metric; the geometer/physicist imposes one.

Through the *Erlangen* program, Felix Klein extended Helmholtz's insights, and made even more trouble for the question of which geometry is physically correct. Klein observed that the differences between Euclidean and non-Euclidean geometries lie in the different definitions of congruence, or the different measurements of distance, angle, and area. Each geometry can be imposed on the same domain of points. Moreover, Klein showed that with appropriate units, the numerical values of the different metrics do not differ appreciably from one another in sufficiently small neighborhoods. Consequently, if the issue of which geometry is correct still makes sense, it cannot be decided in practice. That is, even if we agree on the freely mobile, rigid measuring sticks, intuition and observation are not sufficiently precise to adjudicate the question of whether space is Euclidean. According to Klein, axiomatic systems introduce "exact statements into an inexact situation" (see Nagel [1939, §81–82]). His results show that our inexact intuitions are compatible with the assertion of different, technically incompatible, geometries: "Naive intuition is not exact, while refined intuition is not properly intuition at all, but rises through the logical development of axioms considered as perfectly exact." Thus, "It is just at this point that I regard the non-Euclidean geometries as justified. . . . From this point of view it is a matter of course that of equally justifiable systems of axioms we prefer the simplest, and that we operate with Euclidean geometry just for that reason" (Klein [1921, 381]).

A main theme of the *Erlangen* program was that different geometries can be characterized in terms of their "symmetries," in other words, properties that are left fixed by certain isomorphisms. This focus on interpretation and isomorphism is of a piece with the rise of model theory and the emergence of structuralism.

3 A Tale of Two Debates

Klein joined Helmholtz in adopting an official empiricist philosophy of geometry, although his own work apparently refuted this philosophy. At this point, it seems, even the proponents of non-Euclidean geometry were not ready to drop intuition and perception entirely. Freudenthal [1962, 613] pointed out that the non-Euclidean developments were opposed on the ground that the authors did not take intuition and perception seriously enough. In an amusing historical irony, some opponents dubbed the new science "meta-mathematics" or "meta-geometry." They meant these terms in a strong, pejorative sense: non-Euclidean geometry is to respectable mathematics as metaphysics is to physics. It took some doing to exorcize intuition and perception entirely.

The advocates of non-Euclidean geometry won the war, but Euclidean geometry survived. As Coffa [1986, 8, 17] noted, this posed a problem:

> During the second half of the nineteenth century, through a process still awaiting explanation, the community of geometers reached the conclusion that all geometries were here to stay. . . . [T]his had all the appearance of being the first time that a community of scientists had agreed to accept in a not-merely-provisory way all the members of a set of mutually inconsistent theories about a certain domain. . . . It was now up to philosophers . . . to make epistemological sense of the mathematicians' attitude toward geometry. . . . The challenge was a difficult test for philosophers, a test which (sad to say) they all failed.
>
> For decades professional philosophers had remained largely unmoved by the new developments, watching them from afar or not at all. . . . As the trend toward formalism became stronger and more definite, however, some philosophers concluded that the noble science of geometry was taking too harsh a beating from its practitioners. Perhaps it was time to take a stand on their behalf. In 1899, philosophy and geometry finally stood in eyeball-to-eyeball confrontation. The issue was to determine what, exactly, was going on in the new geometry.

What was going on, I believe, was that geometry was becoming the science of structure. Final blows to a role for intuition and perception in geometry were dealt by Poincaré and Hilbert. Each encountered stiff resistance from a logicist, Russell against Poincaré and Frege against Hilbert.[6] Despite the many differences between Poincaré and Hilbert and between Russell and Frege, the debates have a remarkable resemblance to each other.

3.1 Round one: Poincaré versus Russell

Poincaré [1899], [1900] carried the thread of Helmholtz and Klein to its logical conclusion. With Klein, Poincaré argued that it is impossible to figure out whether physical space is Euclidean by an experiment, such as a series of measurements. We have no access to space other than through configurations of physical objects. Measurements can only be done on physical objects with physical objects. How can we tell whether the instruments themselves conform to Euclidean geometry? For example, how can we tell which edges are really straight? By further measurements? Clearly, there is a vicious regress.

Poincaré then echoed Klein's conclusion that there is no fact about space expressible in Euclidean terms that cannot be stated in any of the standard non-Euclidean systems. The only difference is that things covered by one name (e.g., "straight line") under one system would be covered by different names under the second. Helmholtz could have drawn the same conclusion. Poincaré [1908, 235] wrote: "We know rectilinear triangles the sum of whose angles is equal to two right angles; but equally we know curvilinear triangles the sum of whose angles is less than two right angles. The existence of the one sort is no more doubtful than that of the other. To give the name

---

6. In section 4, I consider a third logicist, Dedekind, who embraced and extended the structuralist insights.

of straights to the sides of the first is to adopt Euclidean geometry; to give the name of straights to the sides of the latter is to adopt the non-Euclidean geometry. So that to ask what geometry is proper to adopt is to ask, to what line is it proper to give the name straight?" Poincaré noted that the various geometries are fully intertranslatable. In the terminology of chapter 3, Euclidean space is structure-equivalent to the non-Euclidean spaces. Thus, the choice of geometry is only a matter of a convention. All that remains is for the theorist to stipulate which objects have straight edges and which do not. Space itself has no metric; we impose one. Poincaré held that it is meaningless to ask whether Euclidean geometry is true:[7] "As well ask whether the metric system is true and the old measures false; whether Cartesian coordinates are true and polar coordinates false. One geometry cannot be more true than another; it can only be more convenient" ([1908, 65]).

In sum, Poincaré's view was that to adopt Euclidean geometry is to accept a series of complicated "disguised definitions" for specifying what sorts of configurations we will call "points," "lines," "triangles," and so forth. These "definitions" have at least a family resemblance to implicit definitions. Strictly speaking, the Euclidean "definitions" do not tell us what a point is. Instead, they specify how points are related to lines, triangles, and so on. For Poincaré, words like "point" and "line" have no independent meaning given by intuition or perception. A point is anything that satisfies the conditions laid down by the axioms of the adopted geometry. In [1900, 78], he wrote, "I do not know whether outside mathematics one can conceive a term independently of relations to other terms; but I know it to be impossible for the objects of mathematics. If one wants to isolate a term and abstract from its relations to other terms, what remains is *nothing*." This looks like a structuralist insight, but we are not there yet. Poincaré held that geometry is about space and that we have intuitions about space. These intuitions, however, do not determine whether we should use Euclidean or non-Euclidean geometry. In modern terms, the intuitions underdetermine the theory. The geometer adopts a *convention* concerning terms like "point" and "line." This convention determines which metric structure we are going to impose on space, and so the convention determines the geometry.[8]

As part of his conventionalism, Poincaré held that the theorems of geometry are not propositions with determinate truth-values. This took some philosophers by surprise. The young Russell took issue with Poincaré's conventionalism and the idea

---

7. Like Klein, Poincaré believed that Euclidean geometry would remain the most convenient theory, no matter what direction science took: "In astronomy 'straight line' means simply 'path of a ray of light.' If therefore negative parallaxes were found . . . two courses would remain open to us; we might either renounce Euclidean geometry, or else modify the laws of optics and suppose that light does not travel rigorously in a straight line. It is needless to add that all the world would regard the latter solution as the most advantageous." From the structuralist perspective, Poincaré and Klein had the options correct even if their predictions were not. The problem is that a Euclidean metric on space-time would require the existence of some strange forces.

8. I am indebted to Janet Folina and Peter Clark for insights into Poincaré's philosophy of mathematics. Much of what follows draws on Coffa [1991, 129–134] and Coffa [1986]. Most of the translations are from Coffa [1991].

that the axioms of geometry are disguised definitions. The debate concerned the metaphysical nature of the basic "undefined" elements of geometry—points, lines, planes, and so on—and our epistemic access to them. Russell thought that Poincaré was confused about the nature of geometry and the nature of definitions.

To set the stage, note that although Russell was no friend of Kantian intuition, he maintained that geometry was about space, and as such he was critical of the introduction of imaginary elements: "If the quantities with which we end are capable of spatial interpretation, then, and only then, our result may be regarded as geometrical. To use geometrical language in any other sense is only a convenient help to the imagination. To speak, for example, of projective properties which refer to the circular points, is a mere *memoria technica* for pure algebraic properties; the circular points are not to be found in space. . . . Whenever, for a moment, we allow our ordinary spatial notions to intrude, the grossest absurdities arise—everyone can see that a circle, being a closed curve, cannot go to infinity" (Russell [1956, 45], first published in 1897). Along similar lines, Russell held that words like "straight" and "distance" have fixed meanings, referring to real properties in real space. Here is where the clash with Poincaré emerges. Russell held that in redefining "line" and "straight" to interpret non-Euclidean geometry, the mathematician does not show that the parallel postulate, say, might be false, but only that the words that we use to express the axioms might mean something different from what they do mean.

Russell's statement prefigured debates over the nature of logical consequence, as part of the development of model theory. Russell came out on the losing side here, as well. To the contemporary logician, in interpreting non-Euclidean geometry, we *do* show that the parallel postulate is not a consequence of the other axioms—and thus that it might be false. In other words, the way we show that a given sentence is not a consequence of other sentences is by reinterpreting the nonlogical terminology.

In a long, critical review of Russell's *An essay on the foundations of geometry*, Poincaré [1899] challenged the philosopher to explain the meanings of primitive terms like "point," "line," and "plane." Russell [1899, 699–700] replied that "one is not entitled to make such a request since everything that is fundamental is necessarily indefinable. . . . I am convinced that this is the only philosophically correct answer. . . . [M]athematicians almost invariably ignore the role of definitions, and . . . M. Poincaré appears to share their disdain."

According to Russell, there are two uses of the word "definition." A mathematical definition, part of knowledge by description, identifies an object as the only one that stands in certain relations to other items, presumed to be known already. Such mathematical definitions are "not definitions in the proper and philosophical sense of the word." To really define a word, one must give its meaning and in at least some cases, this meaning cannot consist in relations to other terms. It is part of Russell's atomism that when it comes to simples, no proper definition is possible—thus his retort to Poincaré's request for a definition of geometric primitives. Russell makes an analogy between defining and spelling. One can spell words, but not letters: "These observations apply manifestly to distance and the straight line. Both belong, one might say, to the geometric alphabet; they can be used to define other terms, but they themselves are indefinable. It follows that any proposition . . . within which these notions occur . . . is not a definition of a word. When I say that the straight line is determined by two points, I assume that *straight line* and *point* are terms already known and understood, and I make a judgement concerning their relations, which will be either true or false" (pp. 701–702).

According to Russell, then, neither Poincaré's conventions nor so-called implicit definitions are definitions in the proper and philosophical sense, and they may not be mathematical definitions either. Russell thus did not see the point of Poincaré's request. The two agree on a conditional: if the axioms have truth-values, then we must be able to detect the meanings of the primitive terms *prior* to the theory. Poincaré adopted conventionalism, because he could not see how to satisfy the consequent. How *do* we determine the meaning of the primitives? Thus, he used the conditional in a modus tollens, rejecting the antecedent. The axioms do not have truth-values.

In contrast, Russell did a modus ponens with the conditional, arguing that we do determine the meanings of the primitives—via the epistemic faculty of acquaintance. In [1904, 593], he wrote that it is "by analysis of perceived objects that we obtain acquaintance with what is *meant* by a straight line in actual space." Here, Poincaré was sarcastic: "I find it difficult to talk to those who claim to have a direct intuition of equality of two distances or of two time lapses; we speak very different languages. I can only admire them, since I am thoroughly deprived of this intuition" ([1899, 274]).

Poincaré also attacked Russell's assertion that the principle of free mobility is an a priori truth. He tried to set Russell straight, echoing the Helmholtz–Klein development: "What is the meaning of 'without distortion'? What is the meaning of 'shape'? Is shape something that we know in advance, or is it, by definition, what does not alter under the envisaged class of motions? Does your axiom mean: in order for measurement to be possible figures must be susceptible of certain motions and there must be a certain thing that will remain invariant through these motions and that we call shape? Or else does it mean: you know full well what shape is. . . . I do not know what it is that Mr. Russell has meant to say; but in my opinion only the first sense is correct" (p. 259). Of course, Russell took the second sense. Coffa [1991, 134] sides with Poincaré: there "is really nothing we can say about the meaning of the geometric primitives beyond what the axioms themselves say." So does the structuralist, for much the same reason.

Of course, the internal mathematical development of geometry favored Poincaré against Russell. Eventually, Russell understood and incorporated the emerging needs and perspectives of mathematics and science. Coffa [1986, 26] noted, "Russell, like Carnap—and unlike Frege and Wittgenstein—always had a hunch that scientists knew what they were doing." In Russell [1993, chapter 6] (first published in 1919), the mathematician is urged to adopt a version of structuralism, even if the philosopher cannot:

> [T]he mathematician need not concern himself with the particular being or intrinsic nature of his points, lines, and planes, ... [A] "point" . . . has to be something that as nearly as possible satisfies our axioms, but it does not have to be "very small" or "without parts." Whether or not it is those things is a matter of indifference, so long as it satisfies the axioms. If we can . . . construct a logical structure, no matter how complicated, which will satisfy our geometrical axioms, that structure may legitimately be called a

> "point". . . . [W]e must . . . say "This object we have constructed is sufficient for the geometer; it may be one of many objects, any of which would be sufficient, but that is no concern of ours . . ." This is only an illustration of the general principle that what matters in mathematics, and to a very great extent in physical science, is not the intrinsic nature of our terms, but the logical nature of their interrelations. . . .
>
> We may say, of two similar relations, that they have the same "structure". For mathematical purposes (though not for those of pure philosophy) the only thing of importance about a relation is the cases in which it holds, not its intrinsic nature. (pp. 59–60)

Except for the parenthetical qualification concerning "pure philosophy," amen.

3.2 Hilbert and the emergence of logic

The program executed in Hilbert's *Grundlagen der Geometrie* [1899] marked both an end to an essential role for intuition in geometry and the beginning of a fruitful era of metamathematics. Structuralism is little more than a corollary to these developments.

Hilbert was aware that at some level, spatial intuition or observation remains the source of the axioms of geometry. In Hilbert's writing, however, the role of intuition is carefully and rigorously limited to motivation and heuristic. Once the axioms have been formulated, intuition is banished. It is no part of mathematics, whether pure or applied. The epigraph of Hilbert [1899] is a quotation from Kant's *Critique of pure reason* (A702/B730), "All human knowledge begins with intuitions, thence passes to concepts and ends with ideas," but the plan executed in that work is far from Kantian. In Hilbert's hands, the slogan "passes to concepts and ends with ideas" comes to something like "is *replaced* by logical relations between ideas." In the short introduction, he wrote, "Geometry, like arithmetic, requires for its logical development only a small number of simple, fundamental principles. These fundamental principles are called the axioms of geometry. The choice of the axioms and the investigation of their relations to one another . . . is tantamount to the logical analysis of our intuition of space."

One result of banishing intuition is that the presented structure is freestanding, in the sense of chapter 3. *Anything at all* can play the role of the undefined primitives of points, lines, planes, and the like, so long as the axioms are satisfied. Otto Blumenthal reports that in a discussion in a Berlin train station in 1891, Hilbert said that in a proper axiomatization of geometry, "one must always be able to say, instead of 'points, straight lines, and planes', 'tables, chairs, and beer mugs.'"[9]

As noted earlier, Pasch and others had previously emphasized that the key to rigor is this requirement that the structure be freestanding. The slogan is that logic is topic-neutral. With Hilbert, however, we see consequences for the essential nature of the very subject matter of mathematics. We also see an emerging metamath-

9. "Lebensgeschichte" in Hilbert [1935, 388–429]; the story is related on p. 403. See Stein [1988, 253]; Coffa [1991, 135]; and Hallett [1990, 201–202].

ematical perspective of "logical analysis," with its own important mathematical questions.

Hilbert [1899] does not contain the phrase "implicit definition," but the book clearly delivers implicit definitions of geometric structures. The early pages contain phrases like "the axioms of this group define the idea expressed by the word 'between'" and "the axioms of this group define the notion of congruence or motion." The idea is summed up as follows: "We think of . . . points, straight lines, and planes as having certain mutual relations, which we indicate by means of such words as 'are situated', 'between', 'parallel', 'congruent', 'continuous', etc. The complete and exact description of these relations follows as a consequence of the *axioms of geometry*" (§1).

To be sure, Hilbert also says that the axioms express "certain related fundamental facts of our intuition," but in the subsequent development of the book, all that remains of the intuitive content is the use of *words* like "point" and "line," and the diagrams that accompany some of the theorems.[10] Bernays [1967, 497] sums up the aims of Hilbert [1899]:

> A main feature of Hilbert's axiomatization of geometry is that the axiomatic method is presented and practiced in the spirit of the abstract conception of mathematics that arose at the end of the nineteenth century and which has generally been adopted in modern mathematics. It consists in abstracting from the intuitive meaning of the terms . . . and in understanding the assertions (theorems) of the axiomatized theory in a hypothetical sense, that is, as holding true for any interpretation . . . for which the axioms are satisfied. Thus, an axiom system is regarded not as a system of statements about a subject matter but as a system of conditions for what might be called a relational structure. . . . [On] this conception of axiomatics, . . . logical reasoning on the basis of the axioms is used not merely as a means of assisting intuition in the study of spatial figures; rather logical dependencies are considered for their own sake, and it is insisted that in reasoning we should rely only on those properties of a figure that either are explicitly assumed or follow logically from the assumptions and axioms.

Interest in metamathematical questions surely grew from the developments in non-Euclidean geometry, as a response to the failure to prove the parallel postulate. In effect (and with hindsight), the axioms of non-Euclidean geometry were shown to be satisfiable. Hilbert [1899] raised and solved questions concerning the satisfiability of sets of geometric axioms. Using techniques from analytic geometry, he employed

---

10. According to the emerging demands of rigor, diagrams may be dangerous. A reader who relies on a diagram in following a demonstration cannot be sure that the conclusion is a logical consequence of the premises. Intuition may have crept back in. Apparently, Hilbert did not want to go so far as to echo Lagrange's boast that his work did not contain a single diagram (see section 2). The presence of diagrams in Hilbert [1899] may be a concession to either the human difficulty in following logical proofs or the continuing need to interpret the results in space, or at least to illustrate the application of the results to space. Hilbert coauthored a second geometry book, whose aim is to "give a presentation of geometry . . . in its visual, intuitive aspects" (Hilbert and Cohn-Vossen [1932]). Hilbert contrasts this "approach through visual intuition" with the tendency toward "abstraction" (in Hilbert [1899]) that "seeks to crystallize the *logical* relations inherent in the maze of material that is being studied."

real numbers to construct a model of all of the axioms, thus showing that the axioms are "compatible," or satisfiable. If spatial intuition were playing a role beyond heuristics, this proof would not be necessary. Intuition alone would assure us that all of the axioms are true (of real space), and thus that they are all compatible with each another. Geometers in Kant's day would wonder what the point of this exercise is. As we will see, Frege also balked at it.

Hilbert then gave a series of models in which one axiom is false, but all the other axioms hold, thus showing that the indicated axiom is independent of the others. The various domains of points, lines, and so forth, of each model are sets of numbers, sets of pairs of numbers, or sets of sets of numbers. Predicates and relations, such as "between" and "congruent," are interpreted over the given domains in the now-familiar manner.

The second of Hilbert's [1900] famous "problems" extends the metamathematical approach to every corner of mathematics: "When we are engaged in investigating the foundations of a science, we must set up a system of axioms which contains an exact and complete description of the relations subsisting between the elementary ideas of that science. The axioms set up are at the same time the definitions of those elementary ideas." Once again, the "definition" is an implicit definition.

The crucial concept employed in this perspective is that of "truth in a model," the central notion of contemporary model theory—and structuralism (see Hodges [1985]). The theorist treats a group of sentences as if they were uninterpreted—even if the sentences were originally suggested by intuition, observation, or whatever. A domain of discourse is stipulated, and the singular terms and predicate and relation symbols are given extensions over this domain. Although the procedure is now commonplace, the insights were hard-won, historically. Once it is allowed that any domain at all can serve as an interpretation of a given formal language—even a set of tables, chairs, and beer mugs—the structures are freestanding. The relationship between this notion of "truth in a model" and "truth" simpliciter is still a central item on the agenda of philosophy (see chapters 2 and 4, and Shapiro [1997]).

One consequence of the axiomatic, metamathematical approach is that isomorphic models are equivalent. If there is a one-to-one function from the domain of one model onto the domain of the other that preserves the relations of the model, then any sentence of the formal language that is true in one model is true in the other. So, if intuition or perception via ostension is banned, then the best a formal theory can do is to fix its interpretation up to isomorphism. Any model of the theory can be changed into another model just by substituting elements of the domain and renaming.[11] The best we can do is a categorical axiomatization, one in which all the models are isomorphic to each other. Weyl [1949, 25–27] put it well:

---

11. This is one of the main motivations for the Quinean doctrines of the inscrutability of reference and the relativity of ontology (see chapters 2 and 4 in this volume). Quine argues that reference is not determinate even if techniques that involve observation are added to the model-theoretic conception.

> [A]n axiom system [is] a *logical mold of possible sciences*. . . . One might have thought of calling an axiom system complete if in order to fix the meanings of the basic concepts present in them it is sufficient to require that the axioms be valid. But this ideal of uniqueness cannot be realized, for the result of an isomorphic mapping of a concrete interpretation is surely again a concrete interpretation. . . . A science can determine its domain of investigation up to an isomorphic mapping. In particular, it remains quite indifferent as to the "essence" of its objects. . . . The idea of isomorphism demarcates the self-evident boundary of cognition. . . . Pure mathematics . . . develops the theory of logical "molds" without binding itself to one or the other among possible concrete interpretations. . . . The axioms become *implicit definitions* of the basic concepts occurring in them.

This is essentially a structuralistic manifesto.

Hilbert's metamathematical approach, of course, did not come out of the blue. His competitors included the Italians Padoa, Pieri, and Peano.[12] Here, there is no need to elaborate the similarities and differences nor to join arguments of priority. Hilbert's influence was due to the clarity and depth of his work. He led by example. Freudenthal [1962, 619, 621] sums things up:

> The father of rigor in geometry is Pasch. The idea of the logical status of geometry occurred at the same time to some Italians. Implicit definition was analyzed much earlier by Gergonne. The proof of independence by counter-example was practiced by the inventors of non-Euclidean geometry, and more consciously by Peano and Padoa. . . . [I]n spite of all these historical facts, we are accustomed to identify the turn of mathematics to axiomatics with Hilbert's *Grundlagen*: This thoroughly and profoundly elaborated piece of axiomatic workmanship was infinitely more persuasive than programmatic and philosophical speculations on space and axioms could ever be.
>
> There is no clearer evidence for the persuasiveness of Hilbert's *Grundlagen*, for the convincing power of a philosophy that is not preached as a program, but that is only the silent background of a masterpiece of workmanship.

In America, the first few decades of the twentieth century saw the emergence of a school of foundational studies. The members, dubbed "American postulate theorists," included Huntington, E. H. Moore, R. L. Moore, Sheffer, and Veblen (see Scanlan [1991]). Their program was to axiomatize various branches of mathematics, such as geometry, arithmetic, and analysis, and then to study the axiomatizations as such. Their perspective was even more metatheoretic than that of their counterparts across the Atlantic. Huntington [1902] established that his axiomatization of analysis is categorical, and Veblen [1904] did the same for his axiomatization of geometry. Each took the result to establish that the theory has "essentially only one" interpretation.

---

12. Padoa [1900], for example, wrote, "[D]uring the period of *elaboration* of any deductive theory we choose the *ideas* to be represented by the undefined symbols and the *facts* to be stated by the unproved propositions; but, when we begin to *formulate* the theory, we can imagine that the undefined symbols are *completely devoid of meaning* and that the unproved propositions . . . are simply *conditions* imposed upon the undefined symbols" (p. 120).

The model-theoretic notions of consequence and satisfiability emerge through this work. In his 1924 presidential address to the American Mathematical Society, Veblen announced that "formal logic has to be taken over by mathematicians. The fact is that there does not exist an adequate logic at the present time, and unless the mathematicians create one, no one else is likely to do so" (Veblen [1925, 141]). Many extraordinary minds, including Veblen's student, Alonzo Church, took up the call. These logicians and their students delivered model theory and proof theory as we know them today. Coffa [1986, 16] wrote that "in the early decades of our century logic evolved, adjusting to the picture of knowledge that has emerged from geometry."

3.3 Round two: Frege versus Hilbert

Although the spirited correspondence between Frege and Hilbert is thoroughly treated in the literature,[13] I briefly recapitulate some of the exchange here, to emphasize the striking resemblance to the clash between Russell and Poincaré. The correspondence is published in Frege [1976] and translated in [1980]. Like Russell did to Poincaré, Frege lectured Hilbert on the nature of definitions and axioms. According to Frege, axioms should express *truths* and definitions should give the *meanings* and *fix the denotations* of certain terms. With an implicit definition, neither job is accomplished. In a letter dated December 27, 1899, Frege complained that Hilbert [1899] does not provide a definition of, say, "between," because the axiomatization "does not give a characteristic mark by which one could recognize whether the relation Between obtains. . . . [T]he meanings of the words 'point', 'line', 'between' are not given, but are assumed to be known in advance. . . . [I]t is also left unclear what you call a point. One first thinks of points in the sense of Euclidean geometry, a thought reinforced by the proposition that the axioms express fundamental facts of our intuition. But afterwards you think of a pair of numbers as a point. . . . Here the axioms are made to carry a burden that belongs to definitions. . . . [B]eside the old meaning of the word 'axiom', . . . there emerges another meaning but one which I cannot grasp" (pp. 35–36). Frege went on to remind Hilbert that a definition should specify the meaning of a single word whose meaning has not yet been given, and the definition should employ other words whose meanings are already known. In contrast to definitions, axioms and theorems "must not contain a word or sign whose sense and meaning, or whose contribution to the expression of a thought, was not already completely laid down, so that there is no doubt about the sense of the proposition and the thought it expresses. The only question can be whether this thought is true and what its truth rests on. Thus axioms and theorems can never try to lay down the meaning of a sign or word that occurs in them, but it must already be laid down" (p. 36). Frege's point is a simple one. If the terms in the proposed axioms do not have meaning beforehand, then the statements cannot be true (or false), and thus they cannot be axioms. If they do have meaning beforehand, then the axioms cannot be definitions. He added

13. See, for example, Resnik [1980]; Coffa [1991, chapter 7]; Demopoulos [1994]; and Hallett [1994].

that from the truth of axioms, "it follows that they do not contradict one another," and so there is no further need to show that the axioms are consistent.

Hilbert replied just two days later, on December 29. He told Frege that the purpose of Hilbert [1899] is to explore logical relations among the principles of geometry, to see why the "parallel axiom is not a consequence of the other axioms" and how the fact that the sum of the angles of a triangle is two right angles is connected with the parallel axiom. Frege, the pioneer in mathematical logic, could surely appreciate *this* project. The key lies in how Hilbert understood the logical relations. Concerning Frege's assertion that the meanings of the words "point," "line," and "plane" are "not given, but are assumed to be known in advance," Hilbert replied, "This is apparently where the cardinal point of the misunderstanding lies. I do not want to assume anything as known in advance. I regard my explanation . . . as the definition of the concepts point, line, plane. . . . If one is looking for other definitions of a 'point', e.g. through paraphrase in terms of extensionless, etc., then I must indeed oppose such attempts in the most decisive way; one is looking for something one can never find because there is nothing there; and everything gets lost and becomes vague and tangled and degenerates into a game of hide and seek" (p. 39). This is an allusion to definitions like Euclid's "a point is that which has no parts." To try to do better than a characterization up to isomorphism is to lapse into "hide and seek." Later in the same letter, when responding to the complaint that his notion of "point" is not "unequivocally fixed," Hilbert wrote,

> [I]t is surely obvious that every theory is only a scaffolding or schema of concepts together with their necessary relations to one another, and that the basic elements can be thought of in any way one likes. If in speaking of my points, I think of some system of things, e.g., the system love, law, chimney-sweep . . . and then assume all my axioms as relations between these things, then my propositions, e.g., Pythagoras' theorem, are also valid for these things. . . . [A]ny theory can always be applied to infinitely many systems of basic elements. One only needs to apply a reversible one-one transformation and lay it down that the axioms shall be correspondingly the same for the transformed things. This circumstance is in fact frequently made use of, e.g. in the principle of duality. . . . [This] . . . can never be a defect in a theory, and it is in any case unavoidable. (pp. 40–41)

Note the similarity with Hilbert's quip in the Berlin train station. Here it is elaborated in terms of isomorphism and logical consequence.

Because Hilbert did "not want to assume anything as known in advance," he rejected Frege's claim that there is no need to worry about the consistency of the axioms, because they are all true: "As long as I have been thinking, writing and lecturing on these things, I have been saying the exact reverse: if the arbitrarily given axioms do not contradict each other with all their consequences, then they are true and the things defined by them exist. This is for me the criterion of truth and existence" (p. 42). He then repeated the role of what is now called "implicit definition," noting that it is impossible to give a definition of "point" in a few lines, because "only the whole structure of axioms yields a complete definition."

Frege's response, dated January 6, 1900, acknowledged Hilbert's project. Frege saw that Hilbert wanted "to detach geometry from spatial intuition and to turn it into a purely logical science like arithmetic"[14] (p. 43). Frege also understood Hilbert's model-theoretic notion of consequence: to show that a sentence *D* is not a consequence of *A*, *B*, and *C* is to show that "the satisfaction of *A*, *B*, and *C* does not contradict the non-satisfaction of *D*." However, these great minds were not to meet. Frege said that the only way to establish consistency is to give a model: "to point to an object that has all those properties, to give a case where all those requirements are satisfied." The later Hilbert program, of course, supplied another way to prove consistency results.

In the same letter, Frege mocked implicit definitions, suggesting that with them, we can solve problems of theology:

> What would you say about the following:
>
> "Explanation. We imagine objects we call Gods.
> Axiom 1. All Gods are omnipotent.
> Axiom 2. All Gods are omnipresent.
> Axiom 3. There is at least one God." (p. 46)

Frege did not elaborate, but the point is clear. Hilbert said that if we establish the consistency of his axiomatization, we thereby establish the existence of points, lines, and planes. If we establish the consistency of the theology axioms, do we thereby establish the existence of a God? In the terminology of chapter 3, the difference is that Euclidean geometry is a freestanding structure, or at least Hilbert saw it that way. Anything at all can occupy its places. Further, the relations of geometry are formal, completely characterized by the axioms. In contrast, the theology structure (if it makes sense to speak this way) is not freestanding. Not just anything can occupy the God office. Even more, the central properties of omnipotence and omnipresence are not formal. A consistency proof does not establish that the nonformal structure is exemplified.

Frege complained that Hilbert's "system of definitions is like a system of equations with several unknowns" (p. 45). This analogy is apt, and Hilbert would surely accept it. A system of equations does characterize a structure, of sorts. Frege wrote, "Given your definitions, I do not know how to decide the question whether my pocket watch is a point." Again, Hilbert would surely agree, but he would add that the attempt to answer this question is to play the game of hide and seek. Frege's question is reminiscent of the so-called Caesar problem, raised in his *Grundlagen der Arithmetik* [1884] (see chapter 3). To state the obvious, Frege did not think in terms of "truth in a model." For Frege, the quantifiers of mathematics range over *everything*, and a concept is a function that takes all objects as arguments. Thus, "my pocket

14. Frege's logicism did not extend to geometry, which he regarded as synthetic a priori. Like Russell, Frege worried about the ideal elements introduced into geometry. In his inaugural dissertation (of 1873), he wrote, "When one considers that the whole of geometry ultimately rests upon axioms, which receive their validity from the nature of our faculty of intuition, then the question of the sense of imaginary figures appears to be well-justified, since we often attribute to such figures properties which contradict our intuition" (Frege [1967, 1]; see Kitcher [1986, 300]). Wilson [1992] provides an insightful account of the connection between Frege's logicism and his concern with the ideal elements of geometry.

watch is a point" must have a truth-value, and our theory must determine this truth-value.

Neither of them budged. On the following September 16, a frustrated Frege wrote that he could not reconcile the claim that axioms are definitions with Hilbert's view that axioms contain a precise and complete statement of the relations among the elementary concepts of a field of study (from Hilbert [1900], quoted above). For Frege, "there can be talk about relations between concepts . . . only after these concepts have been given sharp limits, but not while they are being defined" (p. 49). On September 22, an exasperated Hilbert replied, "[A] concept can be fixed logically only by its relations to other concepts. These relations, formulated in certain statements I call axioms, thus arriving at the view that axioms . . . are the definitions of the concepts. I did not think up this view because I had nothing better to do, but I found myself forced into it by the requirements of strictness in logical inference and in the logical construction of a theory. I have become convinced that the more subtle parts of mathematics . . . can be treated with certainty only in this way; otherwise one is only going around in a circle" (p. 51). I cannot think of a better formulation of structuralism. An implicit definition characterizes a type of structure—a single structure if the axiomatization is satisfiable and categorical.

Hilbert took the rejection of Frege's perspective on concepts to be a major innovation. In a letter to Frege dated November 7, 1903, he wrote, "[T]he most important gap in the traditional structure of logic is the assumption . . . that a concept is already there if one can state of any object whether or not it falls under it. . . . [Instead, what] is decisive is that the axioms that define the concept are free from contradiction" (pp. 51–52). The opening paragraphs of Hilbert [1905] contain a similar point, explicitly directed against Frege.

For his part, Frege used his own logical system to recapitulate much of Hilbert's orientation. In the January 6, 1900 letter, Frege suggests that Hilbert (almost) succeeded in defining "second-level concepts." The axiomatization does not provide an answer "to the question 'What properties must an object have in order to be a point (a line, a plane, etc.)?', but they contain, e.g., second-level relations, e.g., between the concept *point* and the concept *line*" (p. 46). Frege was correct. Hilbert did not give necessary and sufficient conditions for an arbitrary object to be a point. Instead, he showed how points are related to each other and to lines, planes, and the like. He gave necessary and sufficient conditions for *systems* of objects to exemplify the structure of Euclidean geometry.

In subsequent lectures on geometry ([1903a], [1906], translated in Frege [1971]), Frege formulated these second-level relations with characteristic rigor. He replaced the geometric terms—the items we would call nonlogical—with bound, higher-order *variables*. In Frege's hands, Hilbert's implicit definition is transformed into an *explicit definition* of a second-level concept.[15] Frege also used his logical system to recapitulate something like Hilbert's model-theoretic notion of independence.

15. This is similar to a technique in contemporary philosophy, often traced to Ramsey [1925]. See chapter 3, section 6, in this volume.

An assertion in the form "the satisfaction of *A*, *B*, and *C* does not contradict the nonsatisfaction of *D*" is rendered as a formula in Frege's system.

Thus, it is not quite true that model theory was foreign to Frege's logical system. Because his language was designed to be universal, anything could be expressed in it—even statements about models of axiomatizations with noted nonlogical terminology.

In sum, Frege and Hilbert did manage to understand each other, for the most part. Nevertheless, they were at cross-purposes in that neither of them saw much value in the other's point of view. The central feature of Hilbert's orientation is that the main concepts being defined are given different extensions in each model. This is the contemporary notion of a nonlogical term (see Demopoulos [1994]). Frege's own landmark work in logic, the *Begriffsschrift* [1879], contains nothing like nonlogical terminology. Every lexical item has a fixed sense and a fixed denotation or extension. Everything is already fully interpreted, and there is nothing to vary from interpretation to interpretation. This feature, shared by the other major logicist Russell, has been called "logic as language," whereas Hilbert's approach is "logic as calculus" (see, for example, van Heijenoort [1967a] and Goldfarb [1979]).

Demopoulos adds that a related source of the cross-purposes is the Fregean principle that the sense of an expression completely determines its reference. Anyone who grasps the sense of an expression like "point" or "natural number" has the means to determine of anything whatsoever whether it is a point or a natural number. Hallett [1990] (and [1994]) calls this the "fixity of reference." A related Fregean (and Russellian) principle is that every well-formed sentence in a proper mathematical theory makes a fixed assertion about a fixed collection of objects and concepts. Each proposition has a truth-value determined by the nature of the referenced objects and concepts. Coffa [1991] calls this "propositionalism." For geometry, as well as other mathematical theories, Hilbert (and Poincaré) rejected both the fixity of reference and propositionalism.

In the terminology of this book (chapter 3), Hilbert was an eliminative structuralist. To reiterate, eliminative structuralism is a structuralism without structures. Hilbert's rejection of the fixity of reference and propositionalism indicates that he did not join the logicists and take, say, numerals as genuine singular terms. Thus, according to Hilbert, there is no single object denoted by "zero" and there is no single collection of objects that is the extension of "the natural numbers." Instead, "the natural numbers" refers to a type of system. If this interpretation is correct, then when Hilbert speaks of "existence," as entailed by consistency, he is not using the term in the same sense as, say, "Jupiter exists." This is highlighted by Frege's quip about the existence of God. With eliminative structuralism, mathematical assertions are really generalizations over a type of system, and so they do not have the same "deep structure" as ordinary assertions about ordinary physical (or theological) objects.

3.4 Frege

Recall from chapter 3 that *ante rem* structuralism is an attempt to have it both ways. According to this view, numerals are genuine singular terms that denote places in the natural-number structure. The *ante rem* structuralist maintains propositionalism and the fixity of reference.

Although neither Frege nor his interpreters speak in structuralist terms, the development of arithmetic in Frege [1884] goes some way toward structuralism. The structuralism that Frege approached is not the eliminative variety just attributed to Hilbert but full, *ante rem* structuralism. Let there be no mistake, however. I do not mean to attribute structuralism to Frege himself. I propose a starkly un-Fregean modification of the Fregean program.

As noted, Frege held that numerals are genuine singular terms that denote natural numbers. To explicate this, several commentators, such as Dummett [1981], [1991]; Wright [1983]; and Hale [1987], attribute a certain "syntactic-priority thesis" to Frege [1884]: if a sentence *S* is true and if a lexical item *t* in *S* is a singular term, then there is an object that *t* denotes. Once matters of grammar, syntax, and truth are fixed, there can be no further question of deciding which of the singular terms really pick out objects in the world. As Wright put it, "The injunction that we should never ask after the *Bedeutung* of a term in isolation but only in the context of a proposition is to be understood as cautioning us against the temptation to think that . . . *after* we are satisfied that, by syntactic criteria, . . . expressions are functioning as singular terms in sentential contexts, a further genuine question can still remain about whether their role is genuinely denotative" (p. 14). According to the syntactic-priority thesis, all that is needed to establish ontological realism for arithmetic is to show that numerals are singular terms and that sentences containing numerals are true. As part of his argument for ontological realism, Frege cited sentences like "the number of moons of Jupiter is four." If some such sentences are true—and some surely are—and if the numerals in them are genuine singular terms, then the syntactic-priority thesis entails that numerals denote something. Because numbers are whatever the denotata of numerals are, Frege concluded that numbers exist. Again, Wright says, "[I]t must not be coherent to suggest the possibility of some sort of independent, language-unblinkered inspection of the contents of the world, of which the outcome might be to reveal that there was indeed nothing there capable of serving as the referents of what Frege takes to be numerical singular terms" (pp. 13–14).

Frege is thus at odds with both eliminative and modal structuralism, as these programs are articulated in chapter 3. According to the in re structuralisms, a statement that appears to be about numbers is a disguised generalization about all natural-number sequences. Numerals are bound variables, not singular terms. According to Frege, however, for any concept *F*, the locution "the number of *F*" denotes something, a number in particular.

To develop arithmetic, Frege made use of a thesis now called "Hume's principle":[16]

For all concepts *F*, *G*, the number of *F* is identical to
the number of *G* if and only if *F* is equinumerous with *G*.

Frege went on to define 0 as the number of an empty concept, and he defined the successor relation among numbers. Then he defined an object *n* to be a *natural num-*

16. Wright [1983] and Hale [1987] suggest a Fregean program in which equivalence relations indicate the existence of (abstract) objects. In chapter 4, this program is compared to one described in Kraut [1980].

ber if *n* bears the ancestral of the successor relation to 0. The natural numbers are 0, the successor of 0, the successor of the successor of 0, and so on. Frege then showed that the standard Peano postulates hold of the natural numbers so construed. That is, each natural number has a unique successor, the successor function is one-to-one on the natural numbers, and induction holds. In present terms, Frege established that his natural numbers exemplify the natural-number structure. This part of Frege's system is consistent (see Wright [1983] and Boolos [1987]).

Notice that Frege's syntactic-priority thesis looks a bit like the aforementioned pronouncement of Hilbert, directed against Frege (translated in Frege [1980, 42]): "[I]f the arbitrarily given axioms do not contradict each other with all their consequences, then they are true and the things defined by them exist. This is for me the criterion of truth and existence." To be sure, a crucial difference between Hilbert's statement and the syntactic-priority thesis is that Frege insists that the sentences be *true* before one can draw inferences concerning ontology, whereas Hilbert requires only that the axioms be consistent. To identify truth and consistency at this stage would surely beg the question in favor of Hilbert. Moreover, truth and falsity are properties of single sentences, whereas consistency is a (holistic) property of entire theories.

Hallett [1990] attempted to temper this difference between Frege and Hilbert (see also Hallett [1994]). Notice first that in order to sustain the Dummett–Wright–Hale interpretation of Frege, we need to be told how the "truth" clause should be understood. If Frege had something like a model-theoretic picture of truth in mind, then his criterion for existence would be useless. To decide whether a sentence is true, the theorist would have to *first* determine what its singular terms denote. This violates Frege's context principle to never ask for the *Bedeutung* of an expression except in the context of a sentence in which it occurs. It reverses the priority.

The acceptability of a sentence is, of course, not an infallible guarantee of its truth. Humans do err on occasion. However, acceptability in a serious practice is surely *evidence* of truth. In mathematics, at least, error is not rampant. If we note this and allow the slide from single sentences to theories, the Fregean syntactic-priority thesis becomes, "If a body of sentences *S* is acceptable according to the canons of acceptability laid down, then there must be objects to which the singular terms of the *S*-sentences purport to refer" (Hallett [1990, 185]). Let us call this the "acceptability thesis."

As noted in chapter 4, Hallett [1990, 240] also argued that Hilbert's "consistency" requirement was meant to be a mathematically tractable gloss on "acceptability" and not a wholesale replacement for it. If "acceptability" replaces "consistency" in Hilbert's message to Frege, the acceptability thesis results. Thus, there may be a little common ground between Frege and Hilbert.

Whatever the value of this tenuous bridge, there is a more far reaching difference between Frege's version and Hilbert's version of the acceptability thesis. To be precise, there is a difference in the role played by the acceptability thesis (or theses). For Hilbert, once the coherence of a theory is established, there is no need to go on to say what its objects are. As far as ontology goes, there is nothing more to say. As Hilbert emphasized over and over, the axioms of Euclidean geometry give the relations of points, lines, and planes to each other, and the axioms of arithmetic give the relations of the numbers to each other. Those relations are all there are to points, lines, planes, and numbers. What you see is what you get, or, to be precise, what you get in the implicit definition is all there is. This is Hilbert's eliminative structuralism. Any attempt to go on to say what the numbers are or what the Euclidean points are is to play "a game of hide and seek" and to go "around in a circle."

Frege, of course, thought otherwise. For him, once the acceptability thesis, or the syntactic-priority thesis, assures us that, say, numbers exist, there is work to be done. We must then figure out what the numbers *are*. Numerals are singular terms and occur in true sentences. Thus, we know that there are objects denoted by numerals. Now we have to figure out *which objects* those numbers are. This extra burden is indicated by Frege's Caesar problem. Until we have figured out which objects the numbers are, we do not know whether the number 2 is identical to, or different from, Julius Caesar. Until we have figured out which objects the Euclidean points are, we do not know whether a given pocket watch is a point. Dummett [1991, 239] wrote, "Frege never advanced the context principle as having the advantages conceded by Russell to the method of postulation: it merely indicated *what* honest toil was called for."

From the Fregean perspective, Hilbert's method does claim the advantages of theft over toil, but this assumes that Hilbert took numbers to be objects. Earlier, I ventured the barbarous anachronism of interpreting Hilbert as an eliminative structuralist. On that view, as articulated here, numerals are not singular terms, and numbers are not objects. Frege held that the *objects* delivered by his version of the acceptability thesis exist in exactly the same sense as any other objects, including horses, planets, Caesars, and pocket watches. All objects, whether abstract or concrete, belong to a single, all-inclusive domain, and the first-order variables of a proper *Begriffsschrift* range over this domain. Given all this, Frege had to take questions like "Is Julius Caesar = 2?" and "Is my pocket watch a point?" seriously.[17]

Frege held that arithmetic is part of logic, and so he tried to locate the natural numbers among the objects of logic. Numbers are certain extensions. Of course, the discovery that Frege's theory of extensions is inconsistent brought his program down in ruins. Dummett [1991, 225] took Russell's paradox to refute the context principle "as Frege had used it." Frege himself may have come to the same conclusion, because, as Dummett pointed out, after the discovery of the antinomy, logical objects "quietly vanished" from Frege's ontology. Dummett concluded that it "was [Frege's] error and his misfortune . . . to have misconstrued the task, an error for which he paid with the frustration of his life's ambition." Sad.

To allow Frege to quit while he is ahead, suppose that we simply drop the extra ontological burden of Frege's version of the acceptability thesis but maintain the bulk of the rest of the program—including the syntactic-priority thesis. My proposed un-Fregean rescue is to give up the idea of a single, all-inclusive domain that contains all objects of mathematics. The result fits *ante rem* structuralism, and even suggests

17. I am indebted to Crispin Wright and Bob Hale for useful discussion of the Caesar problem.

it. Numerals are genuine singular terms, not disguised variables. We have bona fide reference. Numerals denote numbers. But what are these numbers? All we can say about them is what follows from their characterization via Hume's principle and the various definitions. To repeat what I said about Hilbert, what you see is what you get, or what you get from the characterization is all there is. And what you get from the characterization are the relations of the numbers to each other and to concepts with finite extensions.

I submit that it is natural to think of the numbers, so construed, as an *ante rem* structure. When it comes to mathematics, the Fregean all-inclusive domain gives way to the ontological relativity urged here. Each mathematical object is a place in a particular structure, and statements of identity are determinate only if the terms flanking the identity sign denote places in the same structure. Otherwise, the identity is a matter of decision, based on convenience.

Dummett [1991, 235] himself suggested a similar maneuver in discussing what is left of Frege's project. He wrote that "the referents of the newly introduced terms cannot be thought of in any other way than simply as the referents of those terms." Notice the similarity between this remark and what he elsewhere attributes to the "mystical" (i.e., *ante rem*) structuralist: the zero place of the natural-number structure "has no other properties than those which follow from its being the zero" place of that structure. It is not a set, or anything else whose nature is extrinsic to the structure. As Dummett says on behalf of the mystical structuralist, abstract structures are "distinguished by the fact that their elements have no non-structural properties." Practically speaking, this is what we get from the Fregean program minus the Caesar question.[18]

Dummett concluded, however, that the rescued program is not all that healthy. The notion of reference that emerges from the maneuver is "thin" and falls short of ordinary reference. If numerals are understood in terms of the syntactic-priority thesis and Hume's principle, then they are significantly different from singular terms that denote ordinary objects. When reference is "thin," no genuine objects are picked out. According to Dummett, we must give up the dream, noted earlier, that the objects delivered by the syntactic-priority thesis exist in the same sense as all other objects. The ontological realism vanishes with the single, all-inclusive domain.

But does it? I take the reference of a singular term in mathematics to be genuine reference to a place in a structure. The extent to which this reference is "thin" de-

18. That and a bit more. Notice that even with the fateful reference to extensions eliminated, Frege delivered more than the natural-number structure—more than the relations of the natural numbers to each other (see Wright [1983, 117–129]; Hale [1987, 216–217]; and Hallett [1994, 175–176]). Frege successfully accommodated the *application* of the natural-number structure to finite cardinalities. An implicit definition does not, by itself, show how the defined structure or structures can be applied. Thus, a Hilbert-style treatment of the natural numbers would have to be supplemented with an account of how arithmetic can be used to compare and determine cardinalities. Of course, both accounts have to be supplemented with applications to what I call "finite ordinal structures" in chapter 4. Dedekind accomplished this—stay tuned.

pends on how similar ordinary reference is to mathematical reference. I argue in chapters 3 and 4 that model theory provides the right picture in both cases. Moreover, in ordinary discourse, we do sometimes refer to the places of a structure or pattern, using singular terms, via the places-are-objects perspective. This occurs when we discuss an organization or pattern independently of any exemplifications it may have. Perhaps Dummett would retort that this latter reference is also thin. If so, then I am content to leave the dialogue at this juncture. Mathematical reference is no less robust than at least some ordinary reference.

Notice also that Dummett's conclusion about "thin" reference also depends on the extent to which places in a structure are similar to ordinary objects. Is it clear that there is a single, all-inclusive domain of all nonmathematical objects? Is it even clear that there is a sharp border between the mathematical and the mundane? I briefly return to ordinary objects in chapter 8.[19]

4 Dedekind and *ante rem* Structures

To locate a more direct forerunner of *ante rem* structuralism, I back up a bit and turn to another logicist, Richard Dedekind.[20] His *Stetigkeit und irrationale Zahlen* [1872] contains the celebrated account of continuity and the real numbers. The opening paragraph (§1) sounds the familiar goal of eliminating appeals to intuition: "In discussing the notion of the approach of a variable magnitude to a fixed limiting value . . . I had recourse to geometric evidences. Even now such resort to geometric intuition in a first presentation of the differential calculus I regard as extremely useful. . . . But that this form of introduction into the differential calculus can make no claim to being scientific no one will deny. . . . The statement is so frequently made that the differential calculus deals with continuous magnitude, and yet an explanation of this continuity is nowhere given." In order to facilitate the rigorous deduction of theorems, the mathematician needs a precise formulation of the central notion of "continuous magnitude." How can this be accomplished without appeal to intuition?

Dedekind motivated the problem by noting that the rational numbers can be mapped one-to-one into any straight line, given only a unit and a point on the line as origin. Thus, the arithmetical property of "betweenness" shares many (structural) features with its geometrical counterpart. The ancient lesson of incommensurables is that there are points on the line that do not correspond to any rational number. In other words, there are holes or gaps in the rational numbers. Consequently, the rational numbers are not continuous.

---

19. Dieterle [1994, chapter 3] develops a "neo-Fregean" position that rejects the Caesar problem on structuralist grounds. Her main innovation is that the object/concept (or saturated/unsaturated) distinction be taken as *relative* to a background framework. Perhaps Wright's [1983, 124] own notion of a "relatively sortal" predicate can be put to use here.

20. See Sieg [1990] and Stein [1988] for illuminating accounts of Dedekind's influence on Hilbert. I am also indebted to Kitcher [1986]; Parsons [1990]; Hallett [1990]; McCarty [1995]; and Tait [1997].

According to Dedekind, even if an appeal to spatial or temporal intuition were allowed at this point, it would not help. Do we have intuitions of continuity? Lipschitz wrote to Dedekind that the property of continuity is self-evident and so does not need to be stated, because no one can conceive of a line as discontinuous. Dedekind replied that he, for one (and Cantor, for two) can conceive of discontinuous lines (Dedekind [1932, vol. 3, 478]; see Stein [1988, 244]). In [1872, §3], Dedekind wrote that "if space has at all a real existence it is *not* necessary for it to be continuous," and in the preface to the first edition of his later monograph on the natural numbers [1888], he gave a nowhere-continuous interpretation of Euclidean space. His point is that we do not *see* or intuit the line as continuous; we *attribute* continuity to it. But we have as yet no rigorous formulation of continuity. Thus, Dedekind sought an "axiom by which we attribute to the line its continuity."

Dedekind defined a *cut* to be a division of the rational numbers into two sets (A_1, A_2) such that every member of A_1 is less than every member of A_2. If there is a rational number n such that n is either the largest member of A_1 or the smallest member of A_2, then the cut (A_1, A_2) is "produced" by n. Of course, some cuts are not produced by any rational number—thus the discontinuity of the rational numbers. Dedekind's plan is to fill the gaps: "Whenever, then, we have to do with a cut (A_1, A_2) produced by no rational number, we create a new, an *irrational* number a, which we regard as completely defined by this cut (A_1, A_2); we shall say that the number a corresponds to this cut, or that it produces this cut." The language in [1872] suggests that he did not identify the real numbers with the cuts. Instead, the "created" real numbers "correspond" to the cuts. It is not clear what he meant by the phrase "create a new number," and I will return to this shortly. Of course, a structuralist would say that the cuts exemplify the real-number structure.

The opening paragraphs of Dedekind's later monograph on the natural numbers [1888] continue the rejection of intuition: "In speaking of arithmetic (algebra, analysis) as a part of logic I mean to imply that I consider the number-concept entirely independent of the notions or intuitions of space and time." This echoes a passage in the earlier work [1872, §1]: "I regard the whole of arithmetic as a necessary, or at least natural, consequence of the simplest arithmetic act, that of counting, and counting itself as nothing else than the successive creation of the infinite series of positive integers in which each individual is defined by the one immediately preceding; the simplest act is the passing from an already-formed individual to the consecutive new one to be formed." The "counting" here is not the technique of determining the cardinality of a collection or concept extension. Dedekind's counting is the "creation" of the natural numbers, one after another. Thus, his starting point is different from Frege's.

Dedekind [1888, §5] defined a set S and function s to be a "simply infinite system" if s is one-to-one, there is an element e of S such that e is not in the range of s (thus making S Dedekind-infinite), and the only subset of S that both contains e and is closed under s is S itself.[21] In effect, a simply infinite system is a model of the natural numbers.

---

21. The closure condition is formulated with the notion of a "chain," which is quite similar to Frege's ancestral.

At this point, Dedekind defined the "natural numbers" with language that is music to the ears of the structuralist:

> If in the consideration of a simply infinite system *N* set in order by a transformation ϕ we entirely neglect the special character of the elements, simply retaining their distinguishability and taking into account only the relations to one another in which they are placed by the order-setting transformation ϕ, then are these numbers called *natural numbers* or *ordinal numbers* or simply *numbers*. . . . With reference to this freeing the elements from every other content (abstraction) we are justified in calling numbers a free creation of the human mind. The relations or laws which are derived entirely from the conditions . . . are always the same in all ordered simply infinite systems, whatever names may happen to be given to the individual elements.

Like Frege [1884], Dedekind went on to establish that "the natural numbers" satisfy the usual arithmetic properties, such as the induction principle. He defined addition and multiplication on "the natural numbers" and proved that definitions by primitive recursion do define functions. Finally, he gave straightforward applications of "the natural numbers" to the counting and ordering of finite classes. That is, he characterized the finite cardinal structures and the finite ordinal structures in terms of his "natural numbers."

The sentence just quoted suggests a reading of Dedekind as a protoeliminative structuralist, perhaps like Hilbert. Parsons [1990] proposes, but eventually drops, such an interpretation (see also Kitcher [1986, e.g., 333]). Dedekind's published works, [1872] and [1888], are more or less neutral on the distinction between eliminative and *ante rem* structuralism. Either interpretation fits most of the text. Dedekind did use the letter "N" as a singular term, and there are phrases like "we denote numbers by small italics" (§124), which suggests an *ante rem* reading, but his locution could mean something like "in what follows, we use 'N' to denote an arbitrary (but unspecified), simply infinite set, and we use small italics to denote the members of N." It is similar to a contemporary algebraist saying "let *G* be an arbitrary group, and let *e* be the identity element of *G*."

The theorem in §132 is "All simply infinite systems are similar [i.e., isomorphic] to the number series N and consequently . . . also to one another." On an eliminative reading, this is awkward and redundant. It would be better to say (and prove directly) that all simply infinite systems are isomorphic to each other. The same goes for other passages, like that in §133, but again, this does not rule out an eliminative reading. To further understand Dedekind's views on these matters, we turn to his correspondence.

Dedekind's friend Heinrich Weber had suggested that real numbers be *identified* with Dedekind cuts and that natural numbers be regarded as finite cardinals and identified with classes—along the lines of Frege or Russell and Whitehead. Dedekind replied that there are many properties of cuts that would sound very odd if applied to the corresponding real numbers (January 1988; in Dedekind [1932, vol. 3, 489–490]). Similarly, there are many properties of the class of all triplets that should not be applied to the number 3. This is essentially the same point made in Benacerraf's [1965], much discussed in chapters 3 and 4. Notice how far Dedekind's approach is from that of his fellow logicist Frege. Rather than wonder whether 2 is Julius Caesar, whether $2 \in 4$, and whether my socks are members of 2, Dedekind did not want to even ask the questions.

In the same letter, he again invoked the mind's creativity:

> If one wished to pursue your way—and I would strongly recommend that this be carried out in detail—I should still advise that by number . . . there be understood not the *class* (the system of all mutually similar finite systems), but rather something *new* (corresponding to this class), which the mind *creates*. We are of divine species and without doubt possess creative power not merely in material things . . . but quite specially in intellectual things. This is the same question of which you speak . . . concerning my theory of irrationals, where you say that the irrational number is nothing else than the cut itself, whereas I prefer to create something *new* (different from the cut), which corresponds to the cut. . . . We have the right to claim such a creative power, and besides it is much more suitable, for the sake of the homogeneity of all numbers, to proceed in this manner.

Thus, for Dedekind numbers are objects.[22]

As noted earlier, it is not clear what Dedekind meant by the "creativity" metaphor. Matters of charity preclude attributing some sort of subjectivism to this great mathematician. He surely did not think that the existence of the real numbers began at the moment he had the idea of "creating" them from cuts. Moreover, mathematics is objective if anything is; his numbers are the same as any mathematician's. In addition, he was no constructivist. As noted in chapter 1, Dedekind [1888, §1] explicitly adopted excluded middle.

Some commentators, like McCarty [1995] suggest a Kantian reading of Dedekind's "creativity," relating it to the categories rational beings invoke in order to think. This squares nicely with his logicism. On this interpretation, he was broadly Kantian, but he joined Frege in moving the source of arithmetic and analysis from spatial and temporal intuition to the rational mind's ability to deal with concepts. Stein [1988] relates "free creation" to the mind's ability to open and explore new conceptual possibilities, a theme that occurs often in Dedekind's writings.

The most common interpretation is that Dedekind's "free creation" is an abstraction of sorts, perhaps like the account sketched in chapter 4. Tait, for example, wrote of "Dedekind abstraction." Starting with, say, the finite von Neumann ordinals, "we

---

22. Nevertheless, Dedekind did not insist on thinking of numbers as objects. In a letter to Lipschitz (June 10, 1876, in Dedekind [1932, §65]), he wrote, "[I]f one does not want to introduce new numbers, I have nothing against this; the theorem . . . proved by me will now read: the system of all cuts in the discontinuous domain of all rational numbers forms a continuous manifold." This alternate reading is consonant with eliminative structuralism. Hallett [1990, 233] remarks that Dedekind did not insist on numbers being objects, because the existence of numbers as objects is not part of the meaning of "real number." What does matter is "the holding of the right properties," that is, the structure. It is convenient and perspicuous to take numbers to be new objects, but such a view is not required. This is of a piece with the present thesis that eliminative, modal, and *ante rem* structuralism are equivalent to each other (see chapters 3 and 7).

may . . . abstract from the nature of these ordinals to obtain the system *N* of natural numbers. In other words, we introduce *N* together with an isomorphism between the two systems. In the same way, we can introduce the continuum, for example, by Dedekind abstraction from the system of Dedekind cuts. In this way, the arbitrariness of this or that particular 'construction' of the numbers or the continuum . . . is eliminated" ([1986, 369]). If two systems are isomorphic, then the structures obtained from them by Dedekind abstraction are identical. This highlights the importance of the categoricity proof in Dedekind [1888] and explains why he had no objection to other ways of defining the real and natural numbers, and even encouraged that alternate paths be pursued. In the preface to [1888], he agreed that Cantor and Weierstrass had both given "perfectly rigorous," and so presumably acceptable, accounts of the real numbers, even though they were different from his own and from each other.[23] What the accounts share is the abstract structure.

In a letter to Hans Keferstein (February 1890; translated in van Heijenoort [1967, 99–103]), Dedekind reiterated that he regarded the natural numbers to be a single collection of objects. He wrote, "[The] number sequence N is a *system* of individuals, or elements, called numbers. This leads to the general consideration of systems as such." He called the number sequence N the "abstract type" of a simply infinite system. In response to Keferstein's complaint that he had not given an adequate definition of the number 1, he wrote, "I define the *number* 1 as the basic element of the number sequence without any ambiguity . . . and just as unambiguously, I arrive at the *number* 1 . . . as a consequence of the general definition. . . . Nothing further *may* be added to this at all if the matter is not to be muddled." Note the similarity of the last remark to Hilbert's quip that any attempt to get beyond structural properties is to go around in a circle. Dedekind took the numbers—the places in the natural-number structure—to be objects. As noted in chapters 3–4, one advantage of *ante rem* structuralism is that the language of arithmetic is taken at face value. The mind "creates" the numbers as paradigm instances of the abstract type. The abstraction fixes both the grammar and the subject matter of the branch in question.[24]

Cantor also spoke of a process of abstraction, which produces finite and transfinite cardinal numbers as individuals. His language is similar to Dedekind's: "By the 'power' or 'cardinal number' of *M* we mean the general concept, which arises with the help of our active faculty of thought from the set *M*, in that we abstract from the nature of the particular elements of *M* and from the order they are presented. . . . [E]very single element *m*, if we abstract from its nature, becomes a 'unit'; the cardinal number [of *M*] is a definite aggregate composed of units, and this number has existence in our mind as an intellectual image or projection of the given aggregate *M*" ([1932, 282–283]). Cantor gave a similar account of order types, starting with an ordered system: "[W]e understand the general concept which arises from *M* when

23. However, Dedekind wrote that his approach is "somewhat simpler, I might say quieter" than those of Cantor and Weierstrass.

24. I am indebted to Tait [1997] and McCarty [1995] for this way of putting the point, and to Tait [1997] for pointing to the material that follows.

we abstract only from the nature of the elements of *M*, retaining only the order of precedence among them. . . . Thus the order type . . . is itself an ordered set whose elements are pure units" (p. 297). Cantor thus shared with Dedekind a notion of free creation. Numbers and order types are individual objects, quite similar to the finite cardinal and ordinal structures presented in chapter 4.

Russell [1903, 249] took both Cantor and Dedekind to task:

> [I]t is impossible that the ordinals should be, as Dedekind suggests, nothing but the terms of such relations as constitute a progression. If they are to be anything at all, they must be intrinsically something; they must differ from other entities as points from instants or colours from sounds. . . . Dedekind does not show us what it is that all progressions have in common, nor give any reason for supposing it to be the ordinal numbers, except that all progressions obey the same laws as ordinals do, which would prove equally that *any* assigned progression is what all progressions have in common. . . . [H]is demonstrations nowhere—not even where he comes to cardinals—involve any property distinguishing numbers from other progressions.

Russell's complaint here looks like Frege's Caesar problem. Frege [1903, §§138–147] himself launched a similar attack on Dedekind's and Cantor's "free creation." Unlike the other logicists, Dedekind felt no need to locate the natural numbers and the real numbers among previously defined or located objects. This crucial aspect of "free creation" is shared with the present *ante rem* structuralism.

Although Dedekind was at odds with Russell and Frege, he was still a logicist. The preface to Dedekind [1888] states that arithmetic and analysis are part of logic, claiming that the notions are "an immediate result from the laws of thought." This is a prominent Fregean theme. Dedekind added that numbers "serve as a means of apprehending more easily and more sharply the difference of things." The same idea is sounded in the 1890 letter to Keferstein: "[My essay is] based on a prior analysis of the sequence of natural numbers just as it presents itself in experience. . . . What are the . . . fundamental properties of the sequence N . . . ? And how should we divest these properties of their specifically arithmetic character so that they are subsumed under more general notions and under activities of the understanding *without* which no thinking is possible at all but *with* which a foundation is provided for the reliability and completeness of proofs and for the construction of consistent notions and definitions?" The preface to Dedekind [1888] lists some of the relevant "more general notions": "If we scrutinize what is done in counting an aggregate or number of things, we are led to consider the ability of the mind to relate things to things, to let a thing correspond to a thing, or to represent a thing by a thing, an ability without which no thinking is possible."

The general notions, then, are "object," "identity on objects," and "function from object to object." The first two of these are clearly presupposed in standard first-order logic, and the third is a staple of second-order logic (see Shapiro [1991]). These notions are sufficient to define the notion of "one-to-one mapping" and thus "simply infinite system." Except for the infamous "existence proof" (§66) and the bit on "free creation," Dedekind's treatment of the natural numbers (and the real numbers) can be carried out in a standard second-order logic—with no nonlogical terminology. In other words, once superfluous properties are jettisoned, the only remaining notions are those of logic. The relations of the structure are formal, and the structure is freestanding.

5 Nicholas Bourbaki

Among mathematicians, the name of "structuralism" is often associated with a remarkable group of mathematicians who call themselves Nicholas Bourbaki. I do not pretend to do justice to their accomplishments nor to their underlying philosophy.[25]

Under the influence of Hilbert and the American postulate theorists, Bourbaki focused on the idea of implicit definition. In Bourbaki [1950], a much-quoted informal exposition, they praise the axiomatic method as providing a unifying theme for the diverse branches of mathematics. The axiomatic method makes mathematics intelligible. Contrary to a common belief, the value of the enterprise is not the rigor it affords. Bourbaki wrote that the value of the axiomatic method is that it allows the systematic study of the relations between different mathematical theories. For Bourbaki, "structure" emerges as a central notion of mathematics. In working with axiomatic theories, we see how mathematical structures may be embedded in each other. For Bourbaki, relations among structures are a natural setting for mathematical studies.[26]

Bourbaki uses the word "structure" to mean something like a form characterized by a natural group of axioms: "It can now be made clear what is to be understood, in general, by a mathematical structure. The common character of the different concepts designated by this generic name, is that they can be applied to sets of elements whose nature has not been specified; to define a structure, one takes as given one or several relations, into which these elements enter . . . then one postulates that the given relation, or relations, satisfy certain conditions (which are explicitly stated and which are the axioms of the structure under consideration)" (p. 225–226). In a footnote, they sound a theme we saw in the development of geometry: "It goes without saying that there is no longer any connection between this interpretation of the word 'axiom' and its traditional meaning of 'evident truth.'" Axioms are now more like definitions, implicit definitions in particular.

According to Bourbaki, there are three great types of structures, or "mother structures": algebraic structures, such as group, ring, field; order structures, such as partial order, linear order, and well-order; and topological structures, which "furnish an abstract mathematical formulation of the intuitive concepts of limit, neighborhood and continuity, to which we are led by our idea of space" (p. 227).

---

25. See, for example, Corry [1992] and the references therein. In a number of places, Bourbaki expressed scorn for philosophy.

26. In another popular article [1949, 2], Bourbaki wrote that logic was developed with mathematics and, for the most part, only mathematics in mind: "It serves little purpose to argue that logic exists outside mathematics. Whatever, outside mathematics, is reducible to pure logic is invariably found, on close inspection, to be nothing but a strictly mathematical scheme . . . so devised as to apply to some concrete situation." This is a subtheme of this book; see chapters 1, 2 and 8.

Bourbaki called a theory of a structure in the present sense, such as arithmetic and analysis, "univalent." The theory concerns a single structure, unique up to isomorphism. Such structures lie at the "cross-roads, where several more general mathematical structures meet and react upon one another" (p. 229). They note that the first axiomatic treatments, those that "caused the greatest stir," dealt with these univalent theories. The axiomatizations could not be "applied to any theory except the one from which they had been extracted." If the axiomatic enterprise had not evolved beyond such univalent theories, "the reproach of sterility brought against the axiomatic method, would have been fully justified" (p. 230).

The article closes with the idea that their notion of structure is part of a research program within mathematics (p. 231):

> From the axiomatic point of view, mathematics appears thus as a storehouse of abstract forms—the mathematical structures. . . . Of course, it cannot be denied that most of these forms had originally a very definite intuitive content; but, it is exactly by deliberately throwing out this content, that it has been possible to give these forms all the power they were capable of displaying and to prepare them for new interpretations and for the development of their full power.
>
> It is only in this sense of the word "form" that one can call the axiomatic method a "formalism". The unity which it gives to mathematics is not the armor of formal logic, the unity of a lifeless skeleton; it is the nutritive fluid of an organism at the height of its development, the supple and fertile research instrument to which all the great mathematical thinkers since Gauss have contributed.

Notice the continuing theme of purging intuition from mathematics.

Bourbaki's monumental and influential mathematical work appeared over several decades, illustrating this theme in practice. Although their *Theory of sets* [1968] contains a precise mathematical definition of "structure," Corry [1992] shows that this technical notion plays almost no role in the other mathematical work, and only a minimal role in the book that contains it. The notion of "structure" that underlies the work of Bourbaki and contemporary mathematics, is inherently informal:

> The concept of *mother-structures* and the picture of mathematics as a hierarchy of *structures* are not results obtained within a mathematical theory of any kind. Rather, they belong strictly to Bourbaki's images of mathematics; they appear only in nontechnical, popular articles . . . or in the myths that arose around Bourbaki. (Corry [1992, 340])

> [T]he rise of the structural approach to mathematics should not be conceived in terms of this or that formal concept of structure. Rather, in order to account for this development, the evolution of the nonformal aspects of the structural image of mathematics must be described and explained. (Corry [1992, 342])

This book is a contribution to the program described by that last sentence.

Part III

RAMIFICATIONS AND APPLICATIONS

6

Practice

Construction, Modality, Logic

1 Dynamic Language

At the beginning of chapter 1, I mentioned a gap between the practice of mathematics and its current philosophical and semantic formulations.[1] Mathematicians speak and write as if dynamic operations and constructions are being performed: they draw lines, they move figures, they make choices, they apply functions, they form sets. Taken literally, this language presupposes that mathematicians envision creating their objects, moving them around, and transforming them. This manner of speaking goes back to antiquity. Euclid's *Elements* contains statements that express the capabilities of mathematicians to effect geometrical construction or, in other words, the potentialities of geometrical objects to be created or affected by mathematicians. One of the postulates is "Given any two points, to draw a straight line between them." Taken literally, assertions like this are statements of *permission*, or of what combinations of moves are *possible*.

In contrast to the dynamic picture, the traditional Platonist holds that the subject matter of mathematics is an independent, *static* realm. Accordingly, the practice of mathematics does not change the universe of mathematics. In a deep, metaphysical sense, the universe cannot be affected by operations, constructions, or any other human activity, because the mathematical realm is eternal and immutable. There can be no permission to operate on such a domain.

To belabor the obvious, then, the traditional Platonist does not take dynamic language literally. Euclid's *Elements* also contains static language, statements about existing geometrical objects. One postulate is "All right angles are equal." For the Platonist, this is the philosophically correct way to speak.

The passage from Book VII of the *Republic*, in which Plato scolds his mathematical colleagues, is worth repeating: "[The] science [of geometry] is in direct contra-

1. Much of this chapter is a successor to Shapiro [1989].

diction with the language employed by its adepts. . . . Their language is most ludicrous, . . . for they speak as if they were doing something and as if all their words were directed toward action. . . . [They talk] of squaring and applying and adding and the like . . . whereas in fact the real object of the entire subject is . . . knowledge . . . of what eternally exists, not of anything that comes to be this or that at some time and ceases to be." Of course, the geometers in Plato's time did not take his advice, and neither do contemporary mathematicians.

With the clarity of hindsight, one might think that this dispute over language is, pardon the expression, merely verbal. Hilbert's [1899] treatment of Euclidean geometry is written almost entirely with static language. He makes no reference to the capabilities of mathematicians or to the potentialities of objects to be created and acted on. For example, he says that between any two points *there is* a straight line. One might say that in Hilbert's geometry, there is no geometer. Statements of permission are replaced with statements of existence.

The straightforward interpretation of Hilbert [1899] is that it establishes facts about the geometric realm, and so his language (if not his philosophy) conforms to realism in ontology and to realism in truth-value. Yet, for all of that, Hilbert's treatment is easily recognized as Euclidean. Except for a few logical axioms and inferences, many statements in Hilbert [1899] correspond to those in the *Elements*. A contemporary student of logic can routinely "translate" most of Euclid's *Elements* into the static language of Hilbert [1899].

The situation is not as simple as this, on either mathematical or philosophical grounds. Consider, for example, the long-standing problems of trisecting an angle, squaring a circle, and doubling a cube. One instance of the first problem is: "Is it possible to construct an angle of twenty degrees?" Applying the reformulation of dynamic language, the question becomes "Does there exist an angle of twenty degrees?" Clearly, this last is not what the geometers (or the oracle) wanted to know in posing the original construction problem. They proposed a more or less literal question of whether a given angle *can be drawn*. This problem presupposes another question: "What tools are allowed in constructing our angle?" Each set of construction tools results in a different batch of construction problems. In the case at hand, of course, the original problem was to draw an angle of twenty degrees using only a compass and unmarked straightedge. Ancient geometers knew how to trisect an arbitrary angle using a compass and a *marked* straightedge.

Perhaps the static statements and problems are equivalent to their dynamic counterparts when there are no restrictions on construction tools. To draw an angle of twenty degrees, one need only draw a circle and then draw two lines from the center that cut off exactly one-eighteenth of the circumference. These instructions are determinate, if not effective. We "can" square the circle and double the cube in a similar fashion.

Thus we bring the static and the dynamic together by expanding the dynamic, allowing extensive tools. Another way to connect the discourses is to restrict the static. A constructivist might insist that the way to interpret the *Elements* is that the only objects that *exist* are those that can be constructed from an original set of points, using a compass and straightedge. So an angle of twenty degrees does not exist (assuming that the Abel–Galois result is available to our constructivist). This perspective is also problematic. One Euclidean axiom allows the drawing of a circle given any center and radius. If every point on the circumference of such a circle exists, then there will be two such points exactly twenty degrees apart. Another axiom allows the drawing of a line from each of these points to the center of the circle. There is our angle. Is our constructivist interpreter to hold that a circle exists but that some points on its circumference do not?[2]

The upshot is that it is not obvious how static and dynamic language are to interact. The conflation articulated by Hilbert [1899] is that the locution "one can draw" be understood as "there exists." On making this move, however, one loses the ability to formulate some of the original construction problems. The static language has no resources to speak of what can be constructed with this or that set of tools. The other extreme, the constructive route to conflating the languages, is to fix a set of tools and restrict the ontology to what can be constructed with these tools.

As articulated here, structuralism is a realism in truth-value and a realism in ontology. Thus, a structuralist might join Plato and hold that only the static language is literally correct. The universe of structures is a Platonic heaven (or haven). This would be a mistake, because structuralism should aim for an account of *all* mathematics, not just those parts that dovetail nicely with traditional Platonism.

Some of the construction problems were resolved in the negative only when mathematicians articulated the *structure of Euclidean construction* as such and embedded this structure in a richer one. The situation is somewhat analogous to the development of computability in the early decades of this century. Mathematicians could draw conclusions about what *cannot* be computed only after they articulated the structure of computability and embedded this structure in a richer one (see Shapiro [1983b]; Gandy [1988]; and Sieg [1994]). The problem for the present project is to show how one can meaningfully speak of the *structure* of dynamic systems (like geometric construction and algorithmic computation), and to bring such structures under the banner of structuralism. Along the way, we encounter important issues of ontology and the nature and role of logic.

2 Idealization to the Max

On all accounts, the constructions suggested by dynamic language are idealized. The *Elements*, for example, make no allowances for points that are so far apart one cannot draw a line between them without falling asleep or having the straightedge slip or decay. Euclidean lines have no width, even though such lines cannot be drawn. For the sake of discussion, I propose that we think of the constructions as performed by an imaginary, idealized constructor, obtained in thought by extending the abilities of actual human constructors. Then we can sharpen dynamic language and the

2. This issue is related to the one of the supposed gaps in the *Elements*, a principle of continuity. Euclid gave no axiom to sanction an inference that if a line l goes from the outside of a circle to the inside, then l intersects the circle at some point.

various "construction problems" by articulating exactly what abilities and permissions are attributed to the ideal constructor.

Lakatos [1976] describes the historical development of a proof that in any polyhedron, the number of vertices minus the number of edges plus the number of faces is 2 ($V - E + F = 2$). The reader is invited to take an arbitrary polyhedron, of any size, remove one of the faces, stretch the remaining figure out on flat surface, draw some lines on it, and remove the lines one at a time, keeping certain tallies along the way. With the clarity of hindsight, each of these moves is licensed by a principle of topology. There is thus a connection between construction principles—powers attributed to the ideal constructor—and mathematical principles and logical inferences. As Lakatos shows, however, the path to this retrospective clarity was not smooth. The details of the connection were hard-won.

In the dynamic language of contemporary mathematics, the abilities attributed to the ideal constructor are quite remarkable. Consider the following pair of fanciful film clips:

SCENE 1. Professor A, an expert in analysis, is lecturing to a class in advanced calculus.

> PROFESSOR A: Next I will prove the Bolzano-Weierstrass theorem: every bounded infinite set has at least one cluster point. Let S be an arbitrary, bounded infinite set. To prove the theorem,[3] we must produce a point p with the property that every neighborhood of p contains infinitely many points in S . . . We divide C_0 into four equal squares by intersecting lines. One of these smaller squares must contain infinitely many points of S . . . Choosing such a subsquare, label it C_1. We have $C_0 \subseteq C_1$ and both are closed and bounded. Now repeat this process. Divide C_1 into four squares . . . By continuing this, we generate a sequence of closed squares C_n . . . Appealing to the nested set property, there must be a point p that lies in all the sets C_n. This is the point that will turn out to be a cluster point for S . . .

At this moment, a student with a double major in mathematics and philosophy raises her hand and is recognized.

> STUDENT: You are using a constructional language in this lecture. You do not actually mean that you or some ideal mathematician has done this construction, do you? How can anyone do an infinite number of things, and then after *all* of them—on the basis of all of them—do something else, in this case pick the point p?

> PROFESSOR A: Do not take this lecture literally. Of course, there is no such constructional process. I am *describing* a property of *the plane*. From principles of cardinality, I *infer* the existence of infinitely many points in some square C_1, and then in C_2. The axiom of replacement implies the existence of the whole sequence $<C_n>$. Finally, from the nested-set property, I deduce the existence of a point in all of the C_is. I let p be the name of one such point. Now this is what I *mean* by this lecture; the constructional language makes it easier for you to see.

3. From here until the end of this speech, Professor A has copied from Buck [1956, 38]. The dots indicate the omission of material that is of no concern here. Fortunately for us, Professor A knows some logic and some set theory. In scene 2, most of the opening speech is copied from Crossley et al. [1972, 71–72].

SCENE 2. Professor T is lecturing on set theory. The same student is in this class.

PROFESSOR T: Intuitively, this is what is going on in the formation of the constructible sets. As we pass each ordinal, we assign to it a set M_α in the following way. First, $M_0 = \phi$. At the ordinal $\alpha + 1$, which is next after α, we put for $M_{\alpha+1}$ the immediately preceding M_α together with the set of all sets definable in this preceding M_α . . . When we come to a limit ordinal β, we collect together everything we have obtained so far . . . Now, finally, a set is constructible if . . .

STUDENT: Hold it. What do you mean "finally"? Just like Professor A, you are using constructional language. How do you construct your way to step ω, let alone through all of the ordinals?

PROFESSOR T: Patience. You will soon see that all of this is justified. If there is time (and you still insist), I will do it formally next week, with no operative language. I will describe, by transfinite induction, a series of functions on ordinals. I will use the axioms of union and replacement at each stage. However, if I did that without giving this lecture first, none of you would see what is going on.

If the student in these scenes had encountered constructive or intuitionistic mathematics in her education, she would have questioned Professor A's ability to "pick" the squares C_1, C_2, and so on. The ideal constructor knows that it is not the case that each of the four squares in the divided plane has only finitely many points, but unless he knows *which* square has infinitely many such points, how can he do any picking? How does the constructor know which square to pick for C_1, C_2, and so forth? Professor A would no doubt appeal to the principle of excluded middle (of which more later). Because it is not possible for all four squares to have only finitely many points, then one of them must have infinitely many. The constructor names, or picks, one such square.

The point here is that when the mathematics is reformulated in static language, many of the moves allowed to the ideal constructor correspond to logical inferences. It is not obvious that all such moves should be permitted. This observation, together with the fact that the "translation" to static language is not straightforward, suggests that logic is not as universal and topic-neutral as has been thought. Logic, ontology, and allowed construction are intimately bound up with each other. Here I try to sort things out.

3 Construction, Semantics, and Ontology

The standard formalisms of each branch of mathematics reflect the static orientation to our subject. In model theory, a nonalgebraic branch of mathematics is taken to be about a fixed domain of discourse, and inferences are checked by examining what holds in all (static) models of the premises. Except for some constructivists, most theorists take it as obvious that classical logic is sound for this semantics, and thus for all of mathematics.

Historically, of course, the formalization of a branch of mathematics and the metamathematical study of formal systems have illuminated and even furthered several branches of mathematics. With respect to practice, however, formalization is unnatural. Deduction represents little more than the ultimate standard of justification in mathematics. Were one interested in establishing a theorem beyond the doubts of all but the most obstinate skeptic, one would present it as the result of a deduction from (agreed on) axioms or previously established theorems. Mathematicians at work however, are not usually concerned with ultimate justification but with understanding and explanation. Their aim is to gain insight into the mathematical structure at hand in order to understand *why* a particular theorem is true. Formal deduction often does not contribute to this understanding and sometimes obscures it. I suggest, therefore, that philosophers who are attempting to understand mathematics concentrate less on the little used standards of ultimate justification and more on the actual work of mathematicians. It is at least prima facie plausible that the language and techniques active in mathematical understanding and explanation are good indicators of the nature of mathematics itself. Indeed, how a given subject matter is grasped should have something to do with what it ultimately is. Recalling the advice from chapter 1 about the philosophy-first principle, we should pay some attention to mathematics as practiced.

As we have seen, for a traditional realist in ontology—as for Plato himself—only the static mathematical universe has objective existence. Any literal interpretation of the "constructions" is denied. The only end of the operative language is to provide picturesque ways of expressing corresponding static statements. A realist in ontology might concede that dynamic language is somehow necessary for us mortals to understand and gain insight into the mathematical universe. But even if dynamic language is a psychological or a methodological necessity, no inferences concerning the essential subject matter of mathematics should be drawn. As indicated by the film clips, when classical mathematicians use operative language, they claim to be speaking informally and, if pressed, justify the operative motivation by relying on axioms in a formal (presumably static) language. Constructions correspond to existence statements. According to the received views, mathematical practice is the discussion of, or motivation for, deductions from true axioms.

In contrast, there is philosophy of mathematics that takes dynamic language seriously. For an intuitionist, constructions and operations are the essence of (legitimate) mathematics, and dynamic language is the only literally correct way to depict mathematics. According to the intuitionist, each apparently static statement is to be understood as reporting a *construction*. In arithmetic, for example, the intuitionist employs the same formulas as the Platonist, but the former interprets them as abbreviated forms of speech, which, when fully paraphrased, are operative statements. Heyting [1956, 8], for example, wrote, "'2 + 2 = 3 + 1' is taken as an abbreviation of 'I have effected the mental construction indicated by "2 + 2" and "3 + 1" and I have found that they lead to the same result.'"

The intuitionist is thus committed to reinterpreting the static aspects of mathematical discourse as dynamic. From the present perspective, this is the exact opposite of our Platonist. Moreover, on this view, classical mathematics (if taken more or less at face value) is legitimate only to the extent that it can be straightforwardly interpreted as operative. At this point, battles over logic are joined. In present terms, however, the disputes concern just what sorts of powers are attributed to the ideal constructor.

Classical constructors and their intuitionistic counterparts are both idealized, but classical constructors are idealized more.[4]

One crucial difference between the classical constructive mode of thought and the intuitionistic mode is that the former seems to presuppose that there is a (static) external mathematical world that mirrors the constructs. A traditional Platonist (such as Proclus) might claim that the existence of the mathematical world is what *justifies* or grounds the constructs. However, to stay in line with the modest orientation of this book, I just note that classical construction proceeds *as if* there were an external universe that mirrors it. Classical mathematics does not need the sort of justification that the Platonist attempts.

Whatever its metaphysical status, the supposition of an external world suggests certain inferences, some of which are the nonconstructive parts of mathematical practice rejected by intuitionism. For example, if a classical mathematician proves that *not all* natural numbers *lack* a certain property (i.e., she proves a sentence in the form $\neg\forall x\neg\Phi$), she can infer the *existence* of a natural number with this property ($\exists x\Phi$). Following existential instantiation, she can then give a "name" to some such number and do further constructional operations on it. In the proof of the Bolzano-Weierstrass theorem, in scene 1 earlier, we have a similar instance of excluded middle at work. At each stage, the constructor knows that at least one of four squares has infinitely many points from the given set, but he may not know which one. Nevertheless, the constructor can pick one such square, and go on from there.

Intuitionists demur from such inferences and constructions because they understand the principles as relying on the independent, objective existence of the domain of discourse. For them, every assertion must report (or correspond to) a construction. In the present example, an intuitionist cannot assert the existence of a natural number with the said property, because such a number was not constructed. The intuitionist cannot choose a square with infinitely many points from the given sets, because such a square was not identified. Bishop [1967] understands the law of excluded middle as a principle of omniscience.

Another related difference between the classical constructive mode and the intuitionistic mode is that the former allows the completion of infinite processes. We saw this in the film clips. In scene 1, the constructor produces an infinite sequence of squares and *then* picks one point in each. In scene 2, the constructor runs through the ordinals.[5]

4. There is an ambiguity in mathematical (and metamathematical) discourse, which comes to the fore in the present context. In the mathematics literature, words like "construction" are used for both the classical dynamic processes under study here, as depicted in the film clips, and the intuitionistic use, with its specific restrictions. For example, a classical author may refer to the "Bolzano-Weierstrass construction." For the present, context or explicit reference indicates which use is intended.

5. Consider a man who turns on a light at a given moment. One second later, he turns it off; a half second after that, he turns it back on; a quarter second later, off, and so on. After two seconds, is the light on or off? Compared to the accomplishments of our (classical) constructor, these "supertasks" seem tame.

This language abounds. Crossley et al. [1972, 63] give this gloss on the axiom of choice: "If I have a set of non-empty sets, then I can choose one member from each set and put the chosen elements together into a set." If the set is infinite, then this "axiom" licenses the constructor to do an infinite amount of choosing and then after that, do something else. In discussing the iterative notion of set, Gödel [1944] wrote, "The phrase is meant to include transfinite iteration; i.e., the totality of sets obtained by finite iteration is considered to be itself a set and a basis for the further application of the operation 'set of.'"

The supposition that there is a static mathematical universe that mirrors the dynamic language sanctions these infinitary procedures. Consider, once more, the aforementioned proof of the Bolzano-Weierstrass theorem, in scene 1. On the classical view, each construct refers to a *fact* in the static mathematical world. The construction of the first square C_0 reflects the *existence* of a corresponding square in the plane. A more neutral way to put it is that the construction proceeds as if there were a corresponding square in the plane. Similarly for C_1 and C_2. From the discourse, it becomes clear that the construction *could* be continued as far as one wished. This reflects the existence of a square that corresponds to C_n for any particular natural number n. The crucial supposition is that the corresponding squares exist whether or not the construction is actually performed. The existence of the entire sequence of squares is then deduced. By the nested-set theorem, there is a point p in all of the C_ns. The proof concludes with a demonstration of some facts about p.

In this example at least, the supposition allows one to proceed as if an infinite process had been completed—more precisely, the supposition of a static universe suggests that one can infer the existence of what would be the result of such a process, in this case a point in *every* element of the infinite sequence of squares. In the jargon of mathematics, we pass to the limit.

Because, as noted, intuitionistic construction proceeds without this supposition, its ideal constructors are not endowed with the ability to finish infinite tasks. Beth wrote (Beth and Piaget [1966, 47]), "[I]n classical mathematics, it often happens that, in the course of a demonstration, a construction occurs which requires the introduction of an infinite series of successive operations, whilst the demonstration contains a certain inference depending on the result of this construction, that cannot be judged before the infinite series of operations is completed. . . . Such a manner of reasoning is inadmissible from a strictly constructivist point of view; this viewpoint does not allow the use of the result of a construction unless the construction can be completed; now it is clear that one can never carry out a construction consisting of an infinite series of successive operations." Sort of. As Beth was aware, the intuitionist does accept certain constructions that represent the completion of infinite processes, once the construction is restated in classical terms. The difference lies in the classical and intuitionistic conceptions of items like numbers, sequences, and sets. To speak roughly, an intuitionistic real number *is* a Cauchy sequence of rationals. If one gives a rule for a Cauchy sequence, one has given the number. The constructor does not have to carry out a sequence or pass to the limit to "arrive" at the number. In a sense, the sequence itself *is* the number. In general, intuitionists accept the potential infinite, in the form of procedures for continuing indefinitely, but they do not accept the actual infinite, which for them would be the completion of an infinitary procedure. These differences in ontology provide a challenge to the present attempt to bring intuitionistic mathematics into the fold of structuralism (see section 5).

4 Construction, Logic, and Object

My conclusion, thus far, is that classical and intuitionistic mathematics differ in the different types of "construction" that are countenanced; in other words, they allocate different powers to the ideal constructor. When cast in dynamic terms, the methodological principles that separate the camps become different sets of permissions allowed in construction. This subsumes the difference in logic.

There is thus an intimate, three-way link between the ontology that best fits a given stretch of mathematical discourse, the allowed constructions in that area, and the proper or best logic of the discourse. Never mind, for now, what the proper metaphysical, epistemic, and explanatory priority relations among these three might be. Tait [1986, 600–601] endorses a similar connection:[6] "I believe that it is possible to argue that . . . logical principles of proof are the same as principles of construction or definition of objects. . . . [T]here is no natural distinction between mathematics and the logic of mathematics."

Kitcher [1983] articulates and defends a view of mathematics that focuses on construction-type activity. He suggests that in antiquity (and with children), mathematical knowledge began (begins) by focusing on the human ability to assemble and manipulate collections of objects. From this "protomathematical" start, the enterprise of mathematics evolves through a series of idealizations. The focus shifts from the capabilities of actual people to those of an ideal observer-constructor who lacks certain limitations on humans. As above, the adoption of classical logic can be understood in these terms. As we have seen, the dynamic law of excluded middle is an idealization on the knowledge of the constructor. The current results of this evolving process are the branches of abstract mathematics that we study today.

Although Kitcher does not make the connection, there is an affinity between his view and some work on the origins of mathematical concepts in children. Some of the psychological literature takes such concepts to be both dynamic and structural. Cooper [1984], for example, writes, "Consider number development as learning about the space of number. In this space, one must learn where things are, and how to get from one place to another. For purposes of the analogy, the locations are specific numerosities and the actions (transformations) to get from one place to another are additions and subtractions. . . . [C]hildren learn about this space of numbers by traveling in it. It is through experiences of moving in this space of numbers (by addition

---

6. The conclusions of Tait [1981] are also consonant with the present orientation. Tait proposes that the natural numbers be construed as *constructions*, or iterations of a construction procedure. He then argues that for the Hilbert program, *finitism* is the restriction of reasoning about numbers to those methods implicit in the very construction of numbers. Nonfinitary reasoning is the application of methods derived from other sources to the natural numbers.

and subtraction) that children learn its ordinal structure" (p. 158). See also Piaget's half of the aforementioned Beth and Piaget [1966].

Prima facie, the connection between logic, ontology, and construction runs counter to the widely held view that logic is the study of the canons of correct inference common to all systems, no matter what their "objects" might be. The slogan is "logic is topic-neutral." Accordingly, either the classical mathematician or the intuitionist is using the wrong logic. The classical constructor is too strong or his intuitionistic counterpart is too weak.

A full-scale attack on the "logic-is-topic-neutral" thesis would take us too far afield, but perhaps a statement of my perspective on logic and ontology is in order. This part of this book serves as a partial defense.

There are at least two versions of the topic-neutrality theme. I reject one of them outright and leave the other open but severely qualified. The view to be repudiated is a strong thesis that there is but one correct logic—period. Call this view *logical absolutism.* An advocate of this position may claim that this logic is established through a soundness theorem that concerns the model-theoretic semantics common to all language, or at least all language that is sufficiently regimented. Or perhaps the position may be supported by proof-theoretic studies. The most common version of the view is that standard, classical, first-order logic is the one true logic, but some argue for intuitionistic logic as the one true logic; Tennant [1987] endorses intuitionistic relevant logic. The various arguments presuppose that there is a single model-theoretic-type semantics common to all language or a single notion of logical form or a single notion of objects and reference.

Against this, I urge a more eclectic attitude concerning logic, and that intuitionistic and classical systems both be regarded as legitimate branches of mathematics. The perspective delimited earlier is that some inference rules are tied to permissions and abilities attributed to an ideal constructor. My proposal amounts to a suggestion that different sets of permissions and abilities be investigated. I see no reason, other than philosophical prejudice, to restrict the development of mathematical systems to one sort of logic. Let a thousand flowers (try to) bloom.[7]

In support of this eclectic orientation, notice that classical mathematicians and intuitionistic mathematicians do not ultimately have trouble understanding each other, following each others' proofs, or even contributing to each others' school—keeping in mind, of course, that mutual understanding is not an absolute, all-or-nothing matter. To be sure, classical mathematicians often have trouble following intuitionistic discourse at first, but if prejudice is submerged and an honest effort is made, it is not difficult to get used to it. From the other perspective, many remaining intuitionists find themselves in ordinary departments of mathematics and philosophy, and so are

7. Unlike Brouwer, Heyting (e.g., [1956]) sometimes took an eclectic view toward mathematics. In his conciliatory moments, he urged a place for intuitionistic mathematics "alongside" classical mathematics. In chapter 1, I repudiated the view, called "philosophy-first," that philosophy should determine the proper direction for mathematics. Philosophy should not attempt to restrict the mathematician to classical logic either. The intuitionistic differences between actually infinite, potentially infinite, and unbounded might be helpful in modeling linguistic competence.

required to teach standard courses in mathematics and logic. By and large, they do it in good conscience, and they do a fine job. Occasionally, they come up with new results within classical mathematics.

The point here is that in most cases, classical mathematicians recognize the intuitionist's work *as mathematics*, even if they reject its underlying restrictions. Intuitionists must also see at least a tight family resemblance between their work and that of classical colleagues. Thus, I take it as a desideratum that a philosophy of mathematics accommodate both and not reject either one as incoherent. If the theme of structuralism is to be maintained, the structure of nonclassical systems ought to be countenanced.

A second, weaker version of the logic-is-topic-neutral thesis allows multiple logics in some sense but adopts a version of the Kuhnian doctrine of incommensurability between the logics (see Kuhn [1970]). If different "conceptual schemes" employ different logics, then neither scheme is comprehensible to the other. Logic is still topic-neutral in the sense that it applies universally within its "scheme." On this articulation, then, all systems *within a given conceptual framework* have the same logic, and importantly, any system from a rival framework (with a different logic) is incoherent from the point of view of the first. Classical and intuitionistic systems are each incoherent from the perspective of the other.

On this view, there can be no common framework to compare and relate the schemes and the logics. Indeed, a common framework would undermine the incommensurability, and the neutrality of each logic. There may be a classical-logic framework and, perhaps, an intuitionistic framework (and a quantum framework, a predicative framework, a relevant framework, etc.) but never the twain shall meet. Indeed, if there were a single theory for grasping both classical and intuitionistic systems, what would *its* logic be?

Davidson [1974] argues against the very possibility of different conceptual schemes, especially if they are taken to be incommensurable in an extreme form. In any case, if the incommensurability can be sustained at all, it must be tempered. Notice, first, that on the Kuhnian view, *interpretation* across frameworks is possible. Kuhn himself provides lucid accounts of paradigms that are no longer accepted. If the rival schemes were totally incommensurable, we would not understand Kuhn's own work. In the present case, the various interpretations of classical systems into their intuitionistic counterparts (via double negation, etc.) and the Kripke semantics for intuitionistic systems are two examples (see Kripke [1965]). In each case, the language of one scheme is translated into the language of the other, so that theorems are preserved. To be sure, these interpretations do not exhibit equivalence in any straightforward sense. Logical structure is not preserved, of course—that is the point. In short, the mutual interpretation is partial at best.

I just noted that classical mathematicians and intuitionists can and do understand each other, for the most part. I submit that this mutual understanding is not accomplished by one of the aforementioned interpretations. That would take too long. Each mathematician simply learns directly how to work in the other framework. It is similar to someone who can speak two languages without constantly translating between them, or even without being able to translate between them very well. I do not see anything in the anti-incommensurability arguments to rule this out.

This weak version of the separate-and-incommensurable thesis must be further qualified. The goal of the present articulation of structuralism is to accommodate *all* structures, including those with intuitionistic logic. So the philosophical framework of structuralism must be able to talk about systems with either kind of logic. Structuralism should have its own logic, and *that* logic is not neutral. Presumably, the logic of structuralism is classical and, in any case, the logic of this book is classical.

Thus, I do not embrace what may be called *logical relativism*, a vague claim to the effect that all logics are *equal*. This relativism may be incoherent, because it requires a neutral framework from which to *make* the relativistic assertion—a neutral framework that is supposed to hold good for all the frameworks. I fully accept classical logic and use it (without apology) in discussing structures and structuralism—even structures with nonclassical logic.[8]

In sum, then, we must distinguish the internal logic of a system or structure from the external logic of structuralism itself. By way of analogy, consider the distinction between the logic of an object-language theory and the logic of the metatheory. It is quite common to study intuitionistic systems, such as Heyting's arithmetic, using a classical metatheory. In topos theory, there is also a distinction between the "internal logic" of a topos and the "external logic" of topos theory itself. I take it, then, that it is coherent to speak of different structures with different logics in the same breath. That is to say, there is a philosophical or metatheoretic framework for structuralism that fits both classical and intuitionistic systems and structures. We have at least this much commensurability.

Turning to objects, there are two different views that ontology does not depend on language and logic. One has objects *determining* the logic; the other has objects *independent of* logic. The first orientation is a thesis that the appropriate logic for a given subject matter is *determined by* the objects (and relations) under study. So far, this view is consistent with the present eclectic attitude that classical arithmetic and intuitionistic arithmetic are each legitimate branches of mathematics, provided that we hold that the two arithmetics have *different subject matters*. The former studies classical numbers, and the latter studies intuitionistic numbers (and proofs, constructions, etc.). The different logics are a consequence of this difference in subject matter. We first figure out what it is we are talking about, and this determines how we are to talk about it.

On the other hand, one who accepts this objects-determine-logic thesis and also holds that there is only one "kind" of object will embrace logical absolutism, against the eclectic attitude. Frege and Quine both adopt this combination of views, holding that there is only one kind of object and only one logic (see Parsons [1965], [1983]).

8. I am indebted to the discussion of Carnap's relativism in Friedman [1988]. Carnap attempted to formulate various logical frameworks from a neutral background theory of syntax. Friedman argues that this program failed because there cannot be a neutral framework that accomplishes all of Carnap's aims.

A different orientation to ontology is that subject matter is *independent* of logic. Accordingly, if there are rival logics, they represent different ways of discussing or reasoning about the same objects and relations. Thus, intuitionistic arithmetic and classical arithmetic represent different modes of studying the same things—numbers. Such a view underlies the common suggestion that intuitionistic mathematics is mathematics with an epistemic logic (see, for example, Shapiro [1985]).

To follow the conclusions from the previous sections, I firmly reject the second articulation, the independence of object and logic. As argued in chapters 3–4 (and chapter 8), the mathematical universe does not come to us, nor does it exist, already "divided" into objects, waiting to be studied. If anything, it is the other way around—the type of discourse and its allowed inferences determine (at least in part) the nature of the objects. How we are to talk about the objects partially determines what they are. Thus, I also reject the priority of object over logic. The mutual dependence of object and logic is clearest when one focuses on dynamic mathematical practice, in which the *inferences* allowed in the reasoning of a branch of mathematics are directly related to the sorts of moves available to the ideal constructor.

The conclusion is that intuitionistic numbers, intuitionistic construction, and intuitionistic logic are closely interrelated and interdependent. None are autonomous. Similarly, classical numbers, classical construction, and classical logic are also interrelated and interdependent. The problem is to bring both trinities under the purview of structuralism.

5 Dynamic Language and Structure

The time has come to embed the foregoing conclusions into the main theme of this book, mathematics as the science of structure. I have spoken of dynamic systems as being equivalent to static ones. My goal in this section is to lend some precision to this claim.

Let me briefly review the situations in which two systems exemplify the same structure. As developed in chapter 3, there are two different relations among systems that can be used for "sameness of structure." The first and most common is *isomorphism*. Systems S_1 and S_2 are isomorphic if there is a one-to-one function f from the objects and relations of S_1 onto the objects and relations of S_2 such that f preserves all of the relations of the systems. Notice that two systems are isomorphic only if they have the same number of functions and relations of the same degree. The second "sameness of structure" relaxes that situation. Let R be a system and P a subsystem. Define P to be a *full subsystem* of R if they have the same objects (i.e., every object of R is an object of P) and if every relation of R can be defined in terms of the relations of P. The only difference between P and R is that some definable relations are omitted in P. Let M and N be systems. Define M and N to be *structure-equivalent*, or simply *equivalent*, if there is a system R such that M and N are both isomorphic to full subsystems of R.

We learn from model theory that isomorphic interpretations (of the same language) satisfy the same sentences and corresponding open formulas. Similarly, if two systems are equivalent, then there is a straightforward translation from the language and objects of one to the language and objects of the other that preserves both logical structure and satisfaction.

To return to the matter at hand, the dynamic perspective in mathematics can be brought under the purview of structuralism if "dynamic systems"—and their structures—are countenanced. The idea is that the relationships of a system may include constructions and dynamic operations. Notice, however, that after an operation is performed, new objects may have been created. Thus, the collection of objects associated with a dynamic system may not stay fixed.

The dynamic nature of the collection of objects raises problems for the application of isomorphism and structure equivalence to such systems. Consider the notion of function in the metatheory. In the official model-theoretic formalizations of mathematics, a function is a set of ordered pairs; in other words, a function is a *static* set. Each function has a fixed domain and a range. If the domain of a system does not stay put, then how can there be functions on the system? Recall, however, that the notion of function also has a dynamic interpretation: a function is an *operation* or perhaps a construction. On dynamic construals of arithmetic, for example, the successor function is a procedure for producing a new natural number from a given number. The logical truth that every natural number has a successor ($\forall x \exists y(sx = y)$) corresponds to a power attributed to the ideal constructor: given any number, he can construct its successor.

If our plan is to succeed, we need a way to bridge this gap between static and dynamic functions. If we make two global assumptions about dynamic mathematical systems, then there is a structural equivalence between the two notions of function, so that it does not matter which one is employed. These assumptions concern the stability of systems, and they hold for intuitionist and classical dynamic practice alike.

The first assumption is that the ideal constructor cannot destroy objects: within a dynamic system, no operation *reduces* the collection of objects. The ideal constructor does not have an eraser. According to this assumption, the collection of objects associated with a system is nonreducing over time. The second assumption is one of internal global coherence: within a dynamic system, once the constructor has an operation S available, then no activity can preclude the performance of S. Of course, it may happen that other operations make S superfluous, but in any future scenario, S can still be executed. Suppose, for example, that the ideal Euclidean geometer has points B and C available, so that a line connecting B and C may be drawn. According to the assumption at hand, there are no operations the geometer can perform that prevent this line from being drawable.[9]

---

9. These global assumptions are registered in standard Kripke structures for intuitionistic predicate calculus (see Kripke [1965]). If an object n exists at a node, then n exists at all "future" nodes. Once an object is constructed, it exists forever. Second, if a formula Φ is satisfied at a given node (under a certain assignment to the variables), then Φ is satisfied at all future nodes (under the same assignment). Once true, always true. Notice, incidentally, that the global assumptions fail in systems of quantum mechanics. Suppose that a "quantum constructor" is in a position to determine the momentum of

These stability assumptions provide a straightforward framework for extending structuralism to dynamic mathematics. Roughly, define a *dynamic system* to be a collection of *potential objects*, or *possible objects*, together with certain relations, functions, and operations on them. The functions and operations transfer appropriate sequences of actual and possible objects into other possible objects. Given the global stability assumptions, we can speak of the "collection" of objects associated with a dynamic system. This collection consists of the results of every operation the ideal constructor can perform. It serves as the "domain" of the dynamic system.

At least on an informal level, we thus have a framework to compare and relate dynamic and static systems. The latter have only actual objects in the collection, whereas the former may include "potential objects" or "merely possible objects." Nevertheless, we can coherently speak of a function from a dynamic system to a static one, and vice versa. A function can have both actual and potential objects in its domain and range, and the function itself can be construed as static or dynamic at will. We can then speak of isomorphism and equivalence between dynamic and static systems.

For present purposes, I wish to avoid (as much as possible) questions concerning the ontology of *possibilia*. This metaphysical conundrum may not be as central here as it is elsewhere, because present focus is on the *structure* of (static and dynamic) systems. Structures are at least somewhat independent of the status of the objects in systems—whether they are abstract or concrete, actual or potential.

Notice that if one is a realist concerning *possibilia* (as Lewis [1986], for example) and regards them as static, then under the present treatment, dynamic systems have been reduced to static ones. The static collection of objects associated with a dynamic system is the collection of possible objects the ideal constructor is capable of producing. With this reduction, however, we lose the distinction between the potentially infinite and the actual infinite, and this distinction is essential for intuitionism. In any case, the reduction of dynamic objects is not essential and not desirable here. All we need is a framework for discussing and comparing the structures of static systems and dynamic systems.

Under certain conditions, a dynamic system and a static system can exemplify the same structure—the systems can be structure-equivalent. One example, noted above, is traditional Euclidean geometry and its static reformulation in Hilbert [1899]. The same goes for classical arithmetic construed dynamically and classical arithmetic construed statically. For another example, a Turing machine, viewed as a system, is dynamic. Its tape is operated on in various ways, and the underlying theory is meant to capture at least some of the structure of the *activity* of computation. Turing [1936] contains a detailed argument that common *operations* that a human computist performs with pencil and paper can be simulated on a Turing machine. Nevertheless, one can easily redefine Turing machines in static language. That is, one can define a

---

a given particle. If she does not, but measures its position instead, then she can no longer determine the momentum. Some theorists once claimed that quantum mechanics requires a nonstandard logic, one that differs from both classical and intuitionistic (and relevant) logic. This much is consonant with the present theme that logic is tied to available construction.

different class of mathematical objects with the same structure as Turing machines. Turing himself does not do this, but Rogers [1967, 14] does (if we assume that number theory is static): "It is easy to define *Turing machine* in more orthodox mathematical language. Let $T = \{0, 1\}$ and $S = \{0, 1, 2, 3\}$. Then a Turing machine can be defined as a mapping from a finite subset of $\mathbb{N} \times T$ into $S \times \mathbb{N}$. Here T represents the conditions of a tape cell, S represents operations to be performed, and $\mathbb{N}$ gives the possible labels for internal states." This passage is little more than a gesture toward official rigor. Other than these four sentences, Rogers's book deals with Turing machines only as dynamic systems. It seems that for pedagogical and explanatory purposes, the dynamic model is preferred, even if the official semantics calls for a more static treatment.

There are more mundane examples as well. Deduction is a dynamic activity, because the mathematician writes the lines of a derivation in sequence. Presumably, reasoning is a dynamic activity. Formal deductions are supposed to capture the human activity of correct reasoning. In mathematical logic, by contrast, a derivation is understood to be a sequence of strings. A deduction is thus understood as an abstract object, an item from string theory. The transformation works because the structure of derivations, suitably idealized, is exemplified in string theory. Similarly, a chess game is a dynamic activity, and so the system of possible chess games is also dynamic. But one can "identify" each game with the string that consists of the moves as recorded in a standard notation. Again, we have a dynamic system sharing a structure with a static one.

Suppose that a structure S is exemplified by two systems P_1 and P_2 (so that P_1 and P_2 are either isomorphic or structure-equivalent). Then either system can shed light on the structure S. It does not matter whether mathematicians focus on P_1 or P_2. As long as they limit their language, reference, and so forth, to the elements and relations of P_1 or they limit their language to the elements and relations of P_2, their results hold of S. The choice is only a matter of which system they prefer—which system is more conducive to explaining, grasping, developing insights, and so on. If care is taken, a mathematician may use both systems P_1 and P_2 to study S.

Of course, the cases under study here are those in which one system is static and the other dynamic. The foregoing considerations (and the film clips of section 2, the tone of Rogers [1967], etc.) indicate that for purposes of presentation and explanation, some structures lend themselves to dynamic interpretation. Other structures lend themselves to static interpretation. In set theory, it seems, both sorts of interpretations are used, perhaps for different aspects of the field. Parsons [1977], for example, wrote that one should speak of the elements of a set as prior to the set, and he adds that for "motivation and justification . . . it is important to ask in what this 'priority' consists." But none of this is relevant to mathematics itself: "for the practice of set theory . . . only the abstract structure of the relation matters" (p. 336).[10]

10. Parsons goes on to give a static, modal account of set theory, which he argues is more coherent than the dynamic "genetic" account. On the other hand, Chihara [1984], [1990] develops a dynamic interpretation of simple type theory *as an alternative* to common static interpretations, and he argues that the dynamic interpretation is to be *preferred* on ontological grounds. If, as it appears, Chihara's

We have seen that in the official model-theoretic semantics of mathematical languages, a domain of discourse is a static collection. This suggests that the result of formalizing a branch of (classical) mathematics is always a static system, regardless of whether the branch *as practiced* is dynamic or static. In other words, the metamathematical study of a structure seems to require that it be recast in static terms. A dynamic system is more conducive to developing and communicating insights about the places and relations of a structure, insights *internal* to the structure; but in order to take a broader perspective and develop and communicate insights about the structure itself—how it relates to other structures, what the correct logic for studying it is, and so on—one prefers a static reformulation. The important point is that it is one and the same structure.

The conclusion, thus far, is that some dynamic systems are structure-equivalent to static ones or, in other words, some dynamic systems exemplify the same structure as static ones. However, not every dynamic system shares a structure with a static one. To put it differently, some dynamic systems are not structure-equivalent to any static system, at least not in any obvious or straightforward manner. This conclusion is a corollary of the present thesis that ontology, inference, and construction are intimately related.

Consider, for example, intuitionistic arithmetic, a paradigm example of a dynamic system. To start a reductio, suppose that P were a static system that is structure-equivalent to intuitionistic arithmetic. What would P be like? Presumably, its objects would constitute a countable set (putting the domain of proofs aside). Thus, we might as well identify the domain of P with the natural numbers. The relations of P would have the defining conditions of the ordinary successor, addition and multiplication functions, the less-than relation, and the like. In short, then, P has the same elements and relations as classical arithmetic. It is hard to avoid the conclusion that P is *isomorphic* to the classical natural-number structure, in which case it exemplifies *the same structure*. Yet, one might suggest that, by hypothesis, P does not exemplify the classical natural-number structure, because P has a *different logic*. The law of excluded middle does not apply to P.

There is the rub. In making the identification, we gloss over the distinction between potential infinity and actual infinity. The problem, on the surface, is one internal to structuralism: what is the most natural relationship between logic and the (admittedly relative) structure/system dichotomy? Yet, the situation is a manifestation of a deep issue concerning the status or priority of logic. The crux of the problem concerns the extent to which "objects" are independent of the way they are constructed and studied. As indicated in the previous section, my view is that with dynamic discourse, the sorts of objects studied by a branch of mathematics—the sorts of structures—are determined by the allowed constructions and sanctioned inferences. In short, the logic and the objects share a common source: the permissions and the pos-

---

system has the same structure as the classical one, then on my view there is no relevant difference between the two systems—they exemplify the same structure. This question of ontological commitment is revisited in chapter 7.

sible moves allowed to the ideal constructor. My eclectic attitude toward logic is part and parcel of this thesis. We cannot separate logic and structure.

Recall that when abstracting from a system to its structure, one is to ignore features of the objects that do not concern the interrelations. In dynamic systems, the available constructions are certainly part of the interrelations of the objects. In large part, the allowed moves *constitute* the interrelations. Given the present view that the logic of a system is directly related to those constructions, it follows that in moving from system to structure, logic is not something to be abstracted from. That is, "same structure" implies "same logic." Consequently, we cannot conflate the potentially infinite with the actually infinite.

I conclude, then, that no system that exemplifies intuitionistic arithmetic can satisfy classical logic. As noted earlier, classical logic is appropriate for static systems. Thus, no static system can exemplify the same structure as intuitionistic arithmetic. Brouwer [1948, 90–91] reached a similar conclusion: "[E]ven in those mathematical theories which are covered by a neutral language . . . either school operates with mathematical entities not recognized by the other one: there are intuitionistic structures which cannot be fitted into any classical logical frame, and there are classical arguments not applying to any introspective image."

6 Synthesis

Thus we see that some dynamic systems are equivalent to static ones, whereas others are not, depending on what constructional and knowledge-gathering powers are allotted to the ideal constructor. In this section and the next, I turn to dynamic systems that are equivalent to static ones, and thus systems that satisfy classical logic, the axiom of choice, impredicative definition, and so on. In the prevailing nonconstructive climate, these are the systems that get the bulk of attention from mathematicians.

The plan here is to give a brief account of the conditions under which a given dynamic system satisfies the usual classical, nonconstructive inferences and principles. What must be assumed about a system of possible constructs and their constructor in order to treat them as if they were a bunch of eternal, static objects—at least as far as inference is concerned? Just how much idealizing must we do? This section provides what may be called an *ontic route* to classical logic, impredicative definition, and the like, that focuses on the creative powers allotted to the constructor. The next section turns to the so-called Heyting semantics for intuitionistic logic and provides an epistemic or semantic route to classical logic.

The simplest cases are those in which all of the operations together, iterated indefinitely, produce only finitely many new objects—and the relations are all finite in extent. Consider, for example, a dynamic account of the cyclic group of order four trillion. The constructor starts with an initial object e and keeps on making new objects, using a successor construction. At some point, the constructor produces the first object e again. Because the intuitionist agrees that classical logic is appropriate for such thoroughly finite systems, these cases are apparently unproblematic.

In these finite cases, we simply assume that the constructor has in fact completed all possible constructions and, thus, we treat the system as if it were static. Because everything is decidable, we also assume that the constructor can (or has) determined the truth-value of any sentence about the system. Notice that these easy finite cases still involve an idealization on the abilities of the constructor: no *finite* bounds are placed on his ability to perform the operations. The ideal constructor has unbounded time, space, and materials at his disposal. This idealization is readily made throughout mathematics and is shared by all of the traditional philosophies, save only the strictest finitism. Virtually everyone agrees that there is no difference in principle between small natural numbers and astronomical natural numbers, and no one complains about large triangles in geometry. But we are idealizing. This point is related to the observation, in chapter 4, that many of the traditional problems with the infinite arise when it comes to the large finite. We have no causal contact with systems of four trillion. It takes a thought experiment, stretching the imagination, to comprehend constructing a system that large.[11]

Next, consider cases in which all the operations together, iterated indefinitely, produce a *denumerably* infinite number of new objects. This will occur, for example, if the system has only a finite (or countable) number of operations, each of which takes a finite input. The paradigm case is elementary arithmetic, which can be limited to a single operation (successor) that acts on a single input. The ideal constructor produces the natural numbers with this operation, starting with an initial object 0. At most, countably many natural numbers can be constructed this way. For such a system to be straightforwardly equivalent to a static one, we assume that the constructor can complete the entire process and construct the infinitude of new objects. The constructor has the wherewithal to produce an actual infinity. This assumption allows full reference to the entire set, thus permitting a standard Tarskian semantics, classical logic, and impredicative definitions. Suppose, for example, that one tries to define a number or set of numbers with reference to the entire range of natural numbers. The definition is coherent, because we speak as if the constructor has finished the constructing and can, as it were, survey the range of his creation—all of the natural numbers. He is in position to determine the truth-value of any given sentence of arithmetic.

To belabor the obvious once again, this case is more of an idealization than that of the previous one. There, it was assumed that the constructor has no finite bounds on his ability, but it was not assumed that he can complete an infinite number of constructions. Here, the natural numbers are no longer regarded as a mere potential infinity.

Given the present idealization, the constructor can produce the integers if he forms "differences" among natural numbers and introduces variables that range over these differences. The constructor can produce the rational numbers by forming quotients of integers, and he can produce the algebraic numbers by producing roots of equations. As yet, we have not left the countable.

The next cases are those in which some operations have *infinite* inputs. That is to say, the operations deal with infinite sets, sequences, and so on. Now we encounter

11. Well, we did manage to "construct" the national debt.

the possibility of an *uncountable* number of constructed objects. In real analysis, for example, the ideal constructor can take a Cauchy sequence of rational numbers and construct a limit of the sequence. He can pass to the limit. Or the constructor can take a bounded set of real numbers and construct the least upper bound of the set. A more grandiose, and perhaps far-fetched, example is the dynamic understanding of set theory, in which the constructor can form the union of any collection of objects formed at any stage, and he can form a set whose elements are the range of any function whose domain is (already formed as) a set. We saw this idealization in action in scene 2 of the film clips (section 2). The axiom of replacement is usually motivated in terms much like this. The very name "replacement" calls up an image of a constructor taking a given set and replacing its elements one by one. Because, after some training, many mathematicians find the axiom of replacement compelling, there may be something to the metaphor. Wang [1974, 186] wrote, "Once we adopt the viewpoint that we can in an idealized sense run through all members of a given set, the justification of [replacement] is immediate. That is, if, for each element of the set, we put some other given object there, we are able to run through the resulting multitude as well. In this manner, we are justified in forming new sets by arbitrary replacements. If, however, one does not have this idea of running through all members of a given set, the justification of the replacement axiom is more complex." If unrestricted, our constructor has the potential to produce a proper class of constructed objects.

Admittedly, this metaphor stretches the analogy with everyday, nonmathematical construction in time. Nevertheless, mathematicians do speak that way, and I presume that what they say is coherent, if not literal. On the other hand, one may wonder whether there is anything left of our original intuitions concerning ordinary construction (see Parsons ([1977], [1983, essays 10–11]).

Nevertheless, we will press on for a bit. There are two issues here. One concerns the status of the collections of objects that serve as the *inputs* to the indicated operations; the other concerns the outputs. Starting with the former, the most natural route—the route that is conducive to a static interpretation—is to consider the inputs themselves as completed or actual infinities. If we assume that the constructor can "get at" the whole of any relevant sequence or set, then the operations in question may be applied universally. On this assumption, neither the constructor nor us mortals needs to worry about whether a given Cauchy sequence can be defined (in a finite language) before passing to its limit. The assumption at hand thus accords with the tendency in mathematics to move away from the definable toward the extensional (see Maddy [1993]). Mathematicians typically do not check their constructions for any connections to the language they may be using at the time.

The intuitionists, of course, have the opposite tendency. They do not recognize actual infinities, and they restrict the application of operations to those elements that can be constructed, in their restricted sense of construction. In his early writing, Brouwer took a real number to be a Cauchy sequence determined by a rule. Later, he augmented rule-governed sequences with free-choice sequences, but even then the attitude is that Cauchy sequences are potential, not actual infinities.

Some very nonclassical results follow from the intuitionists' perspective. Brouwer showed, for example, that every function defined on a closed interval is uniformly continuous. For an intuitionist, a function f is a mapping from Cauchy sequences to Cauchy sequences. Let c be any real-number sequence. Remember that c is only a potentially infinite sequence. To pursue the present metaphor, the intuitionistic constructor never sees the entire sequence c all at once. Thus, the constructor has to be able to give initial segments of fc based only on finite, initial segments of c. In particular, to approximate fc within a fixed bound, one should need to know only a fixed, finite, initial segment of c. Technically, the requirement is that for every ε, there is an n such that the approximation of fc to within ε depends only on the first n values of c. The function f would thus give the same approximation to fd, for any real number d that has the same first n values as c. This requirement precludes a discontinuity in f. Indeed, if f were discontinuous, then for some inputs c, the constructor would not be able to approximate fc unless he knew the entire sequence c—and this would be to treat c as an actual, not merely potential infinity. For example, let $gx = 1$ if $x < 1$ and $gx = 0$ otherwise. Let c be the Cauchy sequence $<.9, .99, .999, \ldots>$, which converges to 1. So $gc = 0$. But there is no finite initial segment of c that is sufficient to determine an approximation of gc to within .3. In order to know that $gc > .7$, the ideal constructor must have a look at the *entire* sequence. So for an intuitionist, g is not an acceptable function. Classically, of course, discontinuous functions are easily defined and have proven useful in the sciences.

Our second issue concerns the status of the results of the constructions. Consider the extreme case, set theory.[12] One crucial item sanctioned by static interpretations is impredicative definition. In set theory, this is manifest in the axiom scheme of separation, $\forall x \exists y \forall z (z \in y \equiv z \in x \,\&\, \Phi(z))$. Because Φ might contain quantifiers over all sets, our scheme allows definitions that refer to sets of arbitrarily high rank. Indeed, one can define a set in terms of all sets. With dynamic systems, perhaps one can sanction such inferences by further extending the previous idealization and allowing the constructor to have performed all the possible constructions—a proper class of them. The idea is to treat the entire set-theoretic hierarchy as if it were an actual or completed infinity. In this case, an impredicative definition is treated as it is in static systems. It refers to an "already-constructed" collection in order to characterize one of its elements. In Zermelo-Fraenkel set theory, this "already-constructed" collection is a proper class.

This is a mind-boggling idealization. Suppose we think of the ideal constructor as having somehow completed the work and produced the entire set-theoretic hierarchy, complete in all of its glory. He has before him, as it were, all sets. Then why does the constructor fail to go on to consider collections of sets—proper classes—and then produce classes of such classes? This, after all, would only be for the ideal constructor to

12. There are, of course, other theories worthy of consideration in this regard. Type theory, for example, is often given a dynamic interpretation, at least at the level of heuristics (see Hazen [1983]; Chihara [1984], [1990]). Most accounts of simple type theory employ classical logic, and they allow impredicative definitions without restriction. Ramified type theory, of course, does not sanction impredicative definition, but it employs classical logic. In the ramified theory, impredicative definitions might be allowed if an axiom of reducibility is sanctioned. Construed dynamically, this axiom gives the constructor the power to construct (the extension of) every element of every type using only lower-type objects.

go on as before. Is there any reason the constructor should not just keep going? But the constructor was not to stop until he had completed all possible constructions.

The upshot of this consideration is that, contrary to the suggestion under study, there can be no natural stopping place for the ideal constructor when it comes to set theory. One resolution of this difficulty is to extend the distinction between actual infinity and potential infinity to the transfinite. The idea is to accept the dynamic classical perspective of the ideal constructor producing sets by collecting their elements, and assume that *for any given* ordinal α, the constructor can produce the rank V_α as a completed, actual infinity, but he cannot produce the entire set-theoretic hierarchy (at once). This amounts to taking each rank as an actual infinity but not the entire universe. This is a stretched analogy to the intuitionist concession that because the constructor can produce any given natural number n, there is no harm in taking each number n as if it were an actually existing thing. But the constructor never finishes the natural-number system, so ω is only potential. The idea here is that each V_α can be taken as an actuality. The constructor can get that far, but V itself is only potentially infinite.

Some of Cantor's remarks suggest a picture like this. Although he held that individual cardinals are actual infinities, in his much-quoted [1899], he calls the collection of all sets an "inconsistent multitude," because one cannot conceive of this collection as "one finished thing." Even today, it is common for set theorists to balk at considering the entire hierarchy as one finished thing.[13]

The main methodological issue here is whether impredicative definitions are allowed. Under the Cantorian conception of set theory, is there any reason to think that an impredicative definition succeeds in characterizing something? Notice that the *conception* of the system does not immediately sanction the unrestricted axiom of separation. Without further justification, one is not justified in defining a set with a formula that has variables that range over the entire universe, because the latter does not (i.e., never does) exist as a completed entity. The constructor does not have access to all of V. However, this consideration does not preclude the possibility of impredicative definition. It turns out that the legitimacy of impredicative definition is a substantial set-theoretic principle. In the present framework, the ideal constructor, and thus a human counterpart, can define sets by reference to any fixed rank V_α. That is, a definition is legitimate if its variables are restricted to a fixed rank. This allows some impredicative definitions, and it may allow all. If the underlying language is first-order, then there is a reflection theorem to the effect that if a formula is true at all, then it is true at some sufficiently large rank. Thus, the ideal constructor never has to reach "arbitrarily high" in the hierarchy in order to characterize a set. This sanctions each instance of the axiom of separation and allows full impredicative definition.[14] Wang [1974, 209] reached a similar conclusion: "The concept of an unfinished or . . . unfinishable totality . . . permits the classical interpretation of the

13. The same considerations motivate the static modal accounts of set theory mentioned in note 10.

14. If the background framework is second-order, with standard semantics, then the reflection principle is independent of set theory (if consistent with it). The so-called strong reflection principles imply the existence of (small) large cardinals. See Shapiro [1987] and Shapiro [1991, chapter 6].

quantifiers. . . . We may . . . appeal to the reflection principle to argue that the unbounded quantifiers are not really unbounded."

7 Assertion, Modality, and Truth

Our second path to the equivalence of dynamic systems with static counterparts focuses on semantics. I begin with a review of some global arguments for intuitionistic logic.

As presented in chapter 2, antirealism in truth-value is a type of program to understand and evaluate mathematical statements and theories in terms other than truth. According to these philosophies, good or acceptable mathematics is not necessarily true mathematics. For assertabilism, the alternative to truth is warranted assertability, or proof in the case of mathematics. In place of truth conditions, one provides assertability conditions for each sentence, conditions under which a mathematician is warranted in asserting the sentence.[15] Of course, to be true to antirealism, the assertability conditions should not presuppose too much. It will not do, for example, to propose that an arithmetic sentence Φ is assertable just in case the mathematician has good reason to believe that Φ is true of the natural numbers. To state the obvious, this is not a real alternative to standard realism in ontology and realism in truth-value.

Assertabilism is consonant with the dynamic approach to mathematical practice under study in this chapter. If we are thinking of a mathematical statement as reporting the results of an ideal constructor, then perhaps it is best to understand and evaluate these statements in terms of what the ideal constructor is in a position to assert rather than whether the statement is true of some independent reality. This way of putting the matter highlights the fact that on all accounts, the relevant notion of assertability is an idealized one. There lies the problem.

Like all programs, assertabilism promises progress on philosophical problems associated with mathematics. Perhaps an assertabilist epistemology is tractable. Assertion is, after all, a human endeavor, and it is humans who are the producers of mathematics. We evaluate mathematics in human terms, not in relation to some abstract, detached realm. To fulfill this promise, however, the notion of assertability must be articulated in such a way that assertabilism can approach both human competence and mathematics as practiced. My contention is that if assertabilism is to be true to both of these masters, then the promise is not delivered. Assertabilism is not an improvement over ordinary truth-valued semantics on the epistemic front.

Michael Dummett [1973], [1977] argues that reflections on the role of language in communication, and the learning of language, suggest the development of an assertabilist semantics rather than a truth-valued semantics—in general, not just for mathematics. One who understands a sentence must grasp its meaning, and one who learns a sentence thereby learns its meaning. As Dummett puts it, "a model of mean-

15. Because assertability is a modal notion, we see that with these programs, modality is brought to the aid of antirealism. See chapter 7.

ing is a model of understanding" ([1973, 98]). This at least suggests that the meaning of a statement is to bc somchow determined by its *use*:

> [I]f two individuals agree completely about the use to be made of [a] statement, then they agree about its meaning. The reason is that the meaning of a statement consists solely in its role as an instrument of communication between individuals. . . . An individual cannot communicate what he cannot be observed to communicate: if an individual associated with a mathematical symbol or formula some mental content, where the association did not lie in the use he made of the symbol or formula, then he could not convey that content by means of the symbol or formula, for his audience would be unaware of the association and would have no means of becoming aware of it. ([1973, 98])

> To suppose that there is an ingredient of meaning which transcends the use that is made of that which carries the meaning is to suppose that someone might have learned all that is directly taught when the language of a mathematical theory is taught to him, and might then behave in every way like someone who understood the language, and yet not actually understand it, or understand it only incorrectly. ([1973, 99])

Dummett's point is that one cannot *completely* fake understanding.[16] He concludes that accepting these possibilities makes "meaning ineffable, that is, in principle uncommunicable."

Like many slogans, the phrase "meaning is use" can be misleading. So far, there is no positive view of meaning, no semantics, and not much has been said about "use" either. For present purposes, the important insights concern *understanding*. Dummett identifies an important criterion for any semantics that is to play a role in philosophy: understanding should not be ineffable. One understands the concepts invoked in a language if and only if one knows how to use the language correctly. Call this the *use thesis*. If, as Dummett suggests, a semantics is a model of understanding, then any plausible semantics should accommodate the use thesis—it should not make understanding ineffable. The use thesis is a plausible and important constraint on any account of language. However, to foreshadow my conclusion below, it is real human understanding that must be accommodated, not the understanding of a grossly idealized constructor.

Dummett argues that there is a natural route from the use thesis to assertabilist semantics, and thus to what I am calling "antirealism in truth-value," and he argues that assertabilist semantics leads to the rejection of classical logic, and thus to a demand for major *revisions* in mathematics. There is no need here to challenge the alleged connection between the use thesis and assertabilist semantics. Assertabilism does represent an alternative for philosophical semantics, an option we can afford to keep open for now. However, Dummett's link between assertabilist semantics and classical logic is quite relevant to present concerns. If he is right, then assertabilist

16. A realist might agree that a complete fake, or even a large fake, is impossible in some epistemic sense. Plato's own Socratic method seems to presuppose that if one fails to grasp a concept, then there is a situation in which one cannot give intelligent responses. I owe this observation to Allan Silverman.

semantics leaves no room for mathematics *as practiced.* Assertabilism would be at war with the prevailing antirevisionist theme of this book, and it would not be compatible with what I call "working realism" (see chapters 1 and 2).[17]

Advocates of "meaning is use" are often criticized for leaving "use" vague. Surely, some account is needed if this notion is to have such a central role in philosophy. As Wittgenstein [1978, 366–367] put it, "It all depends [on] *what* settles the sense of a proposition. The use of the signs must settle it; but what do we count as the use?"

Some articulations of "use" make it absurd to motivate the revision of practice by invoking the use thesis. If everything the mathematician does (and gets away with) is considered to be legitimate use, then the law of excluded middle is as legitimate as anything. For better or for worse, classical logic has won the day among mathematicians. That is how the enterprise of mathematics is pursued nowadays. On a view like this, it seems, all use is sacrosanct.

There are at least two orientations toward mathematical language that would suggest an interpretation of "use" along these strongly antirevisionist lines. One is formalism, the idea that correct mathematical practice can be codified into formal deductive systems. If classical logic is an ingredient of the appropriate deductive systems, then the issue of classical logic is settled. I will not add here to the many attacks against formalism. Suffice it to note that when Dummett argues for assertabilist semantics, he explicitly states, in several places, that "proof" does not mean "proof in a fixed formal system." For Dummett, proof is inherently informal. In this, Dummett follows virtually every intuitionist, including Brouwer and Heyting.

A second understanding of "use" that undermines revision is what Dummett calls a "holistic" account of mathematical language: "On such a view it is illegitimate to ask after the content of any single statement. . . . [T]he significance of each statement . . . is modified by the multiple connections it has . . . with other statements in other areas of language taken as a whole, and so there is no adequate way of understanding the statement short of knowing the entire language" ([1973, 100]). In light of Dummett's thesis that to understand a sentence is to grasp its meaning, the view in question is a combination of semantic holism and epistemic holism. Dummett's point is that, on a view like this, there is no way to *criticize* a particular statement, such as an instance of the law of excluded middle, short of criticizing the entire language. This is not quite correct. On a holistic view, mathematics can be criticized, and it can respond to criticism. Quine envisions changes in mathematics as coming from recalcitrant empirical data. Clearly, however, on a holistic view, criticism of practice does not come from *semantics* nor from reflections on meaning and understanding generally. Dummett [1991a] argues that the whole enterprise of semantic theory does not go well with holism.

---

17. There is thus an important difference between Dummett's agenda and the present one. Dummett is looking for a philosophical rationale for *revising* classical mathematics, whereas I am exploring philosophical programs underlying *classical* mathematics, in which working realism—classical logic, impredicative definition, and so on—is assumed. It is a working hypothesis here, if not for Dummett, that most of modern mathematics is correct, whatever this correctness amounts to.

Typically, semantics is *compositional*, or what Dummett calls "molecular." The semantic content of a compound statement is analyzed in terms of the semantic content of its parts. Tarskian semantics, for example, is compositional, because the satisfaction of a complex formula is defined in terms of the satisfaction of its subformulas. Dummett's proposal is that the lessons of the use thesis be incorporated into a compositional semantics. Assertability, or provability, is to replace satisfaction, or truth, as the main constituent of a compositional structure. The semantic content of a formula is its assertability condition.

Dummett's proposal invokes the central theme of Heyting's semantics for intuitionistic logic (see Heyting [1931]; Dummett [1977, 12–26]). Instead of providing truth conditions of each formula, Heyting proposes that *proof* or *computation* conditions be supplied. I sketch three clauses:

A proof of a formula in the form $\Phi \vee \Psi$ is a proof of Φ or a proof of Ψ.
A proof of a formula in the form $\Phi \rightarrow \Psi$ is a procedure that can be proved to transform any proof of Φ into a proof of Ψ.
A proof of a formula in the form $\neg\Phi$ is a procedure that can be proved to transform any proof of Φ into a proof of absurdity; a proof of $\neg\Phi$ is a proof that there can be no proof of Φ.

By consensus, Heyting semantics sanctions the inferences of intuitionistic mathematics, items like $\Phi \& \Psi \vdash \Phi$. Heyting and Dummett both argue that, on a semantics like this, the law of excluded middle is not universally upheld. A proof of a sentence of the form $\Phi \vee \neg\Phi$ consists of a proof of Φ or a proof that there can be no proof of Φ. Heyting and Dummett claim that one cannot maintain, in advance, that for every sentence Φ, there is such a proof. This is the main contention before us now. If Heyting and Dummett are wrong, then a Heyting-type semantics can sanction the inferences of classical logic. We would have a viable antirealist program that is not revisionist.[18]

Notice, first, that the proposed revision of practice depends on what one means by "proof" and "procedure." Who, after all, is doing the proving, and who is doing the computing? As noted, virtually every intuitionist is clear that "proof" does not refer to the derivations in a fixed, formal deductive system. So what do "proof" and "procedure" amount to here?

The problem can be illuminated with a quick look at two extremes. First, consider the possibility that "proof" refers to the correct proof tokens actually written

18. The potential compatibility between classical logic and assertabilist, Heyting-type semantics is elaborated in Barbara Scholz's essay "Assertabilism without revision" (winner of the Fink Prize in Philosophy, Ohio State University, 1987). Many of the foregoing considerations are sketches of ideas originally developed there (although Scholz does not agree with my ultimately negative conclusion). Notice, incidentally, that it is not obvious that assertabilism is, in the end, an antirealism. Heyting semantics has variables ranging over proofs and over procedures, and some of the clauses, such as the ones given above for negation and the conditional, are impredicative. In articulating intuitionism and Heyting semantics, McCarty [1987] proposes what may be called a "realist" interpretation of things like proofs and procedures. He calls them "probjects" or, in later work, "data objects."

down, spoken, or otherwise explicitly considered by mathematicians. On this option, we do not idealize at all, and we take "provable" to be just "proved." If Heyting semantics is interpreted this way, then the law of excluded middle is surely not universally valid. There are many sentences that have been neither proved nor refuted, and there always will be. Some sentences are just too long. But on Heyting semantics so construed, modus ponens is not sanctioned either. At any given time, there is likely to be a conditional such that both it and its antecedent have been proved, but no one has bothered to prove the consequent. The principle of universal instantiation is not sanctioned either, so long as there is some correctly asserted, universally quantified sentence in which the variable ranges over a large domain.

The other extreme is the opposite of this one. The idea is to take a "procedure" to be a function, as understood in *classical* mathematics. That is, a procedure is an *arbitrary* correspondence between members of various collections. As for "proof," notice that in the statement of Heyting semantics, the only unanalyzed uses of this term concern the proofs of atomic formulas and proofs that some "procedures" do certain jobs. The above clauses concerning the conditional and the negation are typical. In cases like this, just read "proof" as "correct statement." Clearly, on this "articulation" of Heyting semantics, provability and classical truth are conflated. Heyting semantics becomes a mere rewording of classical, Tarskian semantics, just using the word "provable" for "true." We do not have a genuine alternative to realism in truth-value.

The first of these extremes highlights the fact that *some* idealization of the notions of "proof" and "procedure" is necessary. This is where modality enters the picture. We are not limited to the actual productions by human mathematicians. Rather, in the theme of this chapter, we look at productions by ideal constructors. Once again, our question becomes what assertion powers are to be attributed to these ideal mathematicians. How much do they know? The second extreme serves as a reminder that the idealization should remain true to assertabilism. Idealized proof and idealized procedure should have something to do with actual human abilities to prove and compute. If they do not, then we will lose the epistemic promise of assertabilism. If the abilities attributed to our ideal constructor are too far removed from the human constructors that we all know and love, then the assertabilist cannot claim any major epistemic gains over realism in truth-value. We must sort out modal notions, and the going gets rough at just this point.

A start on a middle ground between our two extremes would be to construe a "proof" to be a "rationally compelling argument." Suitably trained humans do seem to have some sort of ability to recognize proofs. The issue is whether this can be idealized and controlled in an appropriate manner, without falling back into holism and losing the advantages of compositionality. Remember that, *in practice*, what counts as a correct proof depends heavily on historical context, or on what have already been accepted as correct proofs. What is needed here is an a priori conception of "rationally compelling argument" (perhaps along the lines of Myhill [1960] or Wagner [1987]). Then we would try to square this account with a compositional, Heyting-type semantics. We could then turn to excluded middle. The question would be whether the general principle $\Phi \vee \neg\Phi$ is sanctioned by Heyting semantics so construed.

Some expositors (not intuitionists themselves) have "argued" against excluded middle by citing items like the twin-prime conjecture and the Goldbach conjecture (and, perhaps ironically, Fermat's last theorem and the four-color problem), which have been neither proved nor refuted. These arguments do not work with Heyting semantics on our *idealized* notion of proof. The examples are undecided now, but how do we know that they are undecidable? Suppose the twin-prime conjecture and the Goldbach conjecture were decided next year. The mathematical community seems to be on a roll. The expositor will, of course, pick another example of an undecided problem. There are plenty to choose from. But the new example can be undermined similarly, with luck or hard work. On our idealized notion of proof, the expositor must come up with an example and show that there can be no rationally compelling argument establishing it and no rationally compelling argument refuting it. Such an example, however, is impossible. According to the above clause of Heyting semantics for negations, to establish that there is no rationally compelling argument that establishes a sentence Φ is to refute Φ and establish $\neg\Phi$. In short, to show that Φ cannot be proven is to prove $\neg\Phi$.

Call a sentence Φ *absolutely decidable* if either there is a rationally compelling argument that establishes Φ or one that refutes Φ. That is, Φ is absolutely decidable if the ideal constructor can decide Φ. The point is that Heyting semantics does not give a reason to demur at the law of excluded middle unless one can demur at the prospect that all unambiguous mathematical sentences are absolutely decidable.[19] This demurral is a sort of *pessimism*, transferred to ideal constructors. We acknowledge or impose a limit on their epistemic powers. In sum, Heyting semantics plus pessimism undermine classical logic. This conclusion echoes one in Posy [1984], who argues that Brouwer did embrace this sort of pessimism. Posy also argues that Kant's orientation toward mathematics is best understood in terms of Heyting semantics without the pessimism, and I now turn to that combination.

Define *optimism* to be the belief that every sentence of every unambiguous, nonalgebraic mathematical theory is absolutely decidable—at least by an ideal constructor. I propose that Heyting semantics presently construed plus optimism sanctions excluded middle and classical mathematics. If this is a viable alternative, the result would be what Blackburn [1984] calls "quasi-realism." It is an antirealist program that allows one to speak (in the object language) *as if* one were a realist. In present terms, we would have an antirealist program that sanctions the very inferences and principles challenged by the intuitionists on antirealist grounds. We would have an antirealist program for mathematics as practiced, an antirealism in truth-value that is consistent with the methodological principles of working realism.

If one takes historical data to be relevant to the issue of optimism and pessimism, it goes both ways. The optimist can point out examples of problems, like the four-

19. The word "demur" was chosen carefully. Intuitionists do not actually reject excluded middle, in the sense of accepting a sentence in the form $\neg(\Phi \vee \neg\Phi)$. Such a sentence is a contradiction in intuitionistic logic. The law of excluded middle is rejected by intuitionists only in the sense that they fail to accept it as universally valid.

color result and Fermat's last theorem, which have received generally accepted solutions. The pessimist can point out examples of very old problems, like the Goldbach conjecture, which remain open to this day.

It might be argued, in light of this standoff, that the best course is to remain agnostic on the matter of optimism and pessimism. The law of excluded middle would not be undermined by assertabilism, but neither would it be sanctioned. If so, then some philosophers may urge that mathematicians play it safe and stick to intuitionistic logic—a fragment of classical logic accepted by both camps. That way, we are less likely to go wrong. But mathematics is not always well served by caution. Bold assertions and principles are needed on occasion.

From a bold perspective, one may propose that optimism is something of a *regulative ideal* that underlies the practice of mathematics, and the law of excluded middle codifies this ideal.[20] Why, after all, would anyone pose problems and devote so much energy to their solutions if it is not held that the problems are ultimately solvable? As great a mind as Gödel endorsed optimism (Wang [1974, 324–325]; see also Wang [1987]): "[H]uman reason is [not] utterly irrational by asking questions it cannot answer, while asserting emphatically that only reason can answer them. . . . [T]hose parts of mathematics which have been systematically and completely developed . . . show an amazing degree of beauty and perfection. In those fields, by entirely unsuspected laws and procedures . . . means are provided not only for solving all relevant problems, but also solving them in a most beautiful and perfectly feasible manner. This fact seems to justify what may be called 'rationalistic optimism.'" The opening of Hilbert's celebrated "Mathematical problems" lecture [1900] is also an enthusiastic endorsement of optimism:[21] "However unapproachable these problems may seem to us and however helpless we stand before them, we have, nevertheless, the firm conviction that the solution must follow by . . . logical processes. . . . This conviction of the solvability of every mathematical problem is a powerful incentive to the worker. We hear the perpetual call: There is the problem. Seek its solution. You can find it . . . for in mathematics there is no ignorabimus."

Notice that if the law of excluded middle is accepted in the context of Heyting semantics and if, as assumed here, "provable" means something like "provable by the ideal constructor," then up to Church's thesis, one can rigorously sustain the intuitionistic (and antiformalist) thesis that the notion of "provable sentence" of

20. On this view, perhaps, one might balk at calling the law of excluded middle a principle of *logic*, because the regulative ideal is not analytic, a priori, a matter of form, or obvious. The location and importance of the border between mathematics and logic is beyond my present concern.

21. In the same lecture, Hilbert envisions the possibility of our discovering that a certain problem has no solution *in the sense originally intended*. This seems to bear on excluded middle, vis-à-vis Heyting semantics. I am indebted to Michael Detlefsen for making this point. As for Gödel, one should not forget that he was a realist concerning mathematics, and so, for him, a solution to an unambiguously stated problem is the discovery of a mind-independent fact about the mathematical realm. That is, Gödel's optimism may be grounded in his realism. On the other hand, one might think of realism as an outgrowth—an epiphenomenon—of optimism. If the assertabilist program could be sustained, it would show that the outgrowth is not necessary.

arithmetic cannot be captured by any effective deductive system. Under the present assumptions, it follows from the unsolvability of the halting problem that the extension of "provable sentence" is not recursively enumerable, and thus cannot be formalized.[22]

Thus much, of course, is not surprising. However, the situation concerning the epistemic tractability of the extension of "provable by the ideal constructor" is much worse than this. It can be shown that, under optimism, "provable sentence" of arithmetic has virtually all of the formal properties of classical truth. Indeed, it has the same extension as classical truth. Literal analogues of the Tarski T-sentences,

$$\Phi \text{ is provable iff } \Phi,$$

can be derived from optimism and a reflection principle,

$$\text{if } \Phi \text{ is provable, then } \Phi.$$

Thus, the extension of "provable sentence" of arithmetic cannot occur anywhere in the Kleene hierarchy. According to optimism, the extension of "provable sentence" of arithmetic, analysis, set theory, and the like, is just as complex as the extension of "true sentence" in the corresponding realist theory. In effect, we would be saddled with one of the extreme views noted above, the one in which "procedure" is rendered "function," classically conceived, and "provable" amounts to "true." In that case, Heyting semantics is little more than a rewording of classical, truth-valued semantics.

In sum, then, by moving to Heyting semantics and adopting optimism, we are attributing to the ideal constructor the ability to detect classical truth. Recall that the promise of assertabilism was that there may be gains on the epistemic front. Assertability was supposed to be more tractable than classical truth, because assertion (unlike truth) is a human endeavor. We do not gain anything, however, when we idealize assertability so much that all and only classical truths are assertable by the ideal constructor.

All this is grist for Dummett's revisionism. If "provable sentence" is to be epistemically tractable, or less intractable than classical truth, then optimism seems to be ruled out. If we opt for assertabilism and pessimism, then there is a principled reason to demur at the law of excluded middle and, thus, to demand revisions to mathematics.[23] But one person's modus ponens is another's modus tollens. Things do not look good for assertabilism as a quasi-realism, an antirealist philosophy underlying mathematics *as practiced.*

---

22. For each natural number n, let Φ_n be the statement that the Turing machine with Gödel number n halts when given n as input; and let K be $\{n \mid \Phi_n\}$. Let T be $\{\Phi_n \mid n \in K\} \cup \{\neg\Phi_n \mid n \notin K\}$. Presumably, optimism entails that every member of T is "provable" by the ideal constructor. If there were an effective formal system that codified all and only the "provable" sentences in this language, then the set K would be recursive—which it is not. Gödel draws essentially the same conclusion. He argues that either there are absolutely undecidable propositions—optimism is false—or the human mind is not a machine. This, together with Gödel's optimism, yields his antimechanism.

23. McCarty [1987] provides another, more natural route to the demurral of excluded middle via Heyting semantics. The idea is to accept Church's thesis and identify a procedure with a Turing machine.

Perhaps we should not close the door on this antirealist, nonrevisionist program. It seemed like such a good idea. The challenge is to develop notions of "use," "assertability," and "rationally compelling argument," that can simultaneously support a notion of "provability" that is every bit as rich and noneffective as classical "truth" yet more tractable than "truth" on the epistemic front.

8 Practice, Logic, and Metaphysics

Previous sections dealt with a relationship between mathematical practice and mathematical philosophy. The purpose of this final section is to relate these considerations to the more global perspective concerning mathematics and its philosophy. I further articulate the conclusions of chapter 1.

Once again, the practice of a given branch of mathematics uses either a static language or a dynamic language, many appearing to prefer the latter. By itself, dynamic language is more or less neutral on matters of semantics, ontology, and at least to some extent, logic. Semantics, ontology, and logic can be related to the moves and knowledge attributed to the ideal constructor. Static language is conducive to classical logic and standard Tarskian semantics, and realism in ontology is consonant with classical logic and Tarskian semantics. With dynamic systems, this realism "sanctions" the sorts of constructions sufficient to regard the systems as if they were static. Thus, we have a relationship of mutual support between realism and static language.

The question here is whether realism is the chicken or the egg vis-à-vis the practice. Some hold that realism is what ultimately *grounds* the mathematics—philosophy-first. This view, however, is at odds with the historical development of mathematics. If anything, the roles go the other way around. To extend the metaphor, realism (in ontology and truth-value) is an egg that grew into a chicken. More literally, the sorts of constructions and inferences sanctioned by realism were originally accepted on more or less internal mathematical grounds, in some cases after a struggle. In such contexts, classical logic and Tarskian semantics were formulated, developed, and established. This pragmatic orientation is called "working realism" in chapter 2. Once established, however, the principles and semantics take on a normative role and affect the continued development of mathematical practice, thus producing an advanced, normative stage of working realism.[24]

Today, working realism is well established and has proven fruitful, and I do not wish to suggest that there was or is anything irrational or unjustified in adopting it, either by an individual or by the mathematical community itself. Quite the contrary. I do claim, though, that working realism was not a foregone conclusion in the history of mathematics. It is not a priori. Moreover, there is no guarantee that the mathematical community will always be, or should always be, dominated by working

24. Dictionary writing provides an analogy. At first, the enterprise is descriptive of the accepted norms of spelling and word usage. Once a successful dictionary is in place, however, it can come to *constitute* the norms in the sense that the dictionary becomes the source of correct spelling and word usage.

realism. Conceivably, it may come to be rational to abandon or severely alter the methodological framework.

To be sure, there is no way to substantiate this claim concerning the possible future directions of the mathematical community, unless of course it happens—but even that might not settle the matter. In the futuristic scenario, the philosophical opposition can claim that the mathematical community made a big mistake. I might add that I am not at all anxious for any changes in the current logical/semantical/ontological framework. It is too insightful, fruitful, and compelling. Like Hilbert, I am not looking to leave Cantor's paradise. Nevertheless, the first part of my claim—that the framework of working realism was not a foregone conclusion—does receive some support from mathematical development, both on an individual level and on the community level. Gödel [1964] is correct that the axioms of set theory "force themselves on us as true," but the axioms do not force themselves on a first (or second, or third) reading. For virtually any branch of mathematics, the psychological necessity of the axioms and inferences, and the feeling that the axioms are natural and inevitable, comes only at the end of a process of training in which the student acquires considerable practice working within the given system, under the guidance of teachers.

The fact that commonly accepted axioms and inferences are not psychologically compelling at first is also reflected in mathematics courses at the college level. For many students, even bright ones, the most difficult aspect of mid-level mathematics courses is the emphasis on understanding and writing proofs. Moreover, many professors find teaching proofs to be among their most challenging pedagogical tasks. One explanation is that the students do not find even the most basic inferences as natural and compelling as trained mathematicians do.

In the historical development of mathematics, we find periods of ambivalence followed by near unanimity and certainty. One generation of mathematicians struggles and the next generation of philosophers takes the winners to be expounding self-evident, a priori truths. Today, nearly all mathematicians are thoroughly comfortable with such items as the axiom of choice, impredicative definition, and excluded middle. Most, in fact, are not always aware of the situations in which these items are invoked, and, when they are, they do not seem to care whether the uses are necessary. According to the current framework of working realism, this is as it should be. The most fundamental assumptions are those least likely to be questioned. Yet one does not have to do an extensive review of the relevant histories to remember that the items in question were not always so obvious and natural, and they were accepted by the mathematical community only after long and sometimes hard-fought struggles. Some examples were discussed in chapter 1. Classical mathematics as construed here—excluded middle, impredicative definition, completed infinite processes—is scarcely a century old (see chapters 1 and 5).

Much of Lakatos's historical work concerns disagreements among mathematicians. In a collection of notes entitled "What does a mathematical proof prove?" (published posthumously in Lakatos [1978]), he makes a distinction between the preformal development and the formal development of a branch of mathematics. A preformal proof is *not* best seen as a formal proof with missing steps. Rather, it is a heuristic, explanatory, and *exploratory* device. The examples he develops (e.g., in [1976]) indicate that the preformal proofs are dynamic and often rely on vague analogies with what may be called nonmathematical "construction." His favorite example, attributed to Cauchy, relates the dynamic properties of polyhedra to those of physical objects made from rubber sheets (see section 2). Lakatos shows that during the historical development, it is just not clear what a polyhedron is. The development highlights unclarities and vagueness in the preformal notion. According to Lakatos, the subsequent process of "proofs and refutations" ended with the official definitions we have today—the "formal development" of solid geometry.

Putting the Popperian themes aside, Lakatos is quite correct that the preformal development of a branch of mathematics differs from its formal articulation in significant ways. During the preformal period, mathematical objects are not rigorously defined (in terms of other rigorously defined objects). Instead, objects are characterized by ostension in terms of paradigm cases. In Lakatos's prime example, a "polyhedron" is little more than "something like a cube or tetrahedron" or, perhaps, "something like a polygon, but three dimensional." The series of proofs and refutations deal with borderline cases of this vague characterization. Other examples are not hard to find. In ancient Greece, a "magnitude" is "something like a line segment or an area"; more recently, a "complex number" is "something like a real number, but closed under roots"; and a "continuous function" is "a function whose graph is smooth." Even more recently, a "computable function" is one "that can be executed in something like mechanical fashion." Among other things, the preformal development serves to clear ambiguity and vagueness from the intuitive notions.

In structuralistic terms, I propose that the preformal period of a branch of mathematics is a time when the community is attempting to formulate and study a structure (or group of structures) *before* the exact relations are completely articulated. At the outset, it is neither established nor determined just what the structure is and, in some cases, just what the logic is. Sometimes, the structure is indicated by vague reference to (more or less understood) structures *of* physical reality—the structures of moving objects, stretchable rubber sheets, knots, collections of objects, pencils and paper, and so forth. In these cases (at least), preformal mathematics is not all that different from natural science (see chapter 8 and Shapiro [1983a], [1993]). In other cases, the structures are identified by vague abstraction from other mathematical structures. For example, mathematicians began to ask what happens if the square-root operation is extended to negative numbers, or how multiplication might work on infinite cardinal numbers. In short, preformal proofs are *inherently informal*. The reason, I would suggest, is that it does not become determinate until later what is to count as acceptable construction and, thus, as correct inference. The structure, including the extensions of the concepts and the logic, has not been fully articulated yet.

Preformal structures are usually dynamic. I suggest that the preformal period is best characterized as one of experimenting with the possibilities of various moves within the unarticulated vague structures—that is, moves attributed to the ideal constructor. What are the consequences of allowing such a move? What are the presuppositions of such a move? Which kinds of moves are incompatible with which? One focus of the preformal period is the codification of aspects of the paradigm cases that are essential to the structure whose articulation is evolving.

The ultimate answers *are determined* in part by the nature of the physical or mathematical reality under study (as articulated at a given point), but the answers also serve *to determine* and fix the mathematical structure or structures at hand. In short, the successful moves serve to articulate and (eventually) to rigorously formulate these structures. On the dynamic account, the successful moves determine the relevant logic, semantics and ontology. Ultimately, the successful moves determine the framework in which formalization is to take place and, to reiterate one of the present conclusions, the successful moves determine the logic.

Following Kitcher [1983], I propose that there is no universal, a priori notion of success by which all constructions or attempted idealizations are to be judged. Rather, adjudications are made in response to internal pressures within the evolving mathematical community in response to problems and goals, previously taken to be important. This, in turn, is at least partly determined by the reasons the structures are being articulated and the role they play in the intellectual community at large.

One should not conclude from this, of course, that all practice is sacrosanct. The practice of a given branch can be and has been successfully modified, both on pragmatic and on philosophical grounds. The philosophy that dominates a given period grows from the practice of the period (or significant components of it) as an attempt at explanation and codification. Once established, however, methodological principles, like those of working realism, can take on a normative role and can help to sanction, reject, and even suggest new practices. That is, philosophy is sometimes part of the practice. The criteria relate to the goals of practice taken to be important by a given community.

For philosophers, the overriding concern is to account for the historical data and to put it in perspective. The philosophies that are traditionally called Platonism hold that the ontological and semantic views are primary. Accordingly, philosophical realism (in ontology and truth-value) was a foregone conclusion, and this philosophy is what ultimately grounds the semantics and logic of the various branches of mathematics. Temporary ambivalence, both individual and community, is the result of (some) mathematicians' failure to grasp and adopt the correct philosophy. Our Platonist might concede that the productivity of the relevant inferences played a role in convincing mathematicians and philosophers of the correct philosophy but will add that the inferences, semantics, and so forth, are productive *only because* they are sanctioned by the correct philosophy. The order of subjective justification is not the true epistemic order. The Platonist may even admit that the decisions about the various inferences could have turned out otherwise but would add that if that happened, the mathematical community would have made a serious, wrongheaded move. Again, philosophy-first.

I do not know how to refute this view of the situation. At best, I have only contrasted it with (the sketch of) another view. I endorse Kitcher's [1983] metaphor of an "evolutionary" epistemology. The analogy with biology is apt. To modify Kuhn's well-known remarks concerning science, it is obvious to all—Platonist, intuitionist, formalist, and so on—that both mathematical philosophy and mathematical practice have changed over time. This includes subject matter, standards of rigor, and even logic. Analogously, most people interested in biology do admit that species have changed over time. The theory of evolution, however, regards species change as primarily due to pressures within a changing environment, including the development of other species. Evolutionary mathematical epistemology regards mathematical practice, and mathematical philosophy, as evolving primarily in response to developments within the mathematical community and, to some extent, the intellectual, economic, and political communities at large.

7

Modality, Structure, Ontology

1 Modality

Modal notions, in various guises, were invoked throughout this book.[1] Earlier, one of the formulations of structuralism (chapter 3) involved a notion of "possible structure," as an alternative to both *ante rem* structuralism and eliminative structuralism over a domain of *abstracta* (see section 3 of this chapter). Along these lines, some antirealists have expressed sympathy with structuralism, but they quickly noted that it is possible structures and not actual structures that must be invoked. The purpose of this penultimate chapter is to see what, if anything, is added (or subtracted) with this qualifier. I make good on the claim that there are no real gains on the epistemic front with maneuvers like these. The deep interaction between modality and structure makes an interesting case study for philosophy.

Contemporary philosophy of mathematics and, to some extent, metaphysics and epistemology contain two schools of thought. One group favors a comprehensive mathematical theory, such as set theory. Members of this school hold that set-theoretic assertions, say, should be taken more or less literally. Philosophically, this is the double realism of ontology and truth-value. Members of this group include Quine (e.g., [1981]); Putnam [1971]; Resnik [1981]; Maddy [1990]; and me.

Some, but not all, of these philosophers are skeptical of modality or, at any rate, do not think modal notions can play central roles in philosophical explanations. Quine's influential views are near the extreme: "We should be within our rights in holding that no part of science is definitive so long as it remains couched in idioms of . . . modality. . . . Such good uses as the modalities are ever put to can probably be served in ways that are clearer and already known" ([1986, 33–34]). For many philosophers, the *logical* notions of possibility, necessity, and consequence are notable exceptions to this skepticism. This may be because the logical modalities have

1. Many of the ideas and arguments of this chapter were first published in Shapiro [1993a].

(presumably) been *reduced* to set theory, via model theory—despite the prevailing antireductionist spirit. For example, a sentence or proposition Φ is said to be logically possible if and only if there is a model that satisfies Φ. To paraphrase Quine, model theory via set theory is presumably clearer and already known and can serve the purposes of the logical modalities.

Skepticism toward modality is not universal within the first school. Philosophers use powerful set-theoretic tools in a rich and extensive literature on various modal notions. There is more than a family resemblance between model theory and these possible-worlds analyses of alethic and epistemic modalities. The possible-worlds semantics is an attempt to extend the success of model theory to other modal notions.

The second school in the philosophy of mathematics is the opposite of the first. Its members are skeptical of set theory and other mathematical disciplines, at least if they are taken at face value, and members of this school accept at least some forms of modality. To be precise, these philosophers are less skeptical of modality than they are of set theory. So they set out to reformulate mathematics, or something to play the role of mathematics, in modal terms. Prominent members of this school include Field [1980], [1984]; Hellman [1989]; Chihara [1990]; Dummett [1973], [1977]; Blackburn [1984]; and possibly Putnam [1967].

The purpose of these enterprises is to see how far we can go in mathematics and science without asserting the existence of abstract objects like sets (or categories or structures). The members of this school are thus *antirealists in ontology*. Many of the authors, however, are realists in *truth-value*, holding that statements of mathematics, or close surrogates, have objective truth conditions that hold or fail independently of the conventions, minds, and so on, of mathematicians. In many of the systems, for example, there is a version of the continuum hypothesis that is objectively true or objectively false, although the statement is deductively independent of the axioms and other basic statements of its system. As with the first school, even if no one knows whether the continuum hypothesis is true or false, it is true or false just the same.

Of course, there are also philosophers who are antirealists in truth-value, some of whom are out to revise mathematics on philosophical grounds (see chapters 1 and 6). My present focus, however, is on ontological antirealist programs that agree with the first, realist school that the bulk of contemporary mathematics (or a surrogate) is to be taken seriously, and that most of the assertions of mathematicians are objective and nonvacuously true, when properly understood. In short, my concern in this chapter is with the combination of realism in truth-value and antirealism in ontology. For the programs under study, it is a basic datum that the language of mathematics is understood—somehow—by mathematicians and scientists, and mathematicians get it right most of the time.

Many of the antirealist programs invoke modality in order to reduce ontology. It must be emphasized that the modal notions used by the members of the second school are *primitive*. In part, this means that the modal terminology is taken at face value, not reduced to something nonmodal. The members of the school differ among themselves concerning what the modal primitive is, but none of them envisions a realm of possible worlds, a realm of possibilia, or a model theory to explicate the modality.

This stance is a counterpart to the other orientation that takes mathematics at face value, without envisioning a reduction to something nonmathematical. For the first school, set theory is primitive; for the second modality is primitive. Thus, we are invited to consider a trade-off between a vast ontology and an increased ideology.

Some philosophers favor the reduction of ontology on grounds of economy. Less ontology is better. But one can ponder the purpose of this economy. What, after all, is at stake with the general issue of ontology and ontological commitment? What are we doing when we assert or deny the existence of mathematical objects? More important, how are we to evaluate our competing claims? The prevailing criterion of ontological commitment, due to Quine (e.g., [1969]), is that the ontology of a theory is the range of its bound variables. This criterion is straightforward only if the ideology is held fixed. The programs under study here violate this constraint, and things get obscure at this point. With the ideology in flux, one can wonder whether the Quinean criterion tracks any useful property. Each side proposes to eliminate or reduce the most basic notions *used* by the other. Typical remarks about judging the ontology/ideology trade-offs on some sort of "holistic" grounds are not very helpful unless these grounds are elaborated and defended. The interest of this question goes well beyond structuralism in the philosophy of mathematics, but as shall be seen (section 5), structuralists have something to say about the resolution, at least for the philosophy of mathematics.

The (ontological) antirealist programs under study here do have promising beginnings. The epistemology of the various modal notions may be more tractable than an epistemology of abstract objects like sets. With fewer things to know about, there is less to accommodate, and less can go wrong. However, like the conclusion concerning assertabilism, in chapter 6, the contention here is that the promise is not delivered. The epistemological problems with the antirealist programs are just as serious and troublesome as those of realism in ontology. Moreover, the problems are, in a sense, *equivalent* to those of realism. No gain is posted—and sometimes there is a loss. Perhaps the source of the epistemological difficulties lies in the richness of mathematics itself.

I do not attempt a comprehensive account of all current philosophies of mathematics that fit our scheme of realism in truth-value and antirealism in ontology. The projects are all rich and complex, and there is not sufficient space to give them the attention they deserve. The next, and longest section concerns Hartry Field's modal fictionalism, whereas sections 3 and 4 provide brief accounts of some other programs in the present purview, namely, Hellman [1989]; Chihara [1990]; and Boolos [1984], [1985]. Hellman's account is especially relevant here, because it is a modal structuralism.

In each case, the structure of the argument is the same. I show that there are straightforward, often trivial, translations from the set-theoretic language of the realist to the proposed modal language and vice versa. The translations preserve warranted belief, at least, and probably truth (provided, of course, that both viewpoints are accepted, at least temporarily). Under certain conditions, the regimented languages are definitionally equivalent, in the sense that if one translates a sentence Φ of one language into the other, and then translates the result back into the original language, the end result is equivalent (in the original system) to Φ. The contention is that, because of these translations, neither system has a major epistemological advantage over the other.[2] Any insight that modalists claim for their system can be immediately appropriated by realists and vice versa. Moreover, the epistemological *problems* with realism get "translated" as well. The prima facie intractability of knowledge of abstract objects indicates an intractability concerning knowledge of the modal notions, at least as they are developed in the works in question here. To be sure, the modal notions invoked by our antirealists do have uses in everyday (nonmathematical) language, and competent speakers of the language have some pretheoretic grasp of how they work. By itself, however, this pretheoretic grasp does not support the extensively detailed articulation of the modal notions as they are employed by our antirealists in their explications of mathematics. Our grasp of the detailed articulations of the modal notions is mediated by mathematics, set theory in particular. For example, one item concerns the relationship between model theory and the intuitive notion of logical consequence. Of course, everyone who reasons makes use of logical consequence, and so there is some pretheoretic grasp of this modal notion. How does this pretheoretic notion relate to model-theoretic consequence? The question at hand concerns the extent to which our antirealists are entitled to the hard-won, model-theoretic results.

The truth is that logical consequence and set theory illuminate each other. It is unfair to reject set theory, as our modalists do, and then claim that we have a pretheoretic understanding of the modal notions that, when applied to mathematics, exactly matches the results of the model-theoretic explication. The burden on the antirealist is to show how we come by the detailed articulation.

Where, then, is the burden of proof between the schools? Other things equal, it would be nice to take the languages of mathematics, set theory in particular, literally. Mathematics is, after all, a dignified and vital endeavor, and we would like to think that mathematicians mean what they say and know what they are talking about. This is to take mathematics at face value. Other things equal, it would also be nice to avoid the epistemological problems that seem to dog realism. However, the one thing that everyone agrees on is that other things are not equal.

2 Modal Fictionalism

We have spoken of the program in Field [1980] in several earlier places. The plan of Field's book is to accomplish enough of an eliminativist project to avoid an ontological "commitment" to mathematical entities. Field's goal is to show that science *can* be done without mathematics, albeit in a terribly inconvenient manner. As Field sees it, the role of mathematics is to facilitate inferences from physical premises to physical conclusions, what may be called "nominalistic arguments." His claim is that mathematics is *conservative* over science, in the sense that any nominalistic argument that can be derived with the help of intermediate mathematical statements is

---

2. I am indebted to a referee for suggesting that the point be put this way.

itself logically valid. Thus, the role of mathematics in science is to facilitate the logic. Field points out that conservativeness is not the same thing as truth. So if the fictionalist program succeeds, there is no need to regard the mathematics as literally true. Because mathematics is, in principle, dispensable, its assertions may be regarded as statements about fictional entities, much like what we read in novels. Natural numbers and sets are the same kinds of entities as Oliver Twist and Jean-Luc Picard.

There is, of course, a sizable literature on the semantics of fiction, but, at least prima facie, users of fiction are not committed to a fictionalist ontology. Thus, to belabor the obvious, Field proposes a reduction of ontology. He does hold that statements of mathematics are to be taken literally, at face value, but most of the statements are vacuous. For example, "all natural numbers are prime" comes out true, because, for Field, there are no natural numbers. On Field's view, of course, these vacuous truth-values play no role in determining the acceptability of mathematics or the role of mathematics in science. Mathematicians are not exhorted to assert the truth. In effect, statements of mathematics might as well have no truth-value (see Hale [1987, chapter 5, n. 1]). Thus, Field is allied with antirealists in truth-value, at least in spirit. But, as we will see, Field does develop a close surrogate for much mathematics and there are objective, nonvacuous truth conditions for the surrogate mathematical assertions.

There are two parts of the fictionalist program. The first is to develop a nominalistic version of each worthwhile science. This is needed because typical scientific assertions and laws invoke mathematical entities. The second part of the program is the conservativeness result. Our fictionalist should show that adding mathematics (and "bridge principles" that connect mathematical and physical terms) to a nominalistic theory does not yield any "new" nominalistic theorems. It would follow that even though mathematics is useful and, in fact, practically necessary, it is theoretically dispensable.

To illustrate the first part of the program, Field develops a nominalistic version of Newtonian gravitational theory in some detail, with an admirable level of rigor. The ontology of Field's mechanics includes a continuum of space-time points and even more space-time regions. Nominalistic space-time has the same size and much of the structure of $\mathbb{R}^4$, the set of quadruples of real numbers.[3] So Field is not a finitist, not by a long shot. Nevertheless, he argues that points and regions are concrete, not abstract entities, and so his nominalistic version of space-time is not mathematics. I will not pause to evaluate this argument here.

Part of Field's surrogate mathematics deals directly with space-time. In the nominalistic theory, there are analogues of the continuum hypothesis and various

3. In the theory that Field develops, space-time has no preferred frame of reference and no units on which addition, multiplication, etc, are defined. The difference between Field's space-time and $\mathbb{R}^4$ is similar to the contrast between the synthetic geometry of Euclid's *Elements* and contemporary analytic geometry, done in terms of real numbers. In present terms, Field's work highlights a distinction between the structure actually exemplified by space-time and further structure added for theoretical study. See Burgess [1984].

determinacy principles. These nominalistic surrogates are formulated with reference to space-time points and regions only, but they clearly make the same structural statements as the original mathematical prototypes. For Field, the truth-values of the surrogate statements are not vacuous, and the statements are deductively independent of the nominalistic and mathematical theories combined. The surrogate statements are just as objective as the realist thinks the corresponding mathematical principles are.

The role of nonsurrogate mathematics in Field's nominalistic theory is straightforward. By invoking the resources of set theory, a physicist can define *sets* of points and then apply geometric and physical axioms to the *regions* of space-time constituted by these sets—to be precise, to the regions that would be constituted by these sets if the sets existed. This works because the mathematical theories have much of the structure that Field attributes to physical reality. Mathematicians know how to use set theory to study structure.

So we turn to conservativeness. Let P be a variable for collections of assertions of nominalistic physics, q for single nominalistic assertions, and S for collections of mathematical assertions and bridge principles connecting mathematics and physics. Then, to paraphrase Field, mathematics is conservative over the physics if and only if whenever q is a consequence of the combined $P + S$, q is a consequence of P alone. However, there are at least two articulations of this, depending on how the notion of "consequence" is understood. The mathematics is *deductively conservative* if, whenever q can be deduced from $P + S$, q can be deduced from P alone; and the mathematics is *semantically conservative* if q is true in all models of $P + S$ only if q is true in all models of P.

Because mathematical entities apparently have no interaction with concrete ones, one would expect mathematics to be *semantically* conservative over any nominalistic theory, at least if we ignore the bridge principles. Field gives a proof of semantic conservativeness for his system (including the bridge principles) from plausible premises. The proof is carried out in a mathematical metatheory. So far, so good. However, for the fictionalist program, deductive conservativeness seems to be the more relevant notion. If deductive conservativeness holds, then any conclusion that mathematical physicists *derive* can be obtained by their nominalist counterparts, albeit in a more long-winded fashion. If nominalists know that mathematics is deductively conservative, then they can use mathematics with a clear conscience, knowing it is dispensable in principle.

Field's first exposition of the nominalistic physics is second-order. First-order variables range over points, and monadic predicate variables range over regions, which are taken to be mereological sums of points. Like standard second-order logic, the logic of Field's system is inherently incomplete (see Shapiro [1991, chapter 4]). In another work (Shapiro [1983]), I show that mathematics is not deductively conservative over this physics. One can formulate a Gödel sentence G in the nominalistic language, such that G is not provable in the physics but is provable in the set theory (via the bridge principles). The mathematics is, however, semantically conservative, for the reasons that Field gives. In particular, the sentence G is a model-theoretic consequence of the physics alone. This is not surprising, however, because Field's theory of space-time (i.e., the geometry) is categorical and so is semantically complete: *every* truth about the geometry of space-time is a model-theoretic consequence of the nominalistic physics. In short, the mathematics allows us to derive (semantic) consequences of the physical theory that cannot be deduced without the mathematics.[4] In some cases, mathematics is necessary to see what the consequences are. Is this tolerable to a nominalist?

The proper conclusion here is that mathematics sheds light on the model theory that underlies the semantics of our scientific languages. Mathematics illuminates the relevant relations among structures and thus helps us to determine what is a logical (semantic/model-theoretic) consequence of what. The notion of logical consequence is accurately explicated in terms of model theory. In this way mathematics *indirectly* sheds light on the physical world, against Field. It allows us to see what the consequences of our various theories are.

The problem here is that a fictionalist cannot accept this moral, at least not literally. How does the nominalist determine what follows from what? In particular, what sense can a fictionalist make of the various notions of consequence? Surely, the *semantic*, model-theoretic notion of consequence runs counter to the antirealist theme. If one has doubts about positing a realm of abstract objects as the subject matter of real analysis (and part of the subject matter of physics), one will certainly have qualms about the set theory that lies behind the semantic notions. How can talk about fictional entities be essential for determining consequences about the real, concrete world? This is a trivial point, of course, and Field accepts it. He also points out that even the notion of *deductive* consequence is prima facie troublesome for a nominalist, because the straightforward definition of this notion has variables that range over *deductions*, which are at least prima facie abstract objects. Indeed, if the statement of deductive conservativeness is understood in terms of actual, concrete deduction *tokens*, it is clearly false in virtually every nontrivial case. There are written deductions of physical conclusions from physical and mathematical premises for which no one has bothered to write a deduction of the same conclusion from the same physical premises alone. For an antirealist, an ontology of *abstract* deductions, or of possible physical deductions, is about as dubious as an ontology of numbers.

In sum, the fictionalist thesis of conservativeness is stated in terms of *logical consequence*, and the two best historical explications of consequence are unavailable to our fictionalist. One resolution would be to take the notion of logical consequence as *primitive*. Field's own solution is similar. He takes the notion of *logical possibility* as primitive. So, of course, possibility is not to be explicated in terms of models or deductions. It is not to be explicated at all. Logical consequence can then be defined: q is a consequence of P if the conjunction of the items in P with the negation of q is not possible.[5]

---

4. See Shapiro [1983] and Field's reply [1985] for an account of the first-order versions of Field's system.

5. Presumably, logical possibility is a property of sentences or propositions. Again, our fictionalist cannot restrict attention to actual, concrete sentence tokens, and postulating the existence of types would undermine nominalism. The move to modality helps here, at least for single sentences (and finite sets

Field argues that to account for the applicability of mathematics, we need to assume little more than the possibility of the mathematics, not its truth. If his plan works, then the fictionalist can safely maintain that everything that actually exists is concrete, even while enjoying the benefits of mathematics. In effect, we reduce our ontology by envisioning that we insert possibility operators into our regimented language. I say "envisioning" because no one is to use the nominalistic physics. We are to take comfort in the mere existence, or should I say the mere possibility, of the nominalistic physics.

Let us take stock. Traditional realism has a large ontology, consisting of numbers, sets, and the like. Logical possibility is explicated in set-theoretic terms, either as consistency or, more likely, satisfiability (or both). Field proposes a trade-off between this ontology and an unexplicated ideology. With fictionalism, abstract objects—like sets and numbers—are exchanged for a primitive notion of logical possibility (and uncountably many points and regions). The adjudication between realism and fictionalism, if there is to be one, is to be made on some sort of holistic grounds, yet to be specified.

Field concedes that there are puzzling philosophical questions concerning possibility, epistemological questions in particular. However, he argues that, in this regard, fictionalism is no worse than traditional realism. For example, the fictionalist must believe (with good reason) that the axioms for set theory are jointly possible, whereas the realist must believe (with good reason) that they are true. In either case, we are at a loss to figure out what counts as a "good reason."

This is a fair point. The epistemic problems with possibility are seen as trade-offs of epistemic problems with realism. But trade-offs like this can go both ways. Is the fictionalist any *better* off than the realist? As noted above, it is widely agreed that there are major epistemic problems with realism, whether or not my efforts in chapter 4 bear fruit. The promise of fictionalism is that an epistemology of the concrete may be more tractable than an epistemology of the concrete and abstract. However, we now see that the fictionalist requires an epistemology of the actual and possible, to be secured without the benefits of model theory. It is not clear that the fictionalist has made any progress. Is fictionalism any more promising than the enterprise of realism? I would suggest that fictionalism faces direct counterparts of every epistemic problem with realism, and thus the proposed trade-off undermines the fictionalist program.

To help see this, notice that there are direct, trivial translations from our fictionalist's language into the realist's and vice versa. The translation from realism to fictionalism is surely a matter of inserting modal operators in appropriate places, and perhaps conjoining axioms. For example, the translation of a sentence Φ would be of the form $\diamond(\chi \,\&\, \Phi^*)$, where χ is a conjunction of axioms from the background

---

of sentences). Putting matters of use and mention aside, our fictionalist speaks of *possible* tokens. The problem is aggravated when, in the later [1991], Field invokes substitutional quantifiers as a device for *infinitary* quantification (in order to accommodate consequences of theories that are not finitely axiomatizable).

mathematical theory and Φ^* is a variant of Φ, with possibility operators inserted. This is the main idea behind the program. For the converse, fictionalism to realism, simply replace possibility with satisfiability. A subformula of the form "Ψ is possible" becomes "Ψ is satisfiable."

With any proposed reduction or elimination like this, there is a question of what the translations are supposed to "preserve." I do not claim that the present "translations" preserve meaning. According to Field, the translations also do not preserve ontological commitment—that is the point. If anyone thinks that these facts disqualify the proposed "translations" as translations, another term, like "transformation" or "function," can be used.

It is not easy to say just what the translations do preserve. For a first approximation, let us speak loosely in a joint perspective of the realist and the modalist/fictionalist. Let us temporarily assume that it is meaningful and nonvacuous to speak of "truth" in *both* the realist and the modal languages. Of course, this is not a tolerable end position. The joint perspective takes on both the realist's ontology and the fictionalist's unexplicated modality, and so it assumes the shortcomings of both philosophies and can claim the benefits of neither of them.[6] Still, on the joint perspective, the claims behind the translations are these: (1) For every sentence Φ in the realist's language, if Φ is true, then the translation Φ' of Φ into the fictionalist language is true. (2) For every sentence Ψ in the language of fictionalism, Ψ is true if and only if the translation Ψ' of Ψ into the language of realism is true. Although Field does not speak of "translation," claim (1) is of a piece with the main thrust of his [1980], [1984], and [1991] (ignoring subtle modifications). Claim (2) follows from the (presumed) explication of logical possibility and consequence in terms of model theory.

The reason claim (1) is not a biconditional is that if the realist's theory is not semantically complete, there may be a sentence Φ such that Φ' and $(\neg\Phi)'$ are both true.[7] However, under certain conditions, the translations represent a tighter connection between the theories, what is sometimes called "definitional equivalence." Let Φ be a sentence in the realist's language, and let Φ' be the result of translating Φ into the modal language of the fictionalist. Let Φ'' be the result of translating Φ' back into the realist's language. Is Φ'' equivalent to Φ in the realist's theory? It is if (i) the background mathematics (of the realist) is finitely axiomatized, so that the same χ will do for all the translations; (ii) the background mathematics is semantically complete; and (iii) there is a reflection principle, to the effect that if a sentence Φ is true then Φ is satisfiable.[8] For the other direction, translate a sentence Ψ from the modal lan-

6. I am indebted to Ty Lightner and Ben Theis for pressing this point.

7. Notice that $(\neg\Phi)'$ is $\diamond(\chi \mathbin{\&} (\neg\Phi)^*)$ and $\neg(\Phi')$ is $\neg\diamond(\chi \mathbin{\&} \Phi^*)$. So $(\neg\Phi)'$ may not be equivalent to $\neg(\Phi')$, and $(\neg\Phi')$ may not be contrary to Φ'.

8. In first-order Zermelo-Fraenkel set theory, conditions (i) and (ii) fail, but (iii) holds. In second-order set theory, condition (i) holds. Condition (ii) is independent, unless an axiom that limits the size of the hierarchy is added. Condition (iii) is equivalent to an axiom of infinity, implying the existence of so-called small large cardinals (see Shapiro [1987] or [1991, chapter 6]).

guage of the fictionalist into the realist's language, and then translate the result back to the modal language. Is the resulting sentence, Ψ'', equivalent to Ψ in the modal theory? Yes, if (i) the realist's axioms are jointly possible and (ii) there is a modal-reflection principle to the effect that if Φ is possible then it is possible that Φ is satisfiable. I return to definitional equivalence in section 5.

So much for the unappealing joint perspective of the realist and the modal fictionalist. What can we make of the translations without it? We speak in terms of the two parties separately and what would be reasonable for each to hold concerning the translations. Let us think in terms of a neutral observer who has learned both frameworks and is trying to decide between them.

Notice, first, that the definitional equivalence of the frameworks, if it holds, can be appreciated by either party. The equivalences are formulated in the separate languages and theories: $\Phi \equiv \Phi''$ in the realist's framework and $\Psi \equiv \Psi''$ in the fictionalist's. So each of the disputants has some reason to see the frameworks as equivalent. Communication between them will be smooth.

Whether the theories are definitionally equivalent or not, we can reformulate the central claims (1) and (2) in terms of belief or warranted belief. That much is available to the neutral observer and to the combatants themselves. The new theses are: (1*) The neutral observer sees that if the realist finds good reason to believe Φ, according to his own lights, then the fictionalist can find good reason to believe Φ' in her framework. (2*) The neutral observer sees that the fictionalist has a good reason to believe Ψ if and only if the realist has a good reason to believe Ψ'. Again, claim (1*) is of a piece with Field's own work, and he surely would accept this way of putting the point. Claim (2*) goes along with the presumed model-theoretic explication of the logical modalities.

Claim (1*) might be disputed by the realist. Suppose, for example, that our realist believes that he has an ability to intuit truths about the set-theoretic hierarchy, via a communion with it. He might claim that he just *sees* that Φ is true, without being able to give any other justification. The modal fictionalist has the option of taking the realist's intuition as evidence for Φ', but it is more natural for her to simply deny that the realist has a good reason to believe Φ—if not by the realist's own lights, then by what those lights should be. I presume that the neutral observer will agree that invoking an unexplained intuition is an epistemic nonstarter—and of course most realists also do not invoke unexplained intuition. In justifying their assertions, they speak of proof, definition, heuristic argument, and so on, just as mathematicians do, and those epistemic tools do "translate."

The fact that there are such smooth and straightforward transformations between the ontologically rich language of the realist and the supposedly austere language of the fictionalist indicates that neither of them can claim a major epistemological advantage over the other. There is a positive side to the equivalence and a negative aspect. The fictionalist can (and does) argue that unless the realist invokes some sort of nonnatural direct apprehension of the mathematical realm, any sort of evidence he can cite for believing in a mathematical assertion Φ can be invoked by the fictionalist in defense of belief in Φ', the joint possibility of Φ with some axioms. However, given the other translation—fictionalism to realism—the reverse applies as well.

Any insight claimed by the fictionalist can be appropriated by the realist. If the systems are definitionally equivalent, the standoff is even tighter. Every sentence of either language is equivalent (in its own framework) to a translation of a sentence from the other.

The negative side of the equivalence is even more troublesome. My contention is that with the translations, the major philosophical *problems* with realism get "translated" as well. For example, let Φ be a mathematical assertion. A fundamental problem for the realist is "How do we know Φ?" or, to be philosophically explicit, "How do we know that Φ holds of the highly abstract ontology?" For us structuralists, the problem is "How do we know that Φ holds in the indicated structure?" Under the translation, these questions become "How do we know that Φ is possible?" or "How do we know that the conjunction of Φ with axioms of the background theory is possible?" There is, after all, no acclaimed epistemology for *either* language. In short, it is hard to see how adding primitive possibility operators to the formation of epistemic problems can make them any more tractable, and, consequently, it is hard to see how the fictionalist has made any progress over the realist on the sticky epistemic problems.

Against this, Field claims that the beliefs needed to support fictionalism are weaker than the beliefs needed to support realism. Again, let Φ be a statement of mathematics. The realist must believe Φ, whereas the nominalist need only believe that the conjunction of Φ and the mathematical axioms is possible. The latter is, presumably, weaker. However, our realist and our fictionalist do not understand the locution "is possible" the same way, and there is no neutral framework in which to state and evaluate this claim of relative strength. For the fictionalist, the possibility operator is a primitive and "Φ (and the axioms) is possible" is indeed weaker than Φ itself. However, under the translation, "Φ (and the axioms) is possible" becomes something like "Φ (and the axioms) is satisfiable." This *may* be weaker than Φ, but in the cases at hand, it is not much weaker (and sometimes it may be stronger).[9] It is hard to see how our neutral observer can adjudicate the claim of relative strength. Recall that on the present structuralist view, mathematical knowledge is mediated by knowledge of what is coherent.

So far, I have described the situation between realism and fictionalism as a balanced standoff, but there is an important asymmetry in the positions. The fictionalist proposes that we reject the realist's system—regarding it as no more than a work of

9. If the languages are second-order and "satisfaction" is given its usual reading, in terms of models whose domains are sets, then "Φ is satisfiable" may be stronger than Φ. Let Z be the axioms of second-order Zermelo-Fraenkel set theory. Then "Z is satisfiable" implies the existence of an inaccessible cardinal, whereas Z itself does not. This observation represents a complication in the aforementioned translations between the fictionalist's and the realist's languages. Suppose, for example, that there are no inaccessible cardinals. Field correctly holds that the axioms of second-order set theory are jointly possible, but under the supposition, there is no *set* that is a model of those axioms. That is, the axioms would not be jointly satisfiable. To avoid this problem, the realist can either assume a reflection principle or else define satisfaction in terms of (possibly proper) classes. See Shapiro [1987], [1991, chapter 6].

fiction—whereas the realist accommodates the modal language, via the model-theoretic explications. The important point is that once the model-theoretic explication is in place, the realist has a lot to say about logical possibility and logical consequence. It is a gross understatement to point out that mathematical logic has been a productive enterprise. The challenge to the fictionalist is to show how she can use the results of model theory, as they bear on the *primitive* modal notion.

There is an interesting irony in Field's development of fictionalism. As noted, he shows in rigorous detail that under reasonable assumptions, mathematics is semantically conservative over nominalistic physics. Field's proof uses a substantial amount of set theory, in the spirit of reductio. By *assuming* the mathematics, Field shows that, in a sense, the mathematics is not necessary for science. For the sake of this argument, he also assumes the correctness of the model-theoretic explication of logical consequence. Indeed, the very statement that mathematics is not necessary for science is rendered in set-theoretic terms. The problem, once again, is that fictionalists cannot accept the explication, for they deny the background set theory.

For illustration, consider an imaginary philosopher who is initially skeptical about mathematics, because of its apparent ontology of *abstracta*. However, she sees that mathematics appears to be necessary for science. At this stage, there is no clear explication of this "necessity thesis," but it seems correct. Science is typically done with mathematics and, as far as she knows, there is no alternative. So she reluctantly accepts mathematics and the set-theoretic foundation of it. On doing so (and after taking a course or two), she learns and comes to accept the prevailing model-theoretic explication of logical possibility and logical consequence. Field shows that she now has the resources to further articulate the thesis that mathematics is necessary for science. She can give a precise rendering of this necessity thesis in model-theoretic terms, with all the precision of mathematics. Moreover, after reading Field [1980], our imaginary philosopher can establish that the thesis is *not true*. In short, she can show that mathematics is not necessary for science, once she understands the necessity in model-theoretic terms. Supposedly, this discovery undermines her initial, reluctant acceptance of mathematics. However, if mathematics is rejected, then so is the explication of consequence and, with that, the "refutation" of the necessity of mathematics in science. That is, if the philosopher does reject the mathematics, she is no longer in a position to see the falsity of the thesis that mathematics is necessary for science. Indeed, she no longer has a clear idea of what the necessity amounts to.

The upshot of this is that even if Field's analysis is correct, it is not clear that one should reject mathematics. With mathematics, we get more than some help with science. We also get a reasonable explication of the logical notions, the very items used to articulate the thesis that mathematics is necessary for science. The fictionalist alternative is to accept a primitive notion of possibility, and we are left with very little idea of what this notion comes to.

In later works ([1984], [1991]), Field addresses this situation. His idea is that fictionalists can use model theory to shed light on (logical) possibility for the same reason they can use mathematics in science. The strategy is to restart the fictionalist program at the level of the metatheory, what is sometimes called "metalogic." One uses mathematics to explicate possibility and consequence and then argues that *this* use of mathematics is conservative; so it, too, can be regarded as fictional. A relevant bridge principle is that a sentence is possible just in case it is satisfiable. Of course, one can then wonder what *this* conservativeness amounts to. Presumably, it invokes a notion of consequence as well. If Field's strategy is pursued, then the foregoing standoff between fictionalism and realism becomes a regress, as we ascend through a hierarchy of metalanguages.[10]

3 Modal Structuralism

Chapter 3 contains a modal alternative to *ante rem* structuralism. Instead of speaking of the existence of structures, we speak of the possible existence of them. Hellman [1989] is a detailed and insightful articulation of this program. In that chapter, I indicated that the modal program is equivalent to both the *ante rem* program and an eliminative program applied to an ontology of *abstracta*. The equivalence is delimited here, in much the same terms as that of the previous section.

Unlike Field, Hellman is an avowed realist in truth-value. He holds that the statements of mathematics, properly interpreted, have objective, nonvacuous truth conditions that hold or fail independently of the minds, conventions, and so on, of mathematicians. Hellman and I have much in common. We are both advocates of second-order logic (see Shapiro [1991]), and we are both structuralists. Hellman contrasts his modal structuralism with a more traditional view, called "objects platonism," which holds that arithmetic, say, is about a particular collection of objects, *the* natural numbers. Here he also parts company with the *ante rem* structuralist.

Hellman's plan is to characterize a version of structuralism that does not presuppose the existence of structures, a structuralism without structures. He notes, first, that second-order languages allow categorical characterizations of important mathematical structures. For example, there is a second-order formula $\mathrm{AR}(X, s)$, which has only the variables X and s free (and has no nonlogical terminology), that asserts that $\langle X, s\rangle$ is a model of the natural numbers. Like Field [1980], the ideology of Hellman [1989] consists of operators for logical possibility and logical necessity. Hellman claims that the modal part of the language allows him to avoid ontological commitments. Instead of asserting that *there is* a natural-number structure, he asserts only that it is possible for there to be such a structure:

$$\Diamond\exists X\exists s(\mathrm{AR}(X, s)).$$

So, like Field's modal fictionalism, ontology is somehow reduced by envisioning that we have inserted boxes and diamonds into our regimented language. Moreover, like the Field program, it is not clear that there is a gain over realism. Even if we eschew an ontology of possible objects, we surely need an epistemology of possible objects—just as the traditional realist needs an epistemology of actual abstract ob-

10. See Hale [1987, 256–257, n. 12] for a criticism of Field's strategy.

jects. How do we know what is possible? No reason is given to think that the modal route is any more tractable than the realist one—and there is reason to think it is not more tractable.

Notice that, once again, there are direct, straightforward translations from the realist's language to the language of the modal structuralist and vice versa. Hellman himself provides the translation from realism to modal structuralism, in careful detail. For the other direction, replace possibility with satisfiability and necessity with logical truth, as before. Hellman [1989, 36–37] suggests this translation as well. In the present case, there is enough detail to establish that the translations provide an *equivalence* of sorts between the systems (see section 5). Start with a sentence Φ of, say, arithmetic. Translate Φ into a sentence Φ' in modal structuralism, and then translate Φ' into Φ'', back in the language of the realist. Because of the reference to satisfaction, Φ'' is a sentence of *set theory*, not arithmetic. So, in a sense, the stakes are raised. Nevertheless, in set theory, it can be shown that Φ'' is equivalent to Φ; to be precise, Φ'' is equivalent to the result of restricting the quantifiers of Φ to ω and replacing the arithmetic terminology with set-theoretic counterparts. The equivalence occurs because Hellman (quite correctly) employs a second-order language, and second-order arithmetic is categorical.[11]

A similar exercise can be carried out in the other direction. Translate a given sentence Ψ of the modal structuralist language into the language of realism and then translate the result back to the language of modal structuralism, producing Ψ''. The equivalence of Ψ and Ψ'' depends on a modal-reflection principle that if a sentence is satisfiable then it is possible.

Because of the translations, the earlier remarks against Field apply here. If one accepts the perspectives of both the realist and the modal structuralist, or if one adopts a neutral perspective between the frameworks, then any insight claimed for an advocate of one system is immediately available to the other. Moreover, the traditional epistemological problems with realism also get transferred to the modal structural system. Finally, because Hellman is out to drop the realist perspective, it is not clear why he is entitled to the traditional, model-theoretic explications of the modal operators of logical necessity and logical possibility. For example, the usual way of establishing that a sentence is possible is to show that it has a model. For Hellman, presumably, a sentence is possible if it might have a model (or if, possibly, it has a model). It is not clear what this move brings us.[12]

---

11. If one starts this process with a sentence of set theory, things are not quite this straightforward, because second-order set theory is not categorical. The equivalence between a sentence and its "double translation" depends on a reflection principle, yielding so-called small, large cardinals. See Shapiro [1987], and Shapiro [1991, chapter 6]. See also Hellman [1989, 71].

12. In set theory, there are some important advantages to the Hellman project. Ordinary second-order set theory has quantifiers that range over proper classes, which are collections of sets that are not themselves sets (see Shapiro [1991, chapter 5], and chapter 6 of this volume). Admittedly, this is a rather inelegant way to avoid Russell's paradox. Because Hellman does not envision a realm of possibilia nor a realm of possible worlds, his system has no unrestricted quantifiers, and there are no (absolutely) proper classes. In the metaphor of possible worlds, the classes of one world are sets in another. This, I

4 Other Bargains

This section concerns Chihara's neoconstructivism and Boolos's plural quantification, two other specimens of the trend to reduce ontology by increasing ideology. Like Hellman, and unlike Field, the authors discussed here are avowed realists in truth-value.

Though the approach is different, Chihara [1990] suffers a fate similar to that of Hellman [1989]. Chihara's innovation is another modal primitive, a "constructibility quantifier." Syntactically, it behaves like an ordinary quantifier: if Φ is a formula and x a certain type of variable, then $(Cx)\Phi$ is a formula, which is to be read "it is possible to construct an x such that Φ." It turns out that the semantics and the proof theory of constructibility quantifiers are also much the same as those of ordinary, existential quantifiers.

According to Chihara, ordinary quantifiers ($\forall$, $\exists$) mark ontological commitment, but constructibility quantifiers do not. Common sense supports this—to the extent that the notion of ontological commitment is part of common sense. If I say, for example, that it is possible to construct two new arenas in Columbus, I am not asserting the existence of such arenas, nor of possible arenas, nor of a possible world that contains such arenas. I make a statement only about what it is possible to do—here in this world.

The formal language developed in Chihara [1990] has infinitely many sorts. Level 0 variables range over ordinary (presumably material) objects. These variables can be bound by standard existential and universal quantifiers and not by constructibility quantifiers. Level 1 variables range over *open sentences* satisfied by ordinary objects. For each $n > 1$, level n variables range over open sentences satisfied by the items in the range of level n–1 variables. All of the open-sentence variables can be bound by constructibility quantifiers and not by ordinary quantifiers. The symbol for the semantic notion of satisfaction is another primitive of the system.

Chihara (of [1990]) may be called a "neoconstructivist," because he is not out to revise mathematics. Like Hellman, Chihara's program is an attempt to have the bulk of contemporary mathematics come out true on an ontologically austere reading. He goes to some length to guarantee extensionality and other classical features. For example, like simple type theory, but unlike his earlier [1973], the system here has impredicative comprehension principles at each level: if $\Phi(x)$ is any formula in which the level 1 variable x occurs free, there is an axiom that asserts that it is possible to construct an open sentence (of level 2) that is satisfied by all and only the level 1 open sentences that would satisfy Φ (if only they existed). The formula Φ may contain bound variables of any level. It is this feature of the system that allows classical, nonconstructive mathematics to be developed in it, and it is this feature that puts the system at odds with intuitionism and predicativism.

---

think, is a tidy feature of the modal structural system. One cost, however, is that with Hellman's desire to avoid intensional entities, he is unable to formulate a general notion of isomorphism, one that would apply to structures in "different worlds" (see my review in *Nous* 27 (1993): 522–525). A similar modal set theory is developed in Parsons [1983, essays 10, 11].

All told, then, Chihara's [1990] system is quite similar to that of ordinary, simple type theory or, equivalently, the theory of a noncumulative set-theoretic hierarchy up to level ω, with urelements. Chihara shows how to translate any sentence of type theory into his system: replace variables over sets of type *n* with level *n* variables over open sentences, replace membership (or predication) with satisfaction, and replace quantifiers over variables of level 1 and above with constructibility quantifiers.

With admirable attention to detail, Chihara goes on to develop arithmetic, analysis, functional analysis, and so on, in pretty much the same way as they are developed in simple type theory. For example, there is a theorem that it is possible to construct an open sentence (of level 2) that is satisfied by all and only the level 1 open sentences that are satisfied by exactly four objects. This open sentence plays the role of the number 4 in the account of arithmetic. To be specific, the system of (possible) open sentences that corresponds to the natural numbers exemplifies the structure of the natural numbers. Once again, that is the point.[13]

In Chihara's system, there is a sentence equivalent to the following:

> For every level 3 open sentence α, if α can be satisfied by uncountably many surrogate natural-number open sentences, then α can be satisfied by continuum-many such open sentences.

Such a sentence is obtained by translating a type-theoretic version of the continuum hypothesis into Chihara's language. This sentence is fully objective, and, of course, it is independent of the axioms of the system.

As with Field [1980] and Hellman [1989], there are routine translations between Chihara's neoconstructive language and a formal language for ordinary, simple type theory, which we may call the "language of realism." This time, the translations are immediate. As above, Chihara himself provides the translation from type theory to his system. For the other direction, replace the variables that range over level *n* open sentences with variables that range over type *n* objects, replace the symbol for satisfaction with that of membership (or predication), and, of course, replace constructibility quantifiers with existential quantifiers. That is, to translate from Chihara's language to the standard one, just undo Chihara's own translation. The two systems are definitionally equivalent. So my earlier remarks against modal fictionalism and modal structuralism apply here, as well. From this perspective, Chihara's system is a notational variant of simple type theory. An advocate of one of the systems cannot claim an epistemological advantage over an advocate of the other. Does the ontological advantage matter, and is the ontology offset by added ideology? I return to this matter in section 5.

13. Chihara provides a second, rather interesting and insightful development of real analysis in terms of the possibility of constructing objects of various lengths. Each axiom of real analysis, including the completeness principle, corresponds to a statement of which constructions are possible. Again, the relevant system of possible open sentences exemplifies the real-number structure. The relationship between Chihara's real analysis and the standard one is similar to that between Euclid's geometry and Hilbert's.

Another point against Field [1980] and Hellman [1989] applies here. Is Chihara entitled to the now-standard set-theoretic explications of what sorts of sentences can be constructed? Clearly, constructibility quantifiers are established parts of ordinary language, and competent speakers do have some grasp of how they work. For example, we speak with ease about what someone could have had for breakfast and what a toddler can construct with Lego building blocks. Moreover, there is no acclaimed semantic analysis of these locutions, model-theoretic or otherwise, as they occur in *ordinary language*. These observations seem to underlie Chihara's proposal that the locutions are "primitive." We use them without a fancy model-theoretic analysis. Chihara proposes to use the same locutions to give an ontologically clean rendering of type theory. However, I do not think the everyday notions in question are sufficiently determinate and precise to extend them to mathematics in the way Chihara does, at least not without further argument. For Chihara's construction to proceed to level 4 open sentences, for example, we need to be shown how the structure of the powerset of the continuum is implicit in the ordinary uses of constructibility quantifiers. I submit that we understand how the constructibility locutions work *in Chihara's application to mathematics* only because we have a well-developed theory of logical possibility and satisfiability. Again, this well-developed explication is not primitive or pretheoretic. The articulated understanding is rooted in set theory, via model theory. Set theory is the source of the precision we bring to the modal locutions. Thus, this (partial) account of the modal locutions is not available to an antirealist, not without further ado. In short, we need some reason to believe that, when applied to the reconstruction of mathematics, constructibility quantifiers work exactly as the model-theoretic semantics entails that they do.

Along these lines, both Chihara and Hellman occasionally invoke a possible-worlds semantics, but they regard it as a heuristic, not to be taken literally. Neither of them believes that possible worlds exist. The role of the semantics is to help the reader grasp the intended logic of the formulas and to see what does and does not follow from what. As Chihara puts it, "[T]his whole possible worlds structure is an elaborate myth, useful for clarifying and explaining the modal notions, but a myth just the same" (p. 60). The operative phrase here is "clarifying and explaining." If the structure really is a myth, then I do not see how it explains anything. One cannot, for example, cite a story about Zeus to explain a perplexing feature of the natural world, such as the weather. To be sure, the very notion of explanation represents a deep and complex philosophical problem, one that I do not attempt to resolve here. In correspondence, Chihara pointed out that fictional entities, such as frictionless surfaces and point masses, do occur as part of ordinary scientific explanations of physical phenomena (see, for example, Cartwright [1983] and van Fraassen [1980]).[14] It is not clear that the same goes for philosophy, and philosophical explanation—whatever that is. Intuitively, to explain something is to give a reason for it or, according to Webster's *New twentieth century unabridged dictionary*, to clear from obscurity

14. Chihara also noted that we sometimes tell fictional stories to explain moral principles. I take it that the stories help us to see physical or emotional consequences of various actions.

and make intelligible. In everyday life, a purported explanation must usually be true, or approximately true, in order to successfully explain. I take it that frictionless surfaces, and the like, are parts of respectable scientific explanations of physical phenomena because they approximate actual physical objects. If they did not, then it is hard to see any explaining. It is not clear what, if anything, possible worlds approximate vis-à-vis the modal notions at hand—if not the possibilities themselves or the structure of the possibilities. In any case, I propose that the burden here is on Chihara to tell us more about the modal notions and more about explanation before we can see how possible worlds can clear the modal notions of obscurity and make them intelligible.

To be sure, a myth of possible worlds can *clarify* some things. For example, in modal logic, the structure of possible worlds is often used as a tool to determine which inferences are acceptable and which are not. To show that a given inferences in a modal language is invalid, we describe a system of possible worlds in which the premises hold and the conclusion fails. Chihara and Hellman both make effective use of this tool. From the antirealist perspective, however, the structure does not explain or justify any inferences. How can it? Moreover, the fact that a myth of possible worlds happens to produce the correct modal logic is itself a phenomenon in need of explanation. That is, from the antirealist perspective, the success of possible worlds *adds* to the philosophical puzzle.

I might add that there is something ironic about Chihara's (mythical) possible-worlds semantics. As he describes the system, in a given model, the variable ranges are dutifully distributed into different possible worlds, but this fact plays *no role* in the definition of satisfaction in the modal system he develops. Every object (and every predicate) is rigid and world-bound, and each constructibility quantifier ranges over all objects (of appropriate type) in all worlds. Thus, the worlds themselves do not get used anywhere—just the objects in them. In short, the (mythical) semantics that Chihara develops is just ordinary *model theory*, with some irrelevant structure thrown in. Thus, my contention is that the constructibility quantifier has virtually the same semantics as the ordinary, existential quantifier.

Chihara [1993] begins to address the situation concerning possible worlds. His plan is to paraphrase (or translate) talk of possible worlds into a modal idiom. For example, "there is a possible world in which" becomes "there is a way the world could have been, such that, had the world been that way." On the surface, this locution has a quantifier, "there is a way," and the locution exhibits anaphoric reference, "had the world been that way," which usually signals a bound variable. Chihara, of course, does not understand the terminology this way. He does not believe that "ways the world might be" are objects. Presumably, we have another modal *primitive* at work here, one that is not to be explicated via model theory. I submit that this ploy only pushes the problem up one level—to the philosophical metalanguage. How is the new locution to be understood? What is its logic?

Our last ontology-for-ideology ploy does not involve a modal primitive. George Boolos [1984], [1985] has proposed an alternate way to understand, or interpret, monadic, second-order languages, promising to overcome objections to second-order logic that are based on its presumed ontological commitments (see Quine [1986]).

According to standard semantics (see Shapiro [1991, chapter 3]), a monadic, second-order existential quantifier can be read "there is a class" or "there is a property," in which case, it seems, the locution invokes classes or properties. Against this, Boolos suggests that the quantifier be understood as a plural quantifier, like "there are objects" or "there are people." The following, for example, is sometimes called the Geach–Kaplan sentence:

Some critics admire only one another.

It has a (more or less) straightforward, second-order reading, taking the class of critics as the domain of discourse:[15]

$$\exists X(\exists x Xx \ \& \ \forall x \forall y((Xx \ \& \ Axy) \rightarrow (x \neq y \ \& \ Xy))).$$

According to standard semantics, this formula would correspond to "there is a nonempty *class* X of critics such that for any x in X and any critic y, if x admires y, then $x \neq y$ and y is in X." Notice that this analysis implies the existence of a *class* of critics, whereas the original "Some critics admire only one another" apparently does not.

Natural languages, like English, do allow the plural construction and, in particular, they contain the plural quantifier. So the informal (natural) metalanguage that we *use* to develop formal semantics also contains this quantifier, just as natural language contains the constructibility quantifiers invoked by Chihara. Boolos's proposal is to employ plural quantifiers to interpret monadic, second-order existential quantifiers. Construed this way, he claims, a monadic, second-order language has no ontological commitments beyond those of its first-order sublanguage. To shore up this claim, Boolos [1985] develops a rigorous, model-theoretic semantics along these lines. The metalanguage has second-order quantifiers—read as plural quantifiers—but no terminology for sets (other than what may be in the first-order part of the language).

Although Boolos himself is not an antirealist concerning most mathematical objects, the move to plural quantifiers has been invoked by others to reduce ontology. For example, Lewis [1991], [1993] invokes the plural construction to develop an eliminative structuralism. The idea is to provide second-order characterizations of some structures and thus gain the expressive resources of second-order logic, while maintaining that the second-order variables do not add any ontological commitment.

The situation here is analogous to the previous, modal cases. Here, too, there are straightforward translations between the standard metalanguage, with classes and no plurals, and the classless language with plural quantifiers. Boolos himself provides the translations. As above, I do not see what our antirealist purchases with the ideology. Epistemic qualms about second-order variables become epistemic qualms about plural quantifiers. Moreover, like the situation with Chihara's neoconstructivism, it

15. The given formula holds if there is a single critic who does not admire anybody. This, of course, is not intended by the original "Some critics admire only one another." The problem is that plural quantifiers, like "some critics," imply (or implicate) that there are at least two. This feature can be accommodated in the formalism. Just replace the subformula $\exists x Xx$ with $\exists x \exists y(Xx \ \& \ Xy \ \& \ x \neq y)$.

is not clear a priori that plural quantifiers of ordinary language are sufficiently determinate and precise to be used exactly as second-order quantifiers with standard, model-theoretic semantics.

Resnik [1988a] argues against the Boolos program, suggesting that plural quantifiers of natural language should themselves be understood in terms of classes (or sets). Resnik and Boolos both acknowledge that this sort of dispute leads to a standoff or a regress. Anything that either side says can be reinterpreted by the other, perhaps via the translation. This applies also to the typical epistemic problems with realism and second-order logic, and proposed solutions to the problems. The issue of second-order languages turns on whether we have a serviceable grasp of second-order variables and quantifiers, taken at face value, sufficient for use in foundational studies. With the Boolos program, the issue concerns whether we have a serviceable grasp of plural quantifiers. Boolos claims that we do, citing the prevalence of plurals in ordinary language. Resnik seems to claim that we do not, suggesting that whatever understanding we do have of plural quantifiers is mediated by our understanding of sets and classes (see Shapiro [1991, chapter 9]).

In favor of Resnik, it might be noted that plurals *in general* seem to be rather complex, and there is no consensus among linguists concerning how they are to be understood (see, for example, Landman [1989]). But Boolos does not invoke the full range of plural nouns, only plural *quantifiers*. These are understood reasonably well, about as well as monadic, second-order quantifiers. Resnik, of course, would reiterate that even this is mediated by (first-order) set theory. Once again, I think we have lost our bearings.

5 What Is a Structuralist to Make of All This?

In the preceding tour, we are asked to ponder the legitimacy of our *pretheoretic* grasp of a plural quantifier, a constructibility quantifier, and an operator for logical possibility; then we are to contrast this with ontology or what has been called "ontological commitment." We are not given much guidance in evaluating the trade-offs, or even on what the game is. What are we trying to do in cutting down ontology or cutting down ideology? In this concluding section, I will add some perspective to the question and, hopefully, shed light on the enterprise.

According to the prevailing criterion of ontology and ontological commitment (e.g., Quine [1969]), one's ontology consists of whatever lies in the range of the bound variables in one's envisioned regimented language. The slogan is "to be is to be the value of a variable." Phrases like "you have to quantify over" have become standard, professional philosophical jargon. For his part, Quine does not regard the ontology-via-bound-variables doctrine to be a deep one. It is just an observation that the existential quantifier is a gloss on the ordinary word for existence.

Despite complications like ontological relativity and the inscrutability of reference, the prevailing criterion of ontology is reasonably clear for Quine and his followers, because they insist on an austere ideology. The regimented language is to be first-order, with no modal terminology. Comparing two such systems is more or less straightforward. For example, one might pit a language with variables that range over sets only against a language with variables that range over space-time points, physical objects, and sets.

In contrast, the programs just considered all invite us to expand the ideology of the envisioned, regimented language well beyond Quinean limitations. That is why we lose our bearings. Our antirealists are quite correct that Quine's restrictions on ideology are too severe, and, in any case, the restrictions are not universally accepted on the contemporary philosophical scene. Many philosophers, myself included, pay serious attention to modality and higher-order languages, and we have made significant use of them in philosophical theorizing.

This raises several questions. If we are to maintain some version of the ontology-via-bound-variables doctrine, then what limits should be placed on the ideology to be used? And, once again, how are trade-offs to be adjudicated? If we are to jettison the ontology-via-bound-variables doctrine, then what, if anything, should replace it?

There is one straightforward response from the antirealist camps. For the many philosophers who have not followed Quine in being skeptical of modality, it is not a choice between ideology and ontology. Rather, it is a choice between ontology/ideology and ideology alone, an ideology that both sides accept anyway. In the case of Field [1980], for example, the choice is between a concrete ontology/ideology and an abstract ("platonist") ontology *plus the same ideology* and the same concrete ontology. Clearly, the argument concludes, ideology alone or ideology plus the concrete is preferred.

The rejoinder to this argument goes to the heart of the matter. In philosophy, Occam's razor should not be wielded blindly. The cutter should show that there is some philosophical gain in the proposed reduction of ontology or ideology. The austere theory should explain more or have a more tractable epistemology—something more tangible than cutting for the sake of cutting. As noted earlier, the programs under study here do have promising beginnings on the epistemic front. The most baffling problem with realism is to show how we can know anything substantial about a realm of *abstracta*. An epistemology of the possible, or of the constructible, or of plural quantification, may be more plausible, especially because we manage to invoke such items in everyday conversation (never mind that we also manage to invoke mathematics every day). To recapitulate the conclusions of previous sections, however, there is no real gain in cutting for the cases at hand. The various translations indicate that, when applied to mathematics, the problems with realism have direct counterparts in the reduced theories. Inserting boxes and diamonds into formulas or changing the quantifiers does not, by itself, add epistemic tractability.

The position of the antirealist is even less tenable. If there is to be no philosophical gain, then there should at least be no loss in the reduced theory (to be precise, no loss that is not compensated for by gains in other areas). The antirealist needs this much just to claim a standoff with the realist. In the cases at hand, however, mathematics has come to play a central role in the standard scientific, modal, and linguistic systems. We cannot get by just as well without the *abstracta*. For example, much of what we know about logical modality is obtained by the connection with set theory, via model theory. That knowledge must be preserved in any reduction, but, as far as I can tell, it is lost in the cases at hand. Consider, for example, the following modal assertions:

First-order set theory together with the assertion that there are exactly 124 infinite cardinals less than the continuum is possible.
First-order set theory, together with the assertion that every set is constructible, is possible.
If there is a huge cardinal, then it is possible that there is a supercompact cardinal.

These assertions follow from the standard accounts of logical possibility via set theory. The various theories are possible because they have models. I presume that Field and Hellman would also accept these modal assertions. Otherwise, there is a net epistemic loss in adopting their theories. Dropping mathematics would cost us our ability to negotiate the modalities. However, it is hard to see what grounds our antirealists would use to support the modal assertions, given that they do not believe in models. One can launch a similar attack on Chihara and those who wish to use the Boolos interpretation of second-order logic to reduce ontology, by citing examples in which model theory is used to determine what can be constructed or examples in which it is used to clarify the second-order consequence relation.

At this point, Field and Hellman might reply that they can obtain the modal assertions by using a *modalized* version of set theory (see, for example, Field [1991]). Presumably, the modalized set theory is justified on the same (or analogous) grounds that the realist uses in defense of ordinary set theory, whatever those grounds may be. But this ploy only pushes the problem back one level, and I do not see what it buys.

Notice, incidentally, that because Chihara [1990] only recapitulates simple type theory, ω-order logic, the appeal to modalized set theory is not available.[16] As for plural quantifiers and second-order languages, Shapiro [1991] contains a lengthy defense of the Quinean thesis that there is a fair amount of mathematics underlying second-order logic (but, contra Quine, this is not a defect of second-order logic).

Rather than rely on the model-theoretic explications of the invoked ideology, the antirealist programs under study invite us to consider adopting modal or plural terminology construed as "primitive." What is the force of this word? The proponents claim that the terminology is found in ordinary language, the informal framework in which we do philosophical, logical, and metamathematical work. This is surely correct. Ordinary language is full of possibility talk, constructibility talk, and plural quantification. Our ontological antirealists are not making anything up—they use what they find. Nevertheless, a pressing question concerns how this terminology is understood. What makes the locutions primitive? For one thing, our proponents do not explicate the terminology. They just use it, without apology, as part of the background language—just as the realist uses mathematical terminology, without apology, in model-theoretic semantics. To be sure, there is no acclaimed model-theoretic analysis or reduction of nonlogical modality, nor is there a rigorous analysis of constructibility quantifiers and plural constructions as they appear in ordinary language. Where analyses have been attempted, they are fraught with controversy. Despite this, we do manage to use the notions. As argued earlier, however, our everyday modal

16. Nolt [1978] is a development of full set theory along lines similar to those of Chihara [1990].

and plural notions, by themselves, are too vague to support the detailed applications to surrogates of set theory or type theory, as envisioned by our ontological antirealists. Moreover, in practice, our grasp of modal and plural terminology *as applied to mathematics* is mediated by mathematics, set theory in particular. We inherit the language/framework, with the connections to set theory already forged. Surely, our antirealists do not claim that we still have some sort of pretheoretic intuitions of these notions, intuitions that remain uncorrupted or unmodified by set theory.

When beginning students are first told about logical possibility and logical consequence, most of them seem to have some idea of what is meant, but consider how much their initial intuitions differ from our refined ones. The antirealist owes us some account of how we plausibly could come to understand the notions in question (as applied here) *as we in fact do*, independent of our mathematics. Without such an account, it is empty to use a word like "primitive," and, without an account, we cannot give a positive assessment of progress to the antirealist programs or even a judgment that they have achieved a balanced trade-off.[17] As far as I know, the task at hand has yet to be attempted.

Recall Quine's [1981, 9] claim that ontology and ontological commitment are not notions of everyday natural language (see chapter 2 of this volume): "The common man's ontology is vague and untidy. . . . We must . . . recognize that a fenced ontology is just not implicit in ordinary language. The idea of a boundary between being and nonbeing is . . . an idea of technical science in the broad sense. . . . Ontological concern is . . . foreign to lay culture, though an outgrowth of it." I suggest that the same goes for what may be called "ideological concern."

So we need a new tool to assess the ontology/ideology of a philosophical/scientific/mathematical theory. If the ideology is not held fixed, the Quinean ontology-via-bound-variables doctrine fails. The criterion is useless and outright misleading. What we need instead is a criterion for ontology and ideology combined. If we restrict attention to mathematics, structuralism has the resources for this.

As a first approximation, the proposed criterion of ontology/ideology is this: a theory is committed to at least the structure or structures that it invokes and uses. If two theories involve the same structures or if the systems described by them exemplify the same structures, then, at least as far as mathematics goes, their ontologies/ideologies are equivalent.

To apply this criterion, of course, we need an account of identity among structures or, better, a criterion for when two systems exemplify the same structure. Resnik (e.g., [1981]) suggests that there is no "fact of the matter" whether two structures are identical or whether two systems exemplify the same structure (see chapter 3 of this volume). If he is right and the structure-identity relation is vague, then there is vagueness in ontology/ideology assessment as well. That is, if it is not determinate whether

---

17. This echoes a point made by John Burgess in a recent conference. It is far-fetched for someone who learned about logical possibility in high-powered courses in mathematical logic, using books like Schoenfield [1967], to go on to claim a "primitive," pretheoretic status for this notion. The same goes for constructibility assertions, if not plural quantifiers.

two structures are identical or whether two systems exemplify the same structure, then it also may not be determinate whether two theories have the same ontology/ideology. This is not problematic, because ontology/ideology comparison may be a matter of degree not an absolute, all-or-nothing affair.

We can still make progress. In previous chapters, two criteria for structure identity were formulated: isomorphism and Resnik's structure equivalence. Isomorphism is too tight a relation for present purposes. Two systems are isomorphic only if they have the same number of relations of the same type. It follows that two systems are isomorphic only if they have the same ideology. To briefly review the definition of structure equivalence, let R be a system and P a subsystem. Then P is a *full subsystem* of R if they have the same objects (i.e., every object of R is an object of P) and if every relation of R can be defined in terms of the relations of P. Let M and N be systems. Then M and N are *structure-equivalent*, or simply *equivalent*, if there is a system R such that M and N are each isomorphic to full subsystems of R.

The thesis here, then, is that if two theories invoke equivalent structures, then they are equivalent on the ontology/ideology scale. In other words, if two theories characterize equivalent structures, then neither one is to be preferred to the other on ontological/ideological grounds alone.

I have not explicitly brought modality into the picture yet. Before turning to that, notice that my conclusions thus far echo a theme of Wilson [1981]. Let ZF be ordinary Zermelo-Fraenkel set theory and let ZP be a "Zermelo-type set theory erected over the natural numbers as urelements." According to the original Quinean criterion, ZF and ZP have different ontologies. The former is committed to sets only, whereas the latter has both sets and numbers (not to mention sets of numbers, etc.). Wilson, quite correctly, notes that "many (or most) mathematicians would probably demur, arguing that the ontology of ZF actually does include the numbers, etc. because it includes an ω-sequence and methods for building the needed sets from it" (pp. 413–414). In present terms, the intended structure of ZF is structure-equivalent to the intended structure of ZP. Moreover, *any* model of ZF is structure-equivalent to a model of ZP and vice versa. With Wilson, I take this as an additional reductio against the Quinean criterion. For a second example, compare a theory that identifies real numbers with Dedekind cuts or Cauchy sequences of rational numbers with a theory that does not. The latter theory keeps the real numbers as separate entities, noting only that they are correlated with Dedekind cuts and Cauchy sequences. Intuitively, the theories are the same, at least as far as ontology goes. As we saw in chapter 5, Dedekind himself thought so. Wilson notes, parenthetically, that "any notion that the reals should not be identified with sets represents as great a misunderstanding of mathematical ontology as the claim that they should" (p. 415). Structure is what matters.

We now have the resources to deal, in part, with Field's *Science without numbers* [1980]. The system of space-time, as described in Field's nominalistic physics, is structure-equivalent to that of $\mathbb{R}^4$ under certain synthetic relations (and is equivalent to the full structure of $\mathbb{R}^4$ if an arbitrary frame of reference and unit of measure is added to the physics or if the language has the ε operator for picking them out; see note 3). The situation here is the mirror image of another conclusion of Wilson [1981], that there is no ontological savings in theories that get rid of unwanted physical items—properties, for example—by replacing them with mathematical constructions, like sets. Field does the opposite of this, removing some unwanted mathematical ontology, replacing it with items in the physical ontology. There is no gain, because the structures are equivalent. Despite Field's argument that space-time is concrete, the regimented theories have the same ontology/ideology.

This still does not accommodate Field's use of more-powerful mathematics hidden behind the modality operators, nor does it accommodate the programs of Hellman, Chihara, and those antirealists who use the Boolos proposal.

As noted at the outset of this chapter, there is a trend to understand modality in terms of abstract objects. The model-theoretic explication of logical possibility is a case in point. Let us call *Tarski's thesis* the statement that a set of sentences is logically possible if and only if it is satisfiable. The various possible-worlds semantics are similar explications of modality in terms of ontology. Chapter 6 concerns other, more-straightforward cases, in which the possible existence of some objects is structure-equivalent to the actual existence of abstract objects.

Another example is string theory. Strings are abstract objects that represent *possible* tokens. Each possible token corresponds to a string, and each string represents a class of possible tokens. Call this the *string thesis*. The same goes for possible deduction tokens and abstract deduction types. Notice that an advocate of the string thesis need not claim that the notion of "string" somehow captures the meaning of "possible token." One need not claim that the relevant theory is an analysis of possible string tokens (whatever that would be). The relevant claims are something like claims of extensional equivalence, in the form "every *A* corresponds to a *B* and vice versa." This is all the "reducing" that we want or need.

Another much-discussed example is Church's thesis, which states that recursiveness and Turing computability are extensionally equivalent to computability. As indicated by the suffix "ability," computability is prima facie a modal notion. A function *f* is computable if one *can* compute all of its values or if it is *possible* for a human or machine to compute *f* (ignoring finite limitations on memory and lifetime). Recursiveness and Turing computability, on the other hand, at least appear to be nonmodal. A Turing machine is a set of ordered quadruples with a certain structure, and a function *f* is Turing computable if there is a Turing machine with a given relation to *f*. The relation is defined in terms of sequences of configurations, which are ordered *n*-tuples with a certain structure. No modality here. Similarly, a function *f* is recursive if there is a finite sequence of functions whose last member is *f* and such that each member of the sequence is either one of a certain class of initial functions or bears a certain relation to earlier functions in the sequence. Again, no modality.

The string thesis can hardly be denied, and Tarski's thesis and Church's thesis are widely accepted among mathematicians and logicians today. To be a little more careful, the theses in question would be denied only by a nominalist, who holds that either there are no abstract objects at all (because models, strings, and Turing machines are abstract) or there are not enough tokens to represent every possibility.

All of these theses represent exchanges of modality for ontology. Instead of speaking of what is possible, we speak of what occurs in an abstract but actual mathematical structure. With Church's thesis, for example, the claim is that the possibilities of computation are reflected accurately in a certain arithmetic or set-theoretic structure. This is the direct opposite of the trend of the school of philosophers represented here by Field, Hellman, and Chihara. These thinkers try to eliminate ontology by invoking modality. In response, I claim that the *same structures* are involved in each exchange, and the theme of this book is that, as far as mathematics is concerned, structure is all that matters. As Putnam [1975, 70] notes in a different context, "[m]athematics has . . . got rid of *possibility* by simply assuming that, up to isomorphism anyway, all possibilities are simultaneously *actual*—actual, that is, in the universe of 'sets.'"

The conclusion is that the modal option is structure-equivalent to the ontological one. To sharpen this, we could extend the notion of structure to the modal realm and speak of possible structures and the structure of possible systems. This strategy was pursued in chapter 6, motivated from other quarters. For present purposes, however, it might be better to look directly at theories themselves, and only indirectly at the structures they invoke. Wilson speaks of the "formal strength" of theories. His central notion is "interdefinability" which can be put as follows: "T and T' are *interdefinable* theories if and only if when T and T' are rewritten to share no non-logical vocabulary, a theory T'' exists which represents a definitional extension of both T and T'." For nonmodal theories, at least, there is a straightforward relationship between the interdefinability of theories and the present notion of structure equivalence: If T is interdefinable with T', then for any model of one theory there is an equivalent model of the other.

Wilson proposes that, under certain conditions, interdefinability serves as part of the criteria for having the same (mathematical) ontology:

> The general point is that the ontology of a mathematician is to be determined by the *formal* properties of the structure he postulates. . . . The claim that any two theories meeting the conditions . . . share an ontology, I shall call *structuralism*, since this thesis represents a plausible reading of the jingle "Mathematics is only interested in structure up to isomorphism."
>
> Accepting structuralism provides a different assessment of 'ontological change' . . . than many of us were taught, dazzled as we were by adroit slashings of Occam's razor. . . . [I]f one accepts a theory of a certain *formal strength*, one cannot deny its standard ontology, no matter in what syntactic guise its assertions may appear. (p. 414)

The last passage is the theme of this chapter. Given our broader context, we speak in terms of ontology/ideology, but the point is the same. To encompass our modal and plural cases, the terminology needs to be extended. Again, two theories T and T' are definitionally equivalent if there is a function f_1 from the class of sentences of T into the class of sentences of T', and a function f_2 from the class of sentences of T' into the class of sentences of T, such that (1) f_1 and f_2 both preserve truth (or warranted assertability, or theoremhood) and (2) for any sentence Φ of T, $f_2f_1(\Phi)$ is equivalent to Φ in T, and for any sentence Ψ of T', $f_1f_2(\Psi)$ is equivalent to Ψ in T'. Recall that, under certain conditions, the "translations" of previous sections represent such functions f_1, f_2.

Definitional equivalence is a natural generalization of Wilson's interdefinability. In the nonmodal cases, for example, if *T* is definitionally equivalent to *T*′ and if the translation functions f_1 and f_2 consist of applying explicit definitions to the nonlogical vocabulary of *T* and *T*′, respectively, then *T* is interdefinable with *T*′, and so every model of one is equivalent to a model of the other. Thus, I propose that definitional equivalence serve as a criterion of the formal strength of modal and nonmodal theories and, with Wilson, that this notion be used as an indication that the intended structures, and thus the ontology/ideology of different theories, are the same. If *T* is definitionally equivalent to *T*′, then neither is to be preferred to the other on ontological/ideological grounds.[18]

My conclusion is now concise. I showed in sections 2–4 that the theories developed in Field [1980]; Hellman [1989]; Chihara [1990]; and Boolos [1984], [1985] are each, under certain (plausible) conditions, definitionally equivalent to a standard, "realist" theory. Thus, the intended structure—and the ontology/ideology—of each theory is the same as that of the corresponding realist theory.

The same goes for the various formulations of structuralism itself, from chapter 3 of this volume. *Ante rem* structuralism, eliminative structuralism formulated over a sufficiently large domain of abstract objects, and modal eliminative structuralism are all definitionally equivalent, and neither is to be preferred to any other on ontological/ideological grounds. I hold that the *ante rem* version is more perspicuous and fits mathematical practice more smoothly, but the approaches are equivalent on the crucial ontological/ideological front. They are also indistinguishable on the troublesome epistemological front.

18. When put so baldly, this conclusion is limited to mathematics. Suppose, for example, that a language of ordinary physical objects were definitionally equivalent to one that refers only to sense-data. With this "evidence," a materialist would conclude that there is no need for mental items like sense-data, because those have been "reduced" to material objects. Likewise, an idealist (or phenomenalist) would claim that there is no need for material objects in our ontology. If other things were equal, there would be a standoff, of sorts, at least on the epistemological front. Nevertheless, I would not conclude that the physical and the mental are equivalent on the ontological/ideological front. Part of the difference between this situation and those under discussion here is that, in mathematics, structure (and so theory strength) is all that matters. The next chapter briefly treats extensions of structuralism beyond mathematics.

8

Life Outside Mathematics

Structure and Reality

This concluding chapter is a brief treatment of some extensions of structuralism beyond mathematics, to science and to ordinary discourse. Since at least Benacerraf [1973], one desideratum for philosophy of mathematics is to provide a uniform semantics for mathematical and ordinary discourse. This has guided much of the present book. In the preferred *ante rem* structuralism, the uniform semantics is a truth-valued, model-theoretic framework, one in which mathematical statements are taken at face value. The connection with ordinary or scientific language would be weakened, however, if the target of the semantics were very different in the cases of mathematical, scientific, and ordinary discourse. This may happen if scientific systems, structures, and objects radically differ from their counterparts in mathematics.

One main concern here is a partial account of the applications of mathematics to the sciences. At least some applications consist of incorporating mathematical structures into physical theories, so that physical systems exemplify mathematical structures. What is almost the same thing, in some theories, the structures of physical systems are modeled or described in terms of mathematical structures. The book closes with a brief account of some possible extensions of the structuralist notion of object.

1 Structure and Science—the Problem

To comprehend the nonmathematical universe scientifically, one must master a considerable amount of mathematics.[1] Consider, for example, the amount of mathematics presupposed by virtually any branch of the natural or social sciences. In most universities, mathematics departments are among the largest just because of the use of mathematics throughout the disciplines.[2] At least at the outset, the philosopher

1. Shapiro [1983] is a precursor of some of the ideas presented in this section and the next.

2. In American universities, only English departments are likely to be as large, and this is because English is presupposed even more widely than mathematics is. Both are sometimes called "service disciplines."

must assume that there is a relationship between the subject matter of mathematics and the subject matter of science, and that it is no accident that mathematics applies to material reality. Any philosophy of mathematics or philosophy of science that does not provide an account of this relationship is incomplete at best. The thesis of this book is that the subject matter of mathematics is structure. Thus, the present problem is to delimit the relationship between mathematical structure and the subject matter of science.

A friend once told me that, during an experiment in a physics lab, he noticed a phenomenon that puzzled him. The class was looking at an oscilloscope, and a funny shape kept forming at the end of the screen. When he asked for an explanation, the lab instructor wrote something on the board (probably a differential equation) and said that the phenomenon occurred because a solution has a zero at a particular value. My friend noted that he became even more puzzled that the occurrence of a zero in a function should count as an explanation of a physical event.

This example indicates that much of the theoretical and practical work in science consists of constructing or discovering mathematical models of physical phenomena. Many scientific and engineering problems are tasks of finding the right differential equation, the right formula, or the right function to be associated with a class of phenomena. A scientific explanation of a physical event often amounts to no more than a mathematical description of it—whatever that means.

As mentioned in chapter 7, the philosophical literature on scientific explanation is long, deep, and troubled. Suffice it to note that, strictly speaking, a mathematical structure, description, model, or theory cannot serve as an explanation of a nonmathematical event without some account of the relationship between mathematics per se and scientific reality. Lacking such an account, how can mathematical/scientific explanations succeed in explaining anything? One cannot begin to understand how science contributes to knowledge without some grasp of what mathematical/scientific activity has to do with the reality of which science contributes knowledge.

The problem can occur on several levels. First, one may wonder how it is possible for a particular mathematical fact to serve as an explanation of a particular nonmathematical event. My friend's puzzlement is on this level. How does a zero of a function explain a pattern on an oscilloscope? An adequate response would be a detailed description of the relevant scientific theory that associates a certain class of functions with a class of physical phenomena. The questioner might be advised to take a few courses. A question could then be raised as to what a class of mathematical objects, such as real-valued functions, can have to do with physical phenomena. This takes the query to a different level. Now we wonder about the relevance of the given mathematical/scientific theory as a whole. Why does it work? A possible reply to this second question would be to point out that similar uses of mathematics have an important role in scientific methodology. If the questions persist, we can note the vast success of this methodology in predicting and controlling the world.

This last reply explains why one might engage in mathematical/scientific research, and it provides assurance that the methodology will continue to predict and control, assuming we solve or ignore standard problems with induction. In the spirit of Hume,

I do not wish to question the entire mathematical/scientific enterprise, much less to raise doubts about it (see the rejection of the philosophy-first principle in chapter 1). Surely, there is no more preferred standpoint than the methodology of science. However, the problem of understanding how the enterprise works, in its own terms, is a legitimate philosophical enterprise.

As noted several times previously, a popular argument for realism for mathematics focuses on the connections between mathematics and science (see, for example, Putnam [1971]). One premise is that mathematics is indispensable for science, and another is that the basic principles of science are (more or less) true. From Quinean holism, the argument concludes that mathematics is true as well—realism in truth-value. If we also take the assertions of mathematics at face value, then we are committed to the existence of numbers, sets, and so on, the ontology of mathematics—realism in ontology. However, even if the premises are true and even if the Quine–Putnam indispensability argument is convincing, it is much too cozy to leave things at this stage. To shore up the argument, the realist must provide an account of exactly *how* mathematics is applied in science. The first premise of the argument is itself a mystery. What does the realm of numbers and sets have to do with the physical world studied in science? How can such items shed light on electrons, bridge stability, and market stability? We cannot sustain the conclusion of the Quine–Putnam indispensability argument until we know this. As I concluded in chapter 2, the philosopher should not be content to simply note the apparent indispensability and then draw conclusions that spawn more questions than they answer. By itself, the indispensability argument has the advantages of theft over toil.

Gödel also recognized the importance of the connections with physical reality. In his famous (or infamous) defense of realism, he wrote that a probabilistic "criterion of truth" for a mathematical proposition like the continuum hypothesis (or its negation) is its "fruitfulness in mathematics and . . . *possibly also in physics*" ([1964, 485], my emphasis). Clearly, fruitfulness in physics cannot be a criterion for *mathematical* truth unless the mathematical realm is related somehow to the physical realm. How else can a mathematical proposition be fruitful in physics? Gödel goes on to suggest that "the 'given' underlying mathematics is closely related to the abstract elements contained in our empirical ideas" (p. 484). He adds an intriguing footnote: "[T]here is a close relationship between the concept of set . . . and the categories of pure understanding in Kant's sense. Namely, the function of both is 'synthesis', i.e., the generating of unities out of manifolds" (n. 26). This remark is consonant with the present conclusion that there is no sharp border between the mathematical and the mundane, and that the very notion of "object" is at least partially structural and mathematical.

Until recently, some philosophers and mathematicians might have dismissed the issue of application with a quip that they are concerned only with pure mathematics. Because most mathematics is not directly aimed at understanding the nonmathematical world, it may be possible to account for its activity and goals without even mentioning a relationship with nonmathematical reality. Such remarks seem to presuppose that there is a difference in kind between pure mathematics and what is called "applied mathematics"—mathematics aimed at the nonmathematical world. Applied mathematics would constitute a separate subject and would call for a separate philosophy, if anyone wanted to stoop so low as to work on one.[3]

However, the border between pure and applied mathematics is not sharp. In one sense, the term "applied mathematics" refers to particular branches of mathematics, such as numerical analysis and the theory of differential equations. These branches often form parts of scientific theories and, in some cases, are prerequisites for the study of those theories. This administrative distinction is important to some funding agencies, but philosophically it is artificial. The practice, writings, and textbooks of the various branches show more similarities than differences concerning such basic aspects as aims, techniques, logic, and even subject matter. Moreover, the applied branches are firmly embedded in the pure ones and typically rely on techniques and results from the pure branches.

Although the day-to-day work of the pure mathematician is not directly aimed at understanding nonmathematical reality, the branches of applied mathematics do not have a monopoly on relevance to science. As Nicolas Goodman [1979, 550] put it, "[M]ost branches of mathematics cast light fairly directly on some part of nature. Geometry concerns space. Probability theory teaches us about random processes. Group theory illuminates symmetry. Logic describes rational inference. Many parts of analysis were created to study particular processes and are still indispensable for the study of those processes. . . . It is a practical reality that our best theorems give information about the concrete world." Occasionally, areas of pure mathematics, such as abstract algebra and analysis find unexpected applications. The roads connecting mathematics and science are rich and varied, and traffic goes in both directions (see Polya [1954], [1977] for a wealth of examples).

If we stick to standard formalizations of *any* branch of mathematics—pure or applied—we may conclude that no applications are possible. Official renderings use formal languages in which all nonlogical terminology is mathematical. As such, the mathematical theories have no substantial logical consequences that contain nonlogical terminology (other than logical truths and the like). In general, if there is no overlap between the nonlogical terminologies of two theories, then one can be applied to the other only if certain *bridge principles* are added. The bridge principles connect the terminologies of the two theories.

Once plausible bridge principles are added, virtually every branch of mathematics has some empirical consequences. For example, it follows from Zermelo-Fraenkel set theory (ZFC) that, say, functional analysis is consistent. So far, of course, we are still internal to mathematics. Consistency is a mathematical property of formal deductions. However, the bridge principles connect deduction strings to possible inscriptions written in accordance with certain rules. With these bridge principles, ZFC

3. In unguarded moments, some platonists have taken pride in the separation of the mathematical universe from the mundane world we all live in. To think about connections to the physical world is to soil the lofty enterprise of mathematics. Plato seemed to hold that the material world is the way it is so that we will see its mathematical nature and be led to contemplate the pristine realm of mathematics and then the realm of Forms.

entails that there can be no written deduction of a contradiction in functional analysis. Imagine a computer *C* that is programmed to search through the theorems of an axiomatization of functional analysis looking for a contradiction. If it finds one, the computer prints "oops" and halts. ZFC and the bridge principles predict that *C* will never print "oops" and halt.[4]

A second use of the phrase "applied mathematics" concerns the informal use of quasi-mathematical concepts and methods in science, engineering, and everyday life. This includes such activities as balancing checkbooks and calculating physical constants and tolerances. The methods and concepts in question do not always bear a close resemblance to the methodology of pure mathematics. For example, scientists and engineers often rely on informal curve fitting and an estimation procedures, sometimes derided as "fudge factors." Much of the reasoning is inductive. However, even this "mathematics" is ultimately rooted in pure mathematics. The mathematical concepts involved include natural and real numbers, number-theoretic functions, and related numerical constructions. The estimation procedures are expedients that should ultimately be dispensable in principle or, let us hope, justified by considerations from mainstream mathematics. I conclude that it would be a mistake to adopt a philosophy that divorces even this use of mathematics from the work of the pure mathematician.

2 Application and Structure

First, a pair of disclaimers. I do not claim that the sketch given in this section accounts for every application of mathematics.[5] Indeed, the present account does not illuminate the more interesting and philosophically troubling applications. For example, I have little to say about the use of complex analysis in quantum mechanics. As far as I can determine, there is no satisfactory account of this, unless one considers instrumentalism to be an account. In effect, to adopt instrumentalism and to refuse to wonder why the instrument works is to give up on our problem. In light of the growing contingent of physicists (and philosophers of physics) who have adopted an instrumentalist philosophy for quantum mechanics, perhaps this case is intractable.

I also have little to say about the uncanny ability of mathematicians to come up with structures, concepts, and disciplines that find unexpected application in science.

4. This result is an instance of the fact that powerful mathematical theories have consequences for the nonsolvability of Diophantine equations. Incidentally, I do not mean to claim that we have the makings here of an empirical test for ZFC (whatever that might mean). Suppose that someone did try to implement *C* on a supercomputer, and suppose that after a year of running, it printed "oops" and quit. It would follow that either functional analysis is inconsistent, ZFC is inconsistent, one of the bridge principles is false, there is a bug in the software, some component of the hardware malfunctioned, or the person hit a wrong key while trying to implement *C*. Can anyone who has experience with modern computers rule out the last three disjuncts? Would anyone in their right mind find inconsistency to be the most probable explanation of the result?

5. I do not remember if I intended a grandiose claim for the account in Shapiro [1983a], although Chihara [1990, 125–131, 135–142] plausibly interprets me that way. I am indebted to Chihara's careful analysis of my early, somewhat naive attempt.

The following scene has played itself out repeatedly. Mathematicians study a given structure. They extend it to another structure for their own, internal purposes; then the newly defined structure finds application somewhere in science. As Weinberg [1986, 725] put it, "It is positively spooky how the physicist finds the mathematician has been there before him or her." And Feynman [1967, 171]: "I find it quite amazing that it is possible to predict what will happen by mathematics, which is simply following rules which really have nothing to do with the original thing." From the mathematical camp, the same sentiment was echoed by Bourbaki [1950, 231]: "[M]athematics appears . . . as a storehouse of abstract forms—the mathematical structures; and it so happens—without our knowing why—that certain aspects of empirical reality fit themselves into these forms, as if through a kind of preadaption." I do not venture an explanation of this phenomenon.[6]

Returning to the more modest matter at hand, the present partial account of the relationship between mathematics and science begins with the observation that the contents of the nonmathematical universe exhibit underlying mathematical structures in their interrelations and interactions. According to classical mechanics, for example, a mathematical structure much like the inverse-square variation of real numbers is exemplified in the mutual attraction of physical objects. In general, physical laws expressed in mathematical terms can be construed as proposals that a certain mathematically defined structure is exemplified in a particular area of physical reality.

If this account is combined with the preferred *ante rem* perspective on structuralism, then the relationship between mathematics and material reality is, in part, a special case of the ancient problem of the instantiation of universals. Mathematics is to reality as universal is to instantiated particular. As noted in chapter 3, the "universal" here is a pattern or structure, and the "particular" is a system of related objects. More specifically, mathematics is to reality as pattern is to patterned.[7]

The same account is available to one who adopts an in re view of structures, construed over systems of abstract objects (see chapter 3). The idea is that some structures are exemplified by both systems of *abstracta* and systems of *concreta*. A similar account of application is also available to a traditional Platonist. Instead of speaking of structures that underly physical reality, our Platonist could appeal to *isomorphisms* (or structure equivalences) between systems of physical objects and systems of mathematical objects. Gödel [1953, 353–354, n. 44] himself wrote, "[I]t should be noted that the second reality, although completely separated from the first one, nevertheless might help us considerably in knowing the latter, e.g., if in some respects, or as to certain parts, it happened to be similar or isomorphic to it. In fact this would correspond closely to the manner in which mathematics is applied in theoretical physics."

---

6. Steiner [1989] is an extensive and compelling account of this problem.

7. I do not claim that a pattern explains why a system is patterned a certain way. Incidentally, according to Turnbull [1978], Plato himself may have held a structuralism of sorts. Turnbull proposes that Platonic Forms be understood as "principles of structure," with the Form of X being the mathematical structure of X. Turnbull relies on an interpretation of the later Dialogues, including the geometric description of physical reality in the *Timaeus*.

A first attempt to articulate the present thesis would be that science proceeds by discovering exemplifications of mathematical structures among observable physical objects. Some applications do have this form (see Barbut [1970]). For example, Lévi-Strauss [1949] discovered a certain class structure, the Kariera system. A tribe is divided into four classes, and there is a certain function for determining the class of a child from the classes of his or her parents. There is an "identity class" in the sense that when a member of this class mates with any member of the tribe, the offspring are members of the other's class. If two members of the same class mate, then their offspring are of the identity class. The classes and the function exemplify the Klein group, a well-known finite structure (see Barbut [1970, 381]).

This simple example illustrates a central feature of the application of mathematics via structure exemplification (or isomorphism). The properties of a structure apply to any system that exemplifies it. The Klein group has an identity element, and so the Kariera system has an identity class. Because the Klein group is abelian, we see that in one respect, the Kariera system is egalitarian. If the classes of two parents were switched, the class of the children would remain the same. In general, when the scientist learns that the mathematician "has been there before," the scientist can apply everything the mathematician has figured out about the structure to the physical system under study. In a physics textbook, Geroch [1985, 1] expresses this idea (see Chihara [1990, 128]): "What one often tries to do in mathematics is to isolate some given structure for . . . study: what constructions, what results, what definitions, what relationships are available in the presence of a certain mathematical structure . . . ? But this is exactly the sort of thing that can be useful in physics, for, in a given physical application, some particular mathematical structure . . . arises from the physics of the problem. . . . The idea is to isolate mathematical structures . . . to learn what they are and what they can do. Such a body of knowledge, once established, can then be called upon whenever it makes contact with the physics."

The foregoing sketch more or less fits Frege's [1884] account of the applications of arithmetic to ordinary counting (see chapters 4 and 5 of this volume). Although Frege would not put it this way, Hume's principle shows how the natural-number structure is exemplified by equivalence classes of finite extensions, under equinumerosity.

Most applied structures are not as simple as the Klein group or even the natural numbers. Moreover, many of the applications that are that simple are not very useful in the sciences. Knowledge about simple structures can be reacquired each time. Most of the structures actually invoked in science are uncountably infinite. There is a continuum of space-time points and even more regions. We cannot find exemplifications of such structures among observable physical objects or systems of such objects.

Notice also that, when it comes to physical science, the simple account of direct exemplification works only if all of the places of the exemplified structure are filled with physical objects (or systems). Otherwise, the structures are not fully exemplified. However, some mathematical structures find application even if some places of the structure do not correspond to anything physically real. A case in point is complex analysis. The very name "imaginary number" suggests that no straightforward physical interpretation of these mathematical structure positions is forthcoming. Yet complex analysis is useful in physics and engineering. A related phenomenon is the use of a physically uninterpreted theory, such as set theory, to solve problems that are undecidable in a weaker theory, such as arithmetic. As indicated earlier, some of these results have applications in recursive function theory and thus in computability. One would be hard put to find a physical exemplification of the set-theoretic hierarchy, let alone an exemplification that relates to computability.

In sum, my simple account of application must be extended. First off, I return to a recurring theme of this book: modality. Typically, science is not limited to actual systems. Rather, the scientist studies all possible systems of a certain type. The classical physicist does not directly study the attractions of actual physical objects but rather laws governing the attraction of all possible physical objects or, better, laws governing the attractions of idealized counterparts to physical objects.

This point is forcefully defended in Chihara [1990, 6–15], against Quine's insistence on purely extensional, nonmodal languages. Science could not get by on Quinean austerity. The scientist typically studies how physical objects *would behave* under certain idealized conditions—conditions that are never met in practice (see also Cartwright [1983]). Physical laws have implicit or explicit clauses that "other things are equal" or, better, "other things are absent." In reality, other things are never equal, and other things are never absent. Like it or not, modality is always with us, and the scientist invokes modality, despite Quine's admonition to avoid it. The structures of science match actual physical systems only approximately, hopefully to within experimental error.

Recall that the notion of the structure of possible systems was invoked in previous chapters to accommodate mathematical construction, dynamic language, and the use of modality in philosophical programs. I put this notion to work here as well. Another way that mathematics gets applied is that the theorist describes a class of mathematical objects or structures and claims that this class represents the structures of all possible systems of a certain sort. Relations among the objects or structures represent relations among the possible objects. If the claim is (more or less) correct, then theorems about the class of structures will correspond to facts about the possible systems—about what is and what is not possible. Such claims can be tested in practice, as the ideal conditions are approximated.

Here, then, is one way that large infinities enter the physicist's picture—at least large infinities from the perspective of the physicist. Classical mechanics entails that there is at least a continuum of possible configurations of physical objects. There is even a continuum of possible pairs of point masses. We do not have to reify the "possibilities"; we speak of their structure instead.

As mentioned in chapter 7, some psychologists speak in terms much like this when describing the acquisition of psychological concepts. The natural-number structure represents relations among possible collections of objects. As in Hellman [1989], the principle of infinity is replaced with a principle of possible infinity: there is no limit to the size of possible collections of objects. Chomsky seems to hold a structuralist view like this, in that he believes that the human mind exemplifies structures of transformational grammar. These innate systems account for human linguistic performance, which concerns those sentences that can be recognized as grammatical.

Crowell and Fox [1963] is a mathematics book about knots—twisted pieces of rope. In the introduction, they discuss the problem of using mathematics to study these physical objects or, better, the possible manipulations of these physical objects:

> Definition of a Knot: Almost everyone is familiar with the simplest of the common knots, e.g., the overhand knot . . . and the figure-eight knot. . . . A little experimenting with a piece of rope will convince anyone that these two knots are different: one cannot be transformed into the other without . . . "tying" or "untying". Nevertheless, failure to change the figure-eight into the overhand by hours of patient twisting is no proof that it can't be done. The problem that we shall consider is the problem of showing mathematically that these two knots . . . are distinct from one another. Mathematics never proves anything about anything except mathematics, and a piece of rope is a physical object and not a mathematical one. So before worrying about proofs, we must have a mathematical definition of what a knot is. . . . This problem . . . arises whenever one applies mathematics to a physical situation. The definitions should define mathematical objects that approximate the physical objects under consideration as closely as possible. (p. 3)

When Crowell and Fox suggest that some mathematical objects can resemble or "approximate" physical objects like pieces of rope, they clearly do not mean that some mathematical objects are solid, flexible, and flammable. You cannot twist or burn a number, even approximately. Their claim is that possible relationships and interconnections of pieces of rope formed into knots can be described or modeled in the relationships of a mathematical structure, a topological space in this case. Pieces of rope formed into knots exemplify the structures delimited in the book. The purpose of the enterprise is to prove theorems about possible knots, about what can or cannot be done with rope. If Crowell and Fox are correct in their assumptions about the exemplified structures, then it would follow that particular mathematical theorems represent facts about possible knots. In particular, a theorem that one particular topological configuration cannot be transformed into a second by specified mathematical operations and constructions corresponds to a fact that a figure-eight knot cannot be transformed into an overhand knot without untying and retying it.

Turing's [1936] original argument for the Church–Turing thesis also fits this mold, insofar as he regards the computational process to be subject to mathematical investigation. He makes some straightforward claims that the possible instructions and materials of a human who is following an algorithm have properties that are exemplified by certain mathematical structures. Let us look at one small part of Turing's argument, his proof that a language used in computation must have a finite alphabet. He assumes at the outset that each symbol can be understood or defined to be a set of points in a unit square of a standard metric space. This presupposes that there is some upper bound on the size of a single symbol. There is, after all, a limit to how much space a human can scan at once. Turing also assumes that the set of points that correspond to each symbol is measurable. He then defines a natural metric among the set of possible symbols, so construed. With the induced topology, the symbols form a conditionally compact space. It then follows, by mathematics, that "if we were to allow an infinity of symbols, then there would be symbols differing to an arbitrarily small extent." When we add a premise that there is some limit to the human ability to distinguish among tokens, it follows that an alphabet used in computation must be finite. A structuralist reading of this piece of applied mathematics is straightforward (see Shapiro [1983b] for other, related examples).

The aforementioned physics textbook, Geroch [1985], contains a similar sketch of applications of topology to physics (see Chihara [1990, 140]):

> We imagine some physical system that we wish to study. . . . We introduce the notion of "the state of the system", where we think of the state of the system as a complete description of what every part of the system is like at a given instant of time. (For example, the state of a harmonic oscillator is specified by giving its position and momentum.). . . . By an extended series of manipulations . . . we "discover all the states that are available to [the system]", that is, we introduce a set Γ whose points represent the states of our system. Thus, our mathematical model of the system so far consists simply of a certain set Γ.
>
> We next decide that, more or less, we know what it means physically to say that two states of the system (i.e., two points of Γ) are "nearby". Roughly speaking, two states are "nearby" if "the system does not have to change all that much in passing from the first state to the second". We wish to incorporate this physical idea as mathematical structure on the set Γ. The notion of a topology seems to serve this purpose well. Thus, we suppose that the set of states, Γ, of our system is in fact a topological space, where the topology on Γ reflects "physical closeness" of states.

In present terms, what Geroch calls a "system" is the set of *possible states* of a physical object or configuration of objects. The topological space represents the relevant relationships among these possible states, and theorems about the topological space represent facts about what is and what is not possible for the physical configurations in question.

According to the so-called *semantical* approach (e.g., van Fraassen [1980]), a scientific theory is not a collection of sentences but a collection of models. The models represent possible configurations of physical systems, often idealized. If the models are mathematical structures, then the semantical approach dovetails with the present account (even though the ontology of structuralism may be inconsistent with van Fraassen's other views).[8]

The use of real analysis or geometry in physics makes an interesting case study. The application can be understood in different ways, depending on how points are construed in the *physical* theory. Of course, the status of space points and time points (or space-time points) is an old issue in the interpretation of physics. There are two main positions in the debate. One of them, *substantivalism*, holds that points are physically real and that physical space (or space-time) literally consists of points. A structuralist interpretation of substantivalism is most straightforward, because there is no detour through modality. Applying Euclidean geometry, for example, amounts to postulating uncountably many *physical* points. Classical physics, so construed, entails that Euclidean structure is literally exemplified in space or space-time. The situation might be described as a scientific theory that *incorporates* a mathematical structure.

---

8. I am indebted to Chihara [1990, 139] for pointing out this connection.

Whatever the fate of his ontological theses, Field's [1980] account of the application of mathematics in substantival field theories is correct. Avoiding a technical device that Field employs (of no concern here), let *N* be a nominalistic theory that refers only to space-time points and relations definable on them. Let *S* be a mathematical theory, such as analysis or set theory, together with bridge principles that connect the mathematics to the physics. According to Field, the way mathematics is usually applied is that in the combined theory $N + S$, each nominalistic statement Φ (in the language of *N*) is proved to be equivalent to a statement Φ' in the language of the mathematical theory *S*. Field calls Φ' an *abstract counterpart* to Φ. The mathematics can then be used to derive consequences of nominalistic statements. Suppose that, using the powerful mathematical theory, we see that Φ' follows from the abstract counterparts of the nominalistic theory. We can conclude Φ.

Although Field would not put it this way, the abstract counterparts establish that a system of physical objects exemplifies a mathematical structure. In working with the abstract counterparts, the nominalistic scientist is following the recipe of transferring information about a structure to a system that exemplifies it. In Field's development of Newtonian gravitational theory, abstract counterparts are developed by establishing a structure equivalence between the mathematical structure and the physical system. He proves in the combined theory that there is a structure-preserving one-to-one function *f* from the class of space-time points onto $\mathbb{R}^4$, the set of quadruples of real numbers. For example, if *a*, *b*, *c*, *d* are space-time points, then $\langle a, b\rangle$ is congruent to $\langle c, d\rangle$ if and only if the distance in $\mathbb{R}^4$ between $f(a)$ and $f(b)$ is identical to the distance in $\mathbb{R}^4$ between $f(c)$ and $f(d)$. The function *f* is sometimes called a *representation function*. Because of this isomorphism, one can translate any statement about space-time points into an equivalent statement about real numbers.

The alternative to substantivalism, called *relationalism*, is the view that space (or time or space-time) points are not physically real. The physics only describes relations between (possible) extended physical objects. Proponents of relationalism claim that there is no need to postulate a realm of points that fill space-time in order to describe the relevant relations among actual and possible physical objects. Instead, the relations are directly described in mathematical terms. For example, one might take "distance in meters" as a primitive function from pairs of physical objects to real numbers.

From a relationalist perspective, the account of applications follows the modal route. In describing relations among possible physical objects, the scientist is characterizing a class of systems of physical objects. The mathematics is used to both describe and study the structures of those systems. The relevant feature here is that the relations among the possible physical objects are described in mathematical terms. That is, *physical* relations like distance and force are characterized with the places of standard mathematical structures.

Field's nominalism is related to his substantivalism. In particular, nominalism requires that every structure exemplified in physical reality be completely describable in a language with variables that range over places within that structure, nonempty classes of such places, and perhaps other concrete objects. This guarantees that all the theoretical entities postulated by a given branch of science conform to nominalistic scruples. Both the entities that answer to the places of the structure and entities that are needed to describe the structure are physically real. Typically, relationalists do not meet such scruples. For them, the physical structures are described in terms of mathematical structures. As Field might put it, the substantivalist builds the relevant structure right into the physical systems, by postulating space-time points as physical entities.

The situation with $\mathbb{R}^4$, geometry, and physical space-time illustrates our final case, the application of structures when some of their relations or places do not have physical interpretations. With relationalism, the points themselves are not physically interpreted. With substantivalism, space-time itself has no preferred units and no frame of reference, nor is there addition and multiplication. Moreover, as we saw in chapter 5, space-time may have no metric. These items are supplied by $\mathbb{R}$ and $\mathbb{R}^4$, as applied to space-time.

Recall the embeddings that came up repeatedly in previous chapters, when focus was restricted to mathematics. The natural-number structure is contained in the real-number structure, which is contained in the complex-number structure. Thus, for example, the natural-number structure is illuminated via analytic number theory. Virtually every mathematical structure is contained in the set-theoretic hierarchy, and so set theory illuminates just about every structure. Some scientific applications are like these applications of mathematics to mathematics. A structure that may or may not be physically instantiated is embedded or modeled in a rich mathematical structure. Consideration of the rich structure sheds light on one of its substructures. The rich structure thereby also illuminates any system exemplified by this substructure. If the substructures have physically real, idealized, or modalized physical instantiations, then the rich mathematics indirectly applies to physical reality (see Chihara [1990, 129–130]). As we saw earlier, even set theory has empirical consequences, once suitable bridge principles are added.

The phenomenon of structure embedding helps deflect the following objection. According to structuralism, natural numbers are places within the natural-number structure. This structure is infinite. Thus, according to structuralism, someone who uses small natural numbers in everyday life presupposes an infinite structure. It seems absurd to suggest that every child who learns to count his toes applies an infinite structure to reality, and thus presupposes the structure.

The reply dovetails with an epistemic strategy broached in chapter 4. Simple applications of arithmetic do not require the entire natural-number structure but only various finite cardinal or ordinal structures. In any given case, a sufficiently large finite structure will do. For example, when a child learns to count to twenty and thus learns to count small collections of objects, she is learning a certain finite pattern, the ordinal 20 pattern. She may even learn to enrich this structure with addition and multiplication (as partial functions). Later, the child learns that this pattern can be extended to a larger one, with hundreds of places. If all goes well, she eventually learns that these finite structures themselves have a pattern, and she learns that each of the finite patterns can be embedded in a structure with infinitely many places. Here again, the larger structure is used to shed light on its substructures and on systems that exemplify the substructures.

3 Borders

With Quine, I hold that the boundary between mathematics and science is a blurry one, and there is no good reason to sharpen it. Nevertheless, a blurry border is still a border. This section concerns some fuzzy differences between mathematics and other scientific endeavors.

One difference between the disciplines concerns how structures are presented and studied. In the mature branches of mathematics, structures are described abstractly, independently of anything they may be structures of. Ultimately, mathematical structures are characterized axiomatically (as implicit definitions), or they are defined in set theory. So construed, structures can be studied deductively, allowing full rigor when necessary. In other words, because the mathematical structures are characterized without reference to anything they may be structures of, they can be illuminated in a topic-neutral way, via formal deduction. As we saw in chapter 4, this is one source of the idea that mathematical knowledge is a priori. In contrast to this, scientists are occasionally concerned with systems that are exemplified by the structures, and they may alter the description of a structure accordingly, perhaps in response to empirical data.

Along similar lines, Resnik [1982, 101] distinguishes between types of theories of particular structures: "A pure theory of a pattern is a deductively developed or developable theory which is based upon axioms which purport to characterize the pattern in question. Its assertions do not extend to claims concerning whether, where or how the pattern is instantiated, and they are true of that pattern regardless of its applicability. . . . An applied theory of a pattern [involves] claims stating how the pattern is instantiated." Notice, however, that this provides a rough distinction at best, and it does not separate mathematics from empirical science. First, scientists themselves occasionally invoke deduction in developing a structure, perhaps as part of a hypothetical-deductive inference. Presumably, they are doing mathematics at those moments. Second, as emphasized before, mathematicians often concern themselves with exemplifications of their structures—in other structures. For example, an analyst may study the structure of analytic, complex-valued functionals, or a geometer may study the structure of the conic sections or timelike surfaces or light cones. As emphasized earlier, embedding structures in one another is an essential part of both mathematical and scientific methodology.

Some fields lie on—and challenge—the border between mathematics and science. To return to previous examples, Crowell and Fox [1963] is a book about knots, pieces of rope. Turing's original work [1936] is aimed at human computation ability and, perhaps, mechanical computing devices. So, if librarians were to focus on the intended subject matter, they would classify the works within empirical science. Yet these works are routinely classified as mathematics and used in mathematics courses. Indeed, knot theory and computability are branches of pure mathematics. Of course, I do not mean to quibble with the classification. Turing [1936] and Crowell and Fox [1963] are mathematics par excellence. In each case, all but a few lines are devoted to developing and studying the respective structures as such. The few exceptions concern the connections with physical reality, such as Turing's claim that character tokens exemplify the structure of measurable sets of points. There is little effort to support these empirical claims; they are obvious. In Resnik's terminology, knot theory and computability are, for the most part, *pure theories* of their respective structures. The connection with physical reality may be what originated the inquiry into these structures, but this relationship is of little concern in the study itself.

There are more mainstream examples of fields that straddle the blurry border. Originally, geometry was a theory of physical space, and yet geometry has always been cast as a branch of mathematics (see chapter 5). One would be hard put to distinguish much of the contemporary work in physics—both quantum mechanics and relativity—from mathematics (putting funding agencies aside).

To confirm an observation made at the start of this chapter, branches of pure mathematics and branches of applied mathematics have a common type of subject matter—structures. The interplay between them is a result of modeling one structure within another or, in other words, using one structure to study another structure. The difference between pure mathematics and applied mathematics is, at least in part, an accident of history. Applied mathematics studies structures that traditionally form part of particular scientific theories. There is a continuous grading off from applied mathematics to theoretical science and from there to experimental science. Think of a continuum with, say, set theory at one end, applied mathematics and theoretical science toward the middle, and experimental science at the other end. Applications are found everywhere, and so is the deductive study of structures as such.

4 Maybe It Is Structures All the Way Down

The time has come to speak of ordinary things—of people, stones, cookie crumbs, cards, decks, and events. I will be brief and even more speculative than usual. First, the boundary between mathematics and ordinary discourse is at least as fuzzy as the boundary between mathematics and science. Maddy [1988, 281] correctly notes that one result of the structuralist perspective is a healthy blurring of the distinction between mathematical and ordinary objects:[9] "Are there two realities, one mathematical, one physical, and, if so, why should the theory of the one be relevant to the theory of the other? . . . [S]tructuralism [has] moved away from the more familiar platonistic picture of a realm of mathematical objects completely divorced from the physical world. According to the structuralist, physical configurations often instantiate mathematical patterns. . . . I mention this . . . for the benefit of the naturalistically minded, those for whom any physical/mathematical dualism is a challenge to eliminate all but the physical. I want to suggest that the two might be seen as so interdependent that no separation, let alone elimination, is possible." Natural numbers are places in the natural-number structure, the structure of *counting*. What is it that we count? Objects. Introducing or individuating objects, at any level, is intimately tied up with

9. Maddy is not a structuralist and does not accept the views on ordinary objects sketched in this section. The passages cited here are part of an argument that some virtues of structuralism are shared by her own view, called "set-theoretic realism." See also Maddy [1990, chapter 5].

counting, and this individuation is scarcely distinguishable from the imposition of structure. Maddy continues:

> [I]magine the purely physical world. This would have to be a giant aggregate composed of all the physical stuff in the universe. There is nothing nonphysical in this, but most philosophers, even nominalists, prefer a less amorphous characterization; they begin with all physical objects, or all particles, or all space-time points. . . . [T]o add even this small amount of structure—the differentiation of the amorphous mass into individuals of some kind—is already to broach the mathematical. If we were to consider the mass of physical stuff divided into finite collections, everyone would agree that we have more than the physical. . . . [T]he same is true when the physical stuff is simply divided into individual objects. . . . The only way to confine ourselves to the purely physical is to refrain from any differentiation whatsoever. . . . Could there be a physical world with absolutely no mathematical structure? Whatever our level of success or failure in this shamelessly philosophical imagining . . . it is plain that our world is not one such. Our reality is structured in many ways, into individual objects, into natural kinds, into patterns and structures of many sorts. . . . [T]he idea that physical reality can stand alone, that it comes first and that mathematics is separate, secondary or imposed, must be rejected. (p. 282)

This is a far-reaching idea. To speak of objects at all, we must "structure" the universe, and this structuring is a move toward the mathematical. We find parts of the world structured into continua, into Klein groups, and, to use my favorite example from chapters 3 and 4, baseball defenses. We also find the world structured into objects and systems of objects—into crumbs and cookies.

This theme dovetails with a conclusion from chapter 4. Structuring a chunk of space-time into a system of objects involves mobilizing linguistic and other conceptual machinery. As argued earlier, this mobilization is all that is needed to formulate and discuss structures as such. Along these lines, Frege is known for the idea that we do not have objects without concepts. Without concepts, there is nothing—no things—to count. The use of language involves an ability to manipulate predicates and singular terms, which denote concepts and objects. Frege showed how to construct natural numbers from this raw material.

So far, so good. The temptation, at this point, is to extend the structuralist perspective about mathematical objects to objects generally. This would be a pleasing articulation of my conclusions concerning fuzzy borders. We have seen that Quine [1992] himself embraces a global structuralism, of sorts (see chapter 3 of this volume). Notice, however, that the theses of this book do not require such a bold extension of structuralism. The prevailing desideratum is that there be a uniform semantics for mathematics and sufficiently regimented ordinary discourse. Under structuralism, the relevant semantics for mathematics is model theory. The uniformity condition is satisfied if the nonmathematical discourses enjoy a model-theoretic semantics, provided that we have some grasp of how mathematical structures relate to physical systems. As long as singular terms refer to objects and predicates express relations, the uniformity is assured—even if reference and satisfaction must be understood differently in the various discourses.

Nevertheless, it would be nice if structuralism could be extended beyond mathematics. Thus, I throw caution to the wind and close this book with some tentative remarks on what a global structuralism would involve and how far it can go. I beg the reader's indulgence.

Structuralism entails a relativity of objects. Rather than objects simpliciter, we speak of objects *of a structure*. Each mathematical object is implicitly or explicitly tied to the structure of which it is a place. Structures, in turn, are characterized by coherent theories, via devices like implicit definition (see chapter 4). Recall the structuralist responses to Frege–Benacerraf questions (chapter 3). Is 2 identical to Julius Caesar? Is 2 identical to $\{\{\phi\}\}$ or to $\{\phi, \{\phi\}\}$? Is $3 \in 5$? Some of these questions inquire after the identity between items from different structures. There are no determinate a priori answers to be found. Sometimes answers to such questions not needed and not wanted. In other cases, answers are desirable, and even natural. Consider, for example, identifying the zero of the natural numbers with the zero of the integers. But it is always a matter of decision not discovery. The same goes for the question of whether $3 \in 5$. It is like asking whether 3 is funnier than 5. Membership and relative humor are not relations in arithmetic, and so there is no determinate answer to these questions waiting to be discovered.

Notice that one can raise Frege–Benacerraf questions about ordinary objects. Can we *discover* whether a deck is really identical with its fifty-two cards, or whether a person is identical with her corresponding time-slices, molecules, or space-time points? In philosophy, a lot of ink and toner has gone into figuring out whether an event is or is not identical with a chunk of space-time, or whether the event of Socrates's death is or is not identical with the event of Xanthippe's becoming a widow. Similar puzzles arise with theories of properties, universals, possible worlds, mind states, and brain states. The outlook developed here would suggest that answers to such questions are not always forthcoming. In some cases, answers may be desirable and even natural, but it is a matter of decision based on convenience. The questions are direct analogues of the attempts to determine whether a place within one structure, like the number 2, is or is not identical with a place in another structure, like $\{\{\phi\}\}$, or something else, like Julius Caesar.

Further Frege–Benacerraf questions arise from asking questions about the places of structures by using terminology from outside the theory of the structure. Suppose that during a class on the U.S. Constitution, a student asks about the age of the president (as opposed to the age of Bush or Clinton). Similarly, consider questions about the hair color or batting average of the shortstop (as opposed to the hair color or batting average of a person playing that position on game day). The questions indicate a confusion of the subject matter.

Recall the distinction, in chapters 3 and 4, between the places-are-objects and the places-are-offices perspectives. With places-are-offices, the positions in structures or patterns are taken to be filled with objects or people, drawn from a given ontology. With places-are-objects, the places themselves are treated as bona fide objects. As we saw, this occurred by restricting the language to the theory of the structure, a language limited to terms for the places and relations of the structure in question.

The present plan is to apply the objects/offices dichotomy universally. All objects are places in structures. For each stone, there is at least one pattern such that the stone just *is* a position in that pattern. When that pattern is considered from the places-are-objects perspective, the stone is treated as an object. This perspective presupposes that there is, or could be, another perspective according to which the stone is an office, to be filled with objects drawn from another ontology. Perhaps the stone office can be filled with a collection of space-time points or a collection of molecules.

Dieterle [1994, chapters 4 and 5] pursues a program aimed at providing the details of such a global structuralism, drawing on work by Robert Kraut (e.g., [1980]). She defines a "theory" to be a stretch of discourse (from natural language) that displays some sort of "conceptual homogeneity." The assertions of the theory show how the referents of the terms and predicates of the theory are related to each other. This admittedly vague notion is what plays the role of mathematical axiomatizations in implicit definitions of structures. According to Dieterle, a coherent theory characterizes a structure. The places and relations of the structure are the denotata of the singular terms and predicates of the theory. Each object referred to in a natural language is tied to (one or more) "theories," and so each object is a place in (one or more) structures. Things are admittedly vague here. Structuralism depends on being able to isolate the theories of different structures. This is not as plausible when it comes to scientific or ordinary discourse.

As with mathematics, the system/structure dichotomy is a relative one. What is system from one perspective is structure from another. There is thus a potential regress of theories, each one encompassing the objects of the one before, turning these objects into the places of an exemplified structure. Does the regress end? Is there a natural stopping place, a universal "theory"? If there were, it would be the entire web of natural language, or scientific English—our proverbial mother tongue—suitably regimented. Let us assume that this all-encompassing framework is a (more or less) coherent theory. If so, then the totality of all objects that we talk about with this superframework and all relations that we impose on these objects form a structure. Every object is an object of this theory, and so every object is a place of this gigantic structure. These ultimate structure places, of course, are treated from the places-are-objects perspective, and the envisioned super structure is freestanding (in the sense of chapter 3). Still, this current mother-tongue structure *could* be treated from the places-are-offices point of view, by embedding its "theory"—our entire web of belief—into another theory. If the main Quinean theses are correct, there is more than one way to regiment our full web of belief. Any of several superstructures will do, and so there is a thorough relativity of ontology.

Enough of this speculation. Let us put the envisioned superstructures aside and think in terms of smaller theories. Dieterle, of course, is aware of the differences between mathematical and ordinary objects, but she holds that the differences are not significant. In the terms of this book, mathematical objects are places in freestanding structures, whose relations are formal. Ordinary objects are presumably places in nonfreestanding structures. Thus, the structures characterized by Dieterle's "theories" are exemplified only by certain systems of objects, taken from a background framework. One of her examples is the structure of the U.S. government. Within the envisioned theory, "president" is a singular term that denotes an object, a place in that structure. However, the structure is not freestanding. It can be exemplified only by certain systems of people. Moreover, no one can play the president role unless the person is a native-born citizen, is over thirty-five years of age, and is either elected to the office or succeeds to it.[10]

In other words, with ordinary (pure) mathematics, a given theory is an implicit definition of a freestanding structure or a class of freestanding structures. With the model-theoretic semantics, the formal theory itself determines which systems exemplify the structure. Ordinary objects and ordinary structures are not like this. The theories alone do not determine which systems exemplify the structures, even if the theory is augmented with a general model-theoretic semantics. Thus, beyond mathematics, there is more to reference and satisfaction than is captured in model theory. For ordinary language, model theory must be supplemented with accounts of reference and satisfaction.

On the structuralist perspective, what is an object in a given context depends on what concepts or predicates are in use, and this depends on what concepts or predicates are available. It is through language that we organize the world and divide it into objects. A global consequence of this relativity is that the universe does not come, nor does it exist, divided into objects a priori (so to speak), independent of our language, our framework, or, to use another Wittgensteinian phrase, our form of life. The complex web of beliefs, concepts, and theories that determines how we perceive and understand the world also determines what its objects are and when two of them are the same or different.

Variations on this theme are popular among contemporary philosophers. The idea is part of Quine's views concerning the web of belief and his aforementioned global structuralism. The same thesis is central in Putnam's ([1981], [1987]) internal realism, especially the later versions of it (see chapter 2 of this volume).

Once again, Frege held that ontology is tied to concepts and that linguistic resources track the realms of concepts and objects. Of course, he did not embrace any sort of relativity, because he held that there is a single, absolute realm of concepts and objects. Accordingly, there is, or should be, a single universal language—a *Begriffsschrift*—with variables that range over all objects whatsoever and variables that range over all concepts whatsoever (see Dieterle [1994, chapter 3]). However, a Fregean, or a traditional Platonist for that matter, might admit that what we are able to identify as objects depends on what concepts we manage to grasp and articulate. Dummett

10. There is an interesting issue concerning how the nonfreestanding nature of ordinary structures should be incorporated into Dieterle's framework. Let G be the envisioned "U.S. government theory." As noted, relative to G, "president" is a singular term that denotes an object. What are we to make of the statement that the president must be thirty-five years of age? If this is part of G, then we can speak in the language of G about the age of the president, and so we have an embarrassing matter of determining just what the age of the president is. This is a Frege–Benacerraf question that is formulable in the language of the theory of G. One way out is to insist that the statement that the president must be thirty-five years old is not part of G. The theory G just concerns the *relations* between the officeholders. The age requirement is an extratheoretic statement about which systems can exemplify the structure.

[1973a, 503–505] wrote that "what objects we recognize the world as containing depends upon the structure of our language" and "our apprehension of reality as decomposable into discrete objects is the product of our application to an original unarticulated reality of the conceptual apparatus embodied in our language." Hale [1987, 156] elaborates: "The thought is that we are able to discern objects of any given kind only because we have mastered general, sortal terms whose associated criteria of identity constitute the necessary conceptual apparatus by means of which we can 'slice up' the world in a particular way; and that we can at least conceive the possibility of language users for whom the world would divide up into objects of different kinds, because they employed general sortal terms with quite different criteria of identity." This last surely depends on our language and background "theory."

From a different perspective, a conventionalist about mathematics holds that *mathematical* ontology is dependent on language, but the realm of physical objects is not. The perspective of this book is that there is no sharp boundary between the mathematical and the physical. In both cases, the way the universe is divided into structures and objects—of all kinds—depends on our linguistic resources. Just about everything that relates to ontology is tied to language.

The foregoing thesis is not meant to be an affront to objectivity. Quite the contrary. I remain an unrepentant philosophical realist, in the sense of chapter 2. It is surely correct to maintain that if there had never been any language (or any people), there would be trees, planets, and stars. There would also be numbers, sets of numbers, and Klein groups, if not baseball defenses. Such is the nature of *ante rem* structures. Once a language and a theory impose a structure and sort the universe into objects—be they abstract or concrete—one can sometimes speak objectively about those objects, and we insist (surely correctly) that at least some of the objects were not created by us.[11] Counterfactuals about the ways the world would be must themselves be formulated in our language and form of life—for we know no other. Trivially, had there never been any language, there would be no means of discussing, say, stars, as distinguished from the particles they contain and the galaxies that contain them. However, the lack of sortals available to speakers in a given possible world has nothing to do with which objects that world contains.

11. I am indebted to Michael Resnik for suggestions of how to formulate this point

References

Aspray, W., and P. Kitcher [1988], *History and philosophy of modern mathematics, Minnesota studies in the philosophy of science* 11, Minneapolis, Minnesota University Press.

Azzouni, J. [1994], *Metaphysical myths, mathematical practice*, Cambridge, Cambridge University Press.

Balaguer, M. [1994], "Against (Maddian) naturalized epistemology," *Philosophia Mathematica* (3) 2: 97–108.

Balaguer, M. [1995], "A Platonist epistemology," *Synthese* 103: 303–325.

Barbut, M. [1970], "On the meaning of the word 'structure' in mathematics," in *Introduction to structuralism*, ed. by M. Lane, New York, Basic Books, 367–388.

Barwise, J. [1985], "Model-theoretic logics: Background and aims," in *Model-theoretic logics*, ed. by J. Barwise and S. Feferman, New York, Springer-Verlag, 3–23.

Benacerraf, P. [1965], "What numbers could not be," *Philosophical Review* 74: 47–73; reprinted in Benacerraf and Putnam [1983], 272–294.

Benacerraf, P. [1973], "Mathematical truth," *Journal of Philosophy* 70: 661–679; reprinted in Benacerraf and Putnam [1983], 403–420.

Benacerraf, P., and H. Putnam [1983], *Philosophy of mathematics*, second edition, Cambridge, Cambridge University Press.

Bernays, P. [1935], "Sur le platonisme dans les mathématiques," *L'enseignement mathématique* 34: 52–69; tr. as "Platonism in mathematics," in Benacerraf and Putnam [1983], 258–271.

Bernays, P. [1961], "Zur Frage der Unendlichkeitsschemata in der axiomatischen Mengenlehre," in *Essays on the foundation of mathematics*, ed. by Y. Bar-Hillel et al., Jerusalem, Magnes Press, 3–49.

Bernays, P. [1967], "Hilbert, David," in *The encyclopedia of philosophy,* vol. 3. ed. by P. Edwards, New York, Macmillan and The Free Press, 496–504.

Beth, E. W., and J. Piaget [1966], *Mathematical epistemology and psychology*, Dordrecht, Holland, D. Reidel.

Bishop, E. [1967], *Foundations of constructive analysis*, New York, McGraw-Hill.

Bishop, E. [1975], "The crises in contemporary mathematics," *Historia Mathematica* 2: 505–517.

Blackburn, S. [1984], *Spreading the word*, Oxford, Clarendon Press.

Block, N. (ed.) [1980], *Readings in philosophy of psychology* 1, Cambridge, Harvard University Press.

Boolos, G. [1984], "To be is to be a value of a variable (or to be some values of some variables)," *Journal of Philosophy* 81: 430–449.

Boolos, G. [1985], "Nominalist Platonism," *Philosophical Review* 94: 327–344.

Boolos, G. [1987], "The consistency of Frege's *Foundations of arithmetic*," in *On being and saying: Essays for Richard Cartwright*, ed. by Judith Jarvis Thompson, Cambridge, MIT Press, 3–20.

Bourbaki, N. [1949], "Foundations of mathematics for the working mathematician," *Journal of Symbolic Logic* 14: 1–8.

Bourbaki, N. [1950], "The architecture of mathematics," *American Mathematical Monthly* 57: 221–232.

Bourbaki, N. [1968], *Theory of sets*, Paris, Hermann.

Brouwer, L. E. J. [1948], "Consciousness, philosophy and mathematics," in Benacerraf and Putnam [1983], 90–96.

Buck, R. [1956], *Advanced calculus*, New York, McGraw-Hill.

Burgess, J. [1983], "Why I am not a nominalist," *Notre Dame Journal of Formal Logic* 24: 93–105.

Burgess, J. [1984], "Synthetic mechanics," *Journal of Philosophical Logic* 13: 379–395.

Burgess, J. [1992], "Proofs about proofs: A defense of classical logic," in *Proof, logic and formalization*, ed. by Michael Detlefsen, London, Routledge, 8–23.

Cantor, G. [1899], "Letter to Dedekind," van Heijenoort [1967], 113–117.

Cantor, G. [1932], *Gesammelte Abhandlungen mathematischen und philosophischen Inhalts*, ed. by E. Zermelo, Berlin, Springer.

Carnap, R. [1931], "Die logizistische Grundlegung der Mathematik," *Erkenntnis* 2: 91–105; tr. as "The logicist foundations of mathematics," in Benacerraf and Putnam [1983], 41–52.

Carnap, R. [1934], *Logische Syntax der Sprache*, Vienna, Springer; tr. as *The logical syntax of language*, New York, Harcourt, 1937.

Carnap R. [1942], *Introduction to semantics*, Cambridge, Harvard University Press.

Carnap, R. [1950], "Empiricism, semantics, and ontology," *Revue Internationale de Philosophie* 4: 20–40; reprinted in Benacerraf and Putnam [1983], 241–257.

Cartwright, N. [1983], *How the laws of physics lie*, Oxford, Oxford University Press.

Chihara, C. [1973], *Ontology and the vicious-circle principle*, Ithaca, New York, Cornell University Press.

Chihara, C. [1984], "A simple type theory without Platonic domains," *Journal of Philosophical Logic* 13: 249–283.

Chihara, C. [1990], *Constructibility and mathematical existence*, Oxford, Oxford University Press.

Chihara, C. [1993], "Modality without worlds," in *Philosophy of mathematics: Proceedings of the fifteenth international Wittgenstein symposium* 1, ed. by J. Czermak, Vienna, Verlag-Hölder-Pichler-Tempsky, 253–268.

Coffa, A. [1986], "From geometry to tolerance: Sources of conventionalism in nineteenth-century geometry," in *From quarks to quasars: Philosophical problems of modern physics*, University of Pittsburgh series, vol. 7, Pittsburgh, Pittsburgh University Press, 3–70.

Coffa, A. [1991], *The semantic tradition from Kant to Carnap*, Cambridge, Cambridge University Press.

Cooper, R. [1984], "Early number development: Discovering number space with addition and subtraction," in *Origins of cognitive skills*, ed. by C. Sophian, Hillsdale, New Jersey, Erlbaum, 157–192.

Corry, L. [1992], "Nicolas Bourbaki and the concept of mathematical structure," *Synthese* 92: 315–348.

Crossley, J., C. J. Ash, C. J. Brickhill, J. C. Stillwell, and N. H. Williams [1972], *What is mathematical logic?* New York, Oxford University Press.

Crowell, R., and R. Fox [1963], *Introduction to knot theory*, Boston, Ginn.

Curry, H. [1950], *A theory of formal deducibility*, Notre Dame, Edwards Brothers.

Davidson, D. [1974], "On the very idea of a conceptual scheme," *Proceedings and Addresses of the American Philosophical Association* 47: 5–20.

Dedekind, R. [1872], *Stetigkeit und irrationale Zahlen*, Brunswick, Vieweg; tr. as *Continuity and irrational numbers*, in *Essays on the theory of numbers*, ed. by W. W. Beman, New York, Dover, 1963, 1–27.

Dedekind, R. [1888], *Was sind und was sollen die Zahlen?* Brunswick, Vieweg; tr. as *The nature and meaning of numbers*, in *Essays on the theory of numbers*, ed. by W. W. Beman, New York, Dover, 1963, 31–115.

Dedekind, R. [1932], *Gesammelte mathematische Werke* 3, ed. by R. Fricke, E. Noether, and O. Ore, Brunswick, Vieweg.

Demopoulos, W. [1994], "Frege, Hilbert, and the conceptual structure of model theory," *History and Philosophy of Logic* 15: 211–225.

Dieterle, J. [1994], *Structure and object*, Ph.D. dissertation, The Ohio State University.

Dummett, M. [1973], "The philosophical basis of intuitionistic logic," in *Truth and other enigmas*, by M. Dummett, Cambridge, Harvard University Press, 1978, 215–247; reprinted in Benacerraf and Putnam [1983], 97–129.

Dummett, M. [1973a], *Frege: Philosophy of language*, New York, Harper and Row.

Dummett, M. [1977], *Elements of intuitionism*, Oxford, Oxford University Press.

Dummett, M. [1981], *The interpretation of Frege's philosophy*, Cambridge, Harvard University Press.

Dummett, M. [1991], *Frege: Philosophy of mathematics*, Cambridge, Harvard University Press.

Dummett, M. [1991a], *The logical basis of metaphysics*, Cambridge, Harvard University Press.

Edwards, H. [1988], "Kronecker's place in history," in Aspray and Kitcher [1988], 139–144.

Etchemendy, J. [1988], "Tarski on truth and logical consequence," *Journal of Symbolic Logic* 53: 51–79.

Feynman, R. [1967], *The character of physical law*, Cambridge, MIT Press.

Field, H. [1980], *Science without numbers*, Princeton, Princeton University Press.

Field, H. [1984], "Is mathematical knowledge just logical knowledge?" *Philosophical Review* 93: 509–552; reprinted (with added appendix) in Field [1989], 79–124.

Field, H. [1985], "On conservativeness and incompleteness," *Journal of Philosophy* 82: 239–260; reprinted in Field [1989], 125–146.

Field, H. [1989], *Realism, mathematics and modality*, Oxford, Blackwell.

Field, H. [1991], "Metalogic and modality," *Philosophical Studies* 62: 1–22.

Fine, A. [1986], *The shaky game: Einstein, realism and the quantum theory*, Chicago, University of Chicago Press.

Frege, G. [1879], *Begriffsschrift, eine der arithmetischen nachgebildete Formelsprache des reinen Denkens*, Halle, Louis Nebert; tr. in van Heijenoort [1967], 1–82.

Frege, G. [1884], *Die Grundlagen der Arithmetik*, Breslau, Koebner; *The foundations of arithmetic*, tr. by J. Austin, second edition, New York, Harper, 1960.

Frege, G. [1903], *Grundgesetze der Arithmetik* 2, Olms, Hildescheim.

Frege, G. [1903a], "Über die Grundlagen der Geometrie," *Jahresbericht der Mathematiker-Vereinigung* 12: 319–324, 368–375.

Frege, G. [1906], "Über die Grundlagen der Geometrie," *Jahresbericht der Mathematiker-Vereinigung* 15: 293–309, 377–403, 423–430.

Frege, G. [1967], *Kleine Schriften*, Darmstadt, Wissenschaftlicher Buchgesellschaft (with I. Angelelli).

Frege, G. [1971], *On the foundations of geometry and formal theories of arithmetic*, tr. by Eikee-Henner W. Kluge, New Haven, Yale University Press.

Frege, G. [1976], *Wissenschaftlicher Briefwechsel*, ed. by G. Gabriel, H. Hermes, F. Kambartel, and C. Thiel, Hamburg, Felix Meiner.

Frege, G. [1980], *Philosophical and mathematical correspondence*, Oxford, Blackwell.

Freudenthal, H. [1962], "The main trends in the foundations of geometry in the 19th century," in *Logic, methodology and philosophy of science*, Proceedings of the 1960 Congress, ed. by E. Nagel, P. Suppes, A. Tarski, Stanford, Stanford University Press, 613–621.

Friedman, M. [1983], *Foundations of space-time theories: Relativistic physics and philosophy of science*, Princeton, Princeton University Press.

Friedman, M. [1988], "Logical truth and analyticity in Carnap's *Logical syntax of language,*" in Aspray and Kitcher [1988], 82–94.

Gandy, R. [1988], "The confluence of ideas in 1936," in R. Herken, ed., *The universal Turing machine*, New York, Oxford University Press, 55–111.

Geach, P. [1967], "Identity," *Review of Metaphysics* 21: 3–12.

Geach, P. [1968], *Reference and generality*, Ithaca, New York, Cornell University Press.

Geroch, R. [1985], *Mathematical physics*, Chicago, University of Chicago Press.

Gödel, K. [1944], "Russell's mathematical logic," in Benacerraf and Putnam [1983], 447–469.

Gödel, K. [1951], "Some basic theorems on the foundations of mathematics and their implications," in *Collected Works* 3, Oxford, Oxford University Press, 1995, 304–323.

Gödel, K. [1953], "Is mathematics syntax of language?" in *Collected Works* 3, Oxford, Oxford University Press, 1995, 334–362.

Gödel, K. [1964], "What is Cantor's continuum problem?," in Benacerraf and Putnam [1983], 470–485.

Gödel, K. [1986], *Collected works* 1, Oxford, Oxford University Press.

Goldfarb, W. [1979], "Logic in the twenties: The nature of the quantifier," *Journal of Symbolic Logic* 44: 351–368.

Goldfarb, W. [1988], "Poincaré against the logicists," in Aspray and Kitcher [1988], 61–81.

Goldfarb, W. [1989], "Russell's reasons for ramification," in *Rereading Russell, Minnesota studies in the philosophy of science* 12: 24–40.

Goldman, A. [1986], *Epistemology and cognition*, Cambridge, Harvard University Press.

Goodman, Nicolas [1979], "Mathematics as an objective science," *American Mathematical Monthly* 86: 540–551.

Grassmann, H. [1972], *Gessammelte mathematische und physicalische Werke* 1, ed. by F. Engels, New York, Johnson Reprint Corporation.

Gupta, A., and N. Belnap [1993], *The revision theory of truth*, Cambridge, MIT Press.

Hale, Bob [1987], *Abstract objects*, Oxford, Blackwell.

Hallett, M. [1990], "Physicalism, reductionism and Hilbert," in *Physicalism in mathematics*, ed. by A. D. Irvine, Dordrecht, Holland, Kluwer Academic Publishers, 183–257.

Hallett, M. [1994], "Hilbert's axiomatic method and the laws of thought," in *Mathematics and mind*, ed. by Alexander George, Oxford, Oxford University Press, 158–200.

Hand, M. [1993], "Mathematical structuralism and the third man," *Canadian Journal of Philosophy* 23: 179–192.

Hazen, A. [1983], "Predicative logics," in *Handbook of philosophical logic* 1, ed. by D. Gabbay and F. Guenthner, Dordrecht, Holland, Reidel, 331–407.

Heath, T. [1921], *A history of Greek mathematics*, Oxford, Clarendon Press.

Hellman, G. [1983], "Realist principles," *Philosophy of Science* 50: 227–249.

Hellman, G. [1989], *Mathematics without numbers*, Oxford, Oxford University Press.

Hellman, G. [1996], "Structuralism without structures," *Philosophia Mathematica* (3) 4: 100–123.

Helmholtz, H. von [1921], *Schriften zur Erkenntnistheorie*, ed. by P. Hertz and M. Schlick, Berlin, Springer; tr. as *Epistemological writings*, Dordrecht, Holland Reidel, 1977.

Hersh, R. [1979], "Some proposals for reviving the philosophy of mathematics," *Advances in mathematics* 31: 31–50.

Heyting, A. [1931], "The intuitionistic foundations of mathematics," in Benacerraf and Putnam [1983], 52–61.

Heyting, A. [1956], *Intuitionism, An introduction*, Amsterdam, North Holland.

Hilbert, D. [1899], *Grundlagen der Geometrie*, Leipzig, Teubner; *Foundations of geometry*, tr. by E. Townsend, La Salle, Illinois, Open Court, 1959.

Hilbert, D. [1900], "Mathematische Probleme," *Bulletin of the American Mathematical Society* 8 (1902): 437–479.

Hilbert, D. [1900a], "Über den Zahlbegriff," *Jahresbericht der Deutschen Mathematiker-Vereinigung* 8: 180–194.

Hilbert, D. [1902], *Les principes fondamentaux de la géometrie*, Paris, Gauthier-Villars; the French tr. of Hilbert [1899].

Hilbert, D. [1905], "Über der Grundlagen der Logik und der Arithmetik," in *Verhandlungen des dritten internationalen Mathematiker-Kongresses in Heidelberg vom 8 bis 13 August 1904*, Leipzig, Teubner, 174–185; tr. as "On the foundations of logic and arithmetic," in van Heijenoort [1967], 129–138.

Hilbert, D. [1925], "Über das Unendliche," *Mathematische Annalen* 95: 161–190; tr. as "On the infinite," in van Heijenoort [1967], 369–392; Benacerraf and Putnam [1983], 83–201.

Hilbert, D. [1935], *Gesammelte Abhandlungen,* 3, Berlin, Springer.

Hilbert, D., and S. Cohn-Vossen [1932], *Geometry and the imagination*, tr. by P. Nemenyi, New York, Chelsea Publishing Company, 1952.

Hodes, H. [1984], "Logicism and the ontological commitments of arithmetic," *Journal of Philosophy* 81: 123–149.

Hodges, W. [1985], "Truth in a structure," *Proceedings of the Aristotelian Society* 86 (1985–1986): 135–151.

Horwich, P. [1990], *Truth*, Oxford, Blackwell.

Huntington, E. [1902], "A complete set of postulates for the theory of absolute continuous magnitude," *Transactions of the American Mathematical Society* 3: 264–279.

Kitcher, P. [1983], *The nature of mathematical knowledge*, New York, Oxford University Press.

Kitcher, P. [1986], "Frege, Dedekind, and the philosophy of mathematics," in *Frege synthesized*, ed. by L. Haaparanta and J. Hintikka, Dordrecht, Holland, Reidel, 299–343.

Klein, F. [1921], *Gesammelte mathematische Abhandlungen* 1, Berlin, Springer.

Klein, J. [1968], *Greek mathematical thought and the origin of algebra*, Cambridge, MIT Press.

Kraut, R. [1980], "Indiscernibility and ontology," *Synthese* 44: 113–135.

Kraut, R. [1993], "Robust deflationism," *Philosophical Review* 102: 247–263.

Kreisel, G. [1967], "Informal rigour and completeness proofs," *Problems in the philosophy of mathematics*, ed. by I. Lakatos, Amsterdam, North Holland, 138–186.

Kripke, S. [1965], "Semantical analysis of intuitionistic logic I," in *Formal systems and recursive functions*, ed. by J. Crossley and M. Dummett, Amsterdam, North Holland, 92–130.

Kripke, S. [1975], "Outline of a theory of truth," *Journal of Philosophy* 72: 690–716.

Kuhn, T. [1970], *The structure of scientific revolutions*, second edition, Chicago, University of Chicago Press.

Lakatos, I. [1976], *Proofs and refutations*, ed. by J. Worrall and E. Zahar, Cambridge, Cambridge University Press.

Lakatos, I. [1978], *Mathematics, science and epistemology*, ed. by J. Worrall and G. Currie, Cambridge, Cambridge University Press.

Landman, F. [1989], "Groups," *Linguistics and Philosophy* 12: 559–605, 723–744.

Lawvere, W. [1966], "The category of categories as a foundation for mathematics," in *Proceedings of the conference on categorical algebra in La Jolla, 1965*, ed. by S. Eilenberg et al., New York, Springer, 1–21.

Lebesgue, H. [1971], "A propos de quelques travaux mathematiques recents," *Enseignement mathematique* (2) 17: 1–48.

Lévi-Strauss, C. [1949], *Les Structures élémentaires de la parenté*, Paris, P.U.F.

Levy, A. [1960], "Principles of reflection in axiomatic set theory," *Fundamenta Mathematicae* 49: 1–10.

Lewis, D. [1986], *On the plurality of worlds*, Oxford, Blackwell.

Lewis, D. [1991], *Parts of classes*, Oxford, Blackwell.

Lewis, D. [1993], "Mathematics is megethology," *Philosophia Mathematica* (3) 1: 3–23.

Luce, L. [1988], "Frege on cardinality," *Philosophy and Phenomenological Research* 48: 415–434.

Maddy, P. [1981], "Sets and numbers," *Nous* 11: 495–511.

Maddy, P. [1988], "Mathematical realism," *Midwest Studies in Philosophy* 12: 275–285.

Maddy, P. [1990], *Realism in mathematics*, Oxford, Oxford University Press.

Maddy, P. [1993], "Does V equal L?" *Journal of Symbolic Logic* 58: 15–41.

Maddy, P. [1997], "Naturalizing mathematical methodology," in *Philosophy of mathematics today: Proceedings of an international conference in Munich*, ed. by M. Schirn, The Mind Association, Oxford, Oxford University Press, forthcoming.

McCarty, C. [1987], "Variations on a thesis: Intuitionism and computability," *Notre Dame Journal of Formal Logic* 28: 536–580.

McCarty, C. [1995], "The mysteries of Richard Dedekind," in *From Dedekind to Gödel*, ed. by Jaakko Hintikka, *Synthese Library Series* 251, Dordrecht, The Netherlands, Kluwer Academic Publishers, 53–96.

McLarty, C. [1993], "Numbers can be just what they have to," *Nous* 27: 487–498.

Moore, G. H. [1982], *Zermelo's axiom of choice: Its origins, development, and influence*, New York, Springer-Verlag.

Myhill, J. [1960], "Some remarks on the notion of proof," *Journal of Philosophy* 57: 461–471.

Nagel, E. [1939], "The formation of modern conceptions of formal logic in the development of geometry," *Osiris* 7: 142–224.

Nagel, E. [1979], "Impossible numbers: A chapter in the history of modern logic," in *Teleology revisited and other essays in the philosophy and history of science*, New York, Columbia University Press, 166–194.

Neurath, O. [1932], "Protokollsätze," *Erkenntnis* 3: 204–214.

Nolt, J. [1978], *The language of set theory*, Ph.D. dissertation, The Ohio State University.

Padoa, A. [1900], "Essai d'une théorie algébrique des nombres entiers, précede d'une introduction logique à une théorie déductive qelconque," in *Bibliothèque du Congrès international de philosophie*, Paris; tr. as "Logical introduction to any deductive theory," in van Heijenoort [1967], 118–123.

Parsons, C. [1965], "Frege's theory of number," in *Philosophy in America*, ed. by Max Black, Ithaca, New York, Cornell University Press, 180–203; reprinted in Parsons [1983], 150–175.

Parsons, C. [1977], "What is the iterative conception of set?" in *Logic, foundations of mathematics and computability theory*, ed. by R. Butts and J. Hintikka, Dordrecht, Holland, Reidel, 335–367; reprinted in Benacerraf and Putnam [1983], 503–529; and in Parsons [1983], 268–297.

Parsons, C. [1983], *Mathematics in philosophy*, Ithaca, New York, Cornell University Press.

Parsons, C. [1990], "The structuralist view of mathematical objects," *Synthese* 84: 303–346.

Parsons, C. [1995], "Structuralism and the concept of set," in *Modality, morality, and belief: Essays in honor of Ruth Barcan Marcus*, ed. by W. Sinnott-Armstrong, D. Raffman, N. Ascher, Chicago, University of Chicago Press, 74–92.

Pasch, M. [1926], *Vorlesungen über neuere Geometrie*, Zweite Auflage, Berlin, Springer.

Plücker, J., [1846], *System der Geometrie des Raumes*, Dusseldorf.

Poincaré, H. [1899], "Des fondements de la géométrie," *Revue de Métaphysique et de Morale* 7: 251–279.

Poincaré, H. [1900], "Sur les principes de la géométrie," *Revue de Métaphysique et de Morale* 8: 72–86.

Poincaré, H. [1906], "Les mathématiques et la logique," *Revue de Métaphysique et de Morale* 14: 294–317.

Poincaré, H. [1908], *Science et méthode*, Paris, Flammarion, tr. as *Science and method, Foundation of* science, tr. by G. Halsted, New York, Science Press, 1921, 359–546.

Polya, G. [1954], *Mathematics and plausible reasoning*, Princeton, Princeton University Press.

Polya, G. [1977], *Mathematical methods in science*, Washington, D.C., Mathematical Association of America.

Posy, C. [1984], "Kant's mathematical realism," *Monist* 67: 115–134.

Proclus [485], *Commentary on Euclid's elements I*, tr. by G. Morrow, Princeton, Princeton University Press, 1970.

Putnam, H. [1967], "Mathematics without foundations," *Journal of Philosophy* 64: 5–22; reprinted in Benacerraf and Putnam [1983], 295–311.

Putnam, H. [1971], *Philosophy of logic*, New York, Harper Torchbooks.

Putnam, H. [1975], "What is mathematical truth?" in *Mathematics, matter and method: Philosophical papers*, vol. 1, by Hillary Putnam, Cambridge, Cambridge University Press, 60–78.

Putnam, H. [1980], "Models and reality," *Journal of Symbolic Logic* 45: 464–482; reprinted in Benacerraf and Putnam [1983], 421–444.

Putnam, H. [1981], *Reason, truth and history*, Cambridge, Cambridge University Press.

Putnam, H. [1987], *The many faces of realism*, LaSalle, Illinois, Open Court.

Quine, W. V. O. [1937], "New foundations for mathematical logic," *American Mathematical Monthly* 44: 70–80.

Quine, W. V. O. [1941], "Whitehead and the rise of modern logic," in P. A. Schilpp, *The philosophy of Alfred North Whitehead*, New York, Tudor, 127–163.

Quine, W. V. O. [1951], "Two dogmas of empiricism," *Philosophical Review* 60: 20–43.

Quine, W. V. O. [1960], *Word and object*, Cambridge, MIT Press.

Quine, W. V. O. [1969], *Ontological relativity and other essays*, New York, Columbia University Press.

Quine, W. V. O. [1981], *Theories and things*, Cambridge, Harvard University Press.

Quine, W. V. O. [1986], *Philosophy of logic*, second edition, Englewood Cliffs, New Jersey, Prentice-Hall.

Quine, W. V. O. [1992], "Structure and nature," *Journal of Philosophy* 89: 5–9.

Ramsey, F. [1925], "The foundations of mathematics," *Proceedings of the London Mathematical Society* (2) 25: 338–384.

Resnik, M. [1966], "On Skolem's paradox," *Journal of Philosophy* 63: 425–438.

Resnik, M. [1975], "Mathematical knowledge and pattern cognition," *Canadian Journal of Philosophy* 5: 25–39.

Resnik, M. [1980], *Frege and the philosophy of mathematics*, Ithaca, New York, Cornell University Press.

Resnik, M. [1981], "Mathematics as a science of patterns: Ontology and reference," *Nous* 15: 529–550.

Resnik, M. [1982], "Mathematics as a science of patterns: Epistemology," *Nous* 16: 95–105.

Resnik, M. [1985], "How nominalist is Hartry Field's nominalism?," *Philosophical Studies* 47: 163–181.

Resnik, M. [1988], "Mathematics from the structural point of view," *Revue Internationale de Philosophie* 42: 400–424.

Resnik, M. [1988a], "Second-order logic still wild," *Journal of Philosophy* 85: 75–87.

Resnik, M. [1990] "Beliefs about mathematical objects," in *Physicalism in mathematics*, ed. by A. D. Irvine, Dordrecht, Holland, Kluwer Academic Publishers, 41–71.

Resnik, M. [1992], "A structuralist's involvement with modality" (review of Hellman [1989]), *Mind* 101: 107–122.

Resnik, M. [1996], "Structural relativity," *Philosophia Mathematics* (3) 4: 83–99.

Resnik, M. [1997], "Holistic mathematics," in *Philosophy of mathematics today: Proceedings of an international conference in Munich*, ed. by M. Schirn, The Mind Association, Oxford, Oxford University Press, forthcoming.

Rogers, H. [1967], *Theory of recursive functions and effective computability*, New York, McGraw-Hill.

Russell, B. [1899] "Sur les axiomes de la géométrie," *Revue de Métaphysique et de Morale* 7: 684–707.

Russell, B. [1903], *The principles of mathematics*, London, Allen and Unwin.

Russell, B. [1904], "Non-euclidean geometry," *Athenaeum* 4018: 592–593.

Russell, B. [1956], *An essay on the foundations of geometry*, New York, Dover; first published in 1897.

Russell, B. [1993], *Introduction to mathematical philosophy*, New York, Dover; first published in 1919.

Scanlan, M. [1888] "Beltrami's model and the independence of the parallel postulate," *History and Philosophy of Logic* 9: 13–34.

Scanlan, M. [1991], "Who were the American postulate theorists?" *Journal of Symbolic Logic* 56: 981–1002.

Schoenfield, J. [1967], *Mathematical logic*, Reading, Massachusetts, Addison Wesley.

Shapiro, S. [1980], "On the notion of effectiveness," *History and Philosophy of Logic* 1: 209–230.

Shapiro, S. [1983], "Conservativeness and incompleteness," *Journal of Philosophy* 80: 521–531.

Shapiro, S. [1983a], "Mathematics and reality," *Philosophy of Science* 50: 523–548.

Shapiro, S. [1983b], "Remarks on the development of computability," *History and Philosophy of Logic* 4: 203–220.

Shapiro, S. [1985], "Epistemic and intuitionistic arithmetic," in *Intensional mathematics*, ed. by S. Shapiro, Amsterdam, North Holland, 11–46.

Shapiro, S. [1987], "Principles of reflection and second-order logic," *Journal of Philosophical Logic* 16: 309–333.

Shapiro, S. [1989], "Logic, ontology, mathematical practice," *Synthese* 79: 13–50.

Shapiro, S. [1991], *Foundations without foundationalism: A case for second-order logic*, Oxford, Oxford University Press.

Shapiro, S. [1993], "Understanding Church's thesis, again," *Acta Analytica* 11: 59–77.

Shapiro, S. [1993a], "Modality and ontology," *Mind* 102: 455–481.

Shapiro, S. [1995], "Skolem paradox," in *The Oxford companion to philosophy*, ed. by T. Honderich, Oxford, Oxford University Press, 827.

Shapiro, S. [1997], "Logical consequence: Models and modality," *Philosophy of mathematics today: Proceedings of an international conference in Munich*, ed. by M. Schirn, The Mind Association, Oxford, Oxford University Press, forthcoming.

Sher, G. [1991], *The bounds of logic*, Cambridge, MIT Press.

Sieg, W. [1990], "Physicalism, reductionism and Hilbert," in *Physicalism in mathematics*, ed. by A. D. Irvine, Dordrecht, Holland, Kluwer Academic Publishers, 183–257.

Sieg, W. [1994], "Mechanical procedures and mathematical experience," in *Mathematics and mind*, ed. by Alexander George, Oxford, Oxford University Press, 71–140.

Skolem, T. [1922], "Einige Bemerkungen zur axiomatischen Begründung der Mengenlehre," *Matematikerkongressen i Helsingfors den 4–7 Juli 1922*, Helsinki, Akademiska Bokhandeln, 217–232; tr. as "Some remarks on axiomatized set theory," in van Heijenoort [1967], 291–301.

Stein, H. [1988], "*Logos*, logic, and *Logistiké*: Some philosophical remarks on the nineteenth century transformation of mathematics," in Aspray and Kitcher [1988], 238–259.

Steiner, M. [1989], "The application of mathematics to natural science," *Journal of Philosophy* 86: 449–480.

Tait, W. [1981], "Finitism," *Journal of Philosophy* 78: 524–546.

Tait, W. [1986], "Truth and proof: The Platonism of mathematics," *Synthese* 69: 341–370.

Tait, W. [1986a], "Critical notice: Charles Parsons' *Mathematics in philosophy*," *Philosophy of Science* 53: 588–607.

Tait, W. [1997], "Frege versus Cantor and Dedekind on the concept of number," in *Early analytic philosophy: Frege, Russell, Wittgenstein: Essays in honor of Leonard Linsky*, ed. by W. Tait, Chicago, Open Court, 213–248.

Tarski, A. [1933], "Der Warheitsbegriff in dem formalisierten Sprachen," *Studia Philosophica* 1: 261–405; tr. as "The concept of truth in formalized languages," in Tarski [1956], 152–278.

Tarski, A. [1935], "On the concept of logical consequence," in Tarski [1956], 417–429.

Tarski, A. [1944], "The semantic conception of truth and the foundations of semantics," *Philosophy and Phenomenological Research* 4: 341–376.

Tarski, A. [1956], *Logic, semantics and metamathematics*, Oxford, Clarendon Press; second edition, ed. by John Corcoran, Indianapolis, Hackett, 1983.

Tarski, A. [1986], "What are logical notions?" (ed. by John Corcoran), *History and Philosophy of Logic* 7: 143–154.

Taylor, R. [1993], "Zermelo, reductionism, and the philosophy of mathematics," *Notre Dame Journal of Formal Logic* 34: 539–563.

Tennant, N. [1987], *Anti-realism and logic*, Oxford, Oxford University Press.

Tennant, N. [1997], "On the necessary existence of numbers," *Nous*, forthcoming.

Turing, A. [1936], "On computable numbers, with an application to the *Entscheidungsproblem*," *Proceedings of the London Mathematical Society* 42: 230–265; reprinted in *The Undecidable*, ed. by M. Davis, Hewlett, New York, The Raven Press, 1965, 116–153.

Turnbull, R. [1978], "Knowledge of the Forms in the later Platonic Dialogues," *Proceedings and Addresses of the American Philosophical Association* 51: 735–758.

Vaihinger, H. [1913], *Die Philosophie des Als Ob*, Berlin, Verlag von Reuther & Reichard; tr. as *The philosophy of "as if,"* by C. K. Ogden, London, Routledge and Kegan Paul, 1935.

Van Fraassen, B. [1980], *The scientific image*, Oxford, Oxford University Press.

Van Heijenoort, J. [1967], *From Frege to Gödel*, Cambridge, Harvard University Press.

Van Heijenoort, J. [1967a], "Logic as calculus and logic as language," *Synthese* 17: 324–330.

Veblen, O. [1904], "A system of axioms for geometry," *Transactions of the American Mathematical Society* 5: 343–384.

Veblen, O. [1925], "Remarks on the foundations of geometry," *Bulletin of the American Mathematical Society* 31: 121–141.

Wagner, S. [1987], "The rationalist conception of logic," *Notre Dame Journal of Formal Logic* 28: 3–35.

Wang, H. [1974], *From mathematics to philosophy*, London, Routledge and Kegan Paul.

Wang, H. [1987], *Reflections on Kurt Gödel*, Cambridge, MIT Press.

Weinberg, S. [1986], "Lecture on the applicability of mathematics," *Notices of the American Mathematical Society* 33.5.

Weyl, H. [1949], *Philosophy of mathematics and natural science*, Princeton, Princeton University Press; revised and augmented edition, New York, Athenaeum Press, 1963.

Wilson, M. [1981], "The double standard in ontology," *Philosophical Studies* 39: 409–427.

Wilson, M. [1992], "Frege: The royal road from geometry," *Nous* 26: 149–180.

Wilson, M. [1993], "There's a hole and a bucket, dear Leibniz," *Midwest Studies in Philosophy* 18: 202–241.

Wilson, M. [1993a], "Honorable intensions," in *Naturalism: A critical appraisal*, ed. by S. Wagner and R. Warner, Notre Dame, University of Notre Dame Press, 53–94.

Wittgenstein, L. [1953], *Philosophical investigations*, tr. by G. E. M. Anscombe, New York, Macmillan.

Wittgenstein, L. [1978], *Remarks on the foundations of mathematics*, tr. by G. E. M. Anscombe, Cambridge, MIT Press.

Wright, C. [1983], *Frege's conception of numbers as objects*, Aberdeen University Press.

Wright, C. [1992], *Truth and objectivity*, Cambridge, Harvard University Press.

Yaqub, A. [1993], *The liar speaks the truth: A defense of the revision theory of truth*, New York, Oxford University Press, 1993.

Zalta, E. [1983], *Abstract objects: An introduction to axiomatic metaphysics*, Dordrecht, Holland, Reidel.

Zermelo, E. [1904], "Beweis, dass jede Menge wohlgeordnet werden kann," *Mathematische Annalen* 59: 514–516; tr. in van Heijenoort [1967], 139–141.

Index